Pre-Calculus Workbook FOR DUMMIES®

2ND EDITION

by Dr. Yang Kuang and Michelle Rose Gilman,

with Elleyne Kase

John Wiley & Sons, Inc.

Pre-Calculus Workbook For Dummies®, 2nd Edition

Published by
John Wiley & Sons, Inc.
111 River St.
Hoboken, NJ 07030-5774
www.wiley.com

Published simultaneously in Canada

Library of Congress Control Number: 2011921773

ISBN: 978-0-470-92322-1

ISBN 978-0-470-92322-1 (pbk); ISBN 978-1-118-03416-3 (ebk); ISBN 978-1-118-03417-0 (ebk)

Manufactured in the United States of America

V10002473_071918

About the Authors

Dr. Yang Kuang is a proud father of three: Youny, Foris, and Belany. He has been a professor of mathematics at Arizona State University (ASU) since 1988. He entered the University of Science and Technology of China in 1980, went to the University of Oxford in 1984 for a Master's degree and completed a PhD in mathematics at University of Alberta (Canada) in 1988, becoming a mathematics professor at ASU in the same year. Yang Kuang has published over 120 refereed publications and authored or edited many books. He has directed a dozen PhD dissertations and many funded research projects in mathematical biology and medicine. He founded and continues to edit the journal *Mathematical Biosciences and Engineering.* He has given mathematical research talks all over the world and has organized many international mathematical biology and medicine meetings. He is well-known for his pioneer work in the application of delay differential equations in models of biology and medicine and his ongoing work in establishing a solid foundation and framework for building accurate, dynamic, rich population models that explicitly include resource quality dynamics.

His more recent interest is to formulate scientifically well-grounded and computationally tractable mathematical models to describe the rich and intriguing dynamics of various within-host diseases and their treatments, including various types of cancer, diabetic diseases, and virus-induced diseases such as influenza and hepatitis B virus (HBV) infection. He hopes that these models will help speed up the much-needed development of personalized medicine.

Yang Kuang teaches and uses calculus all the time. He is keenly aware of the many challenges facing students and instructors. He hopes this book will be helpful to many of those who want to confront those challenges.

Michelle Rose Gilman is the founder and CEO of Fusion Learning Center and Fusion Academy, a private school and tutoring/test prep facility in Solana Beach, California. She is the author of *Pre-Calculus For Dummies, The ACT For Dummies,* and other books on self-esteem, writing, and motivational topics.

Elleyne Kase is a multidisciplined artist and graphics professional whose creative expertise has been applied commercially in many different avenues, such as fine art designs for the commercial interior design industry, Wayfinding programs for large hospital systems, 3D displays, Web site design, and national brand identity. A bachelor's degree in graphic design from San Jose State University began her professional career in the San Francisco Bay Area, which continued in San Diego design agencies as her family grew.

Book design, a favorite of hers, began early on in her career and has now been applied to two of her own exam prep publications, *Visual Quick Notes for Life Insurance* and *Visual Quick Notes for Life and Health.* Both use her new mind mapping system, an evolution of her study on how the mind absorbs data, to explain simple terms and concepts in an organized visual matrix of graphic elements and text that improves memory retention.

Authors' Acknowledgments

From Yang:

I would like to thank Elleyne Kase for inviting me to take part in this unusual book project and all the people involved in the process. It was a joy working with all of you. I would also like to thank Professor Jamie Whittimore McGill and Dr. Scott Parsell for providing a detailed technical review of this edition.

And last, but not least, I am forever grateful to my wife, Aijun, for her beautiful creation of and the great attention she devotes to the family, so that I can have time to do my work, including carrying out this project.

From Elleyne:

Throughout my varied career I have had the great pleasure of working with professionals of many disciplines. The *For Dummies* series combines the skills of many very talented people. I want to express my gratitude to Bill Gladstone who gave me this rare opportunity to participate; to the editors at Wiley — Lindsay Lefevere, Tracy Boggier, Elizabeth Rea, and Christy Pingleton — for their professionalism, understanding, and skill in bringing such a complex product to market; and to Yang Kuang who juggled his demanding teaching schedule to accommodate this project.

Above all I want to thank my dear mother, who has always supported me in all my varied interests, and my three children, Seth, Shane, and Mimi, who have grown to become wonderful, nurturing, caring adults and my best friends.

Publisher's Acknowledgments

We're proud of this book; please send us your comments at `http://dummies.custhelp.com`. For other comments, please contact our Customer Care Department within the U.S. at 877-762-2974, outside the U.S. at 317-572-3993, or fax 317-572-4002.

Some of the people who helped bring this book to market include the following:

Acquisitions, Editorial, and Media Development

Project Editor: Elizabeth Rea

Acquisitions Editor: Tracy Boggier

Copy Editors: Christine Pingleton, Jennifer Tebbe

Assistant Editor: David Lutton

Technical Editors: Jamie W. McGill, Scott Parsell, Ph.D.

Editorial Manager: Michelle Hacker

Editorial Assistants: Rachelle Amick, Jennette ElNaggar, Alexa Koschier

Cartoons: Rich Tennant (`www.the5thwave.com`)

Composition Services

Project Coordinators: Nikki Gee, Patrick Redmond

Layout and Graphics: Carl Byers, Carrie A. Cesavice, Joyce Haughey, Mark Pinto, Corrie Socolovitch

Proofreaders: Betty Kish, Jessica Kramer

Indexer: Glassman Indexing Services

Publishing and Editorial for Consumer Dummies

Kathleen Nebenhaus, Vice President and Executive Publisher

Kristin Ferguson-Wagstaffe, Product Development Director

David Palmer, Associate Publisher

Publishing for Technology Dummies

Andy Cummings, Vice President and Publisher

Composition Services

Debbie Stailey, Director of Composition Services

Contents at a Glance

Table of Contents

Introduction

You've just discovered the best workbook ever to help you with pre-calculus, if we do say so ourselves. If you've gotten this far in your math career, congratulations! Many students choose to stop their math education after they complete Algebra II, but not you!

If you've picked up this book (and obviously you have, given that you're reading this sentence!), maybe some of the concepts in pre-calc are giving you a hard time, or perhaps you just want more practice. Maybe you're deciding whether you even want to take pre-calc at all. This book fits the bill for all those reasons. And we're here to encourage you on your pre-calc adventure.

We know that you'll find this workbook chock-full of valuable practice problems and explanations. In instances where you feel you may need a more thorough explanation, please refer to *Pre-Calculus For Dummies* by Krystle Rose Forseth, Christopher Burger, and Michelle Rose Gilman (Wiley). This book, however, is a great stand-alone workbook if you need extra practice or want to just brush up in certain areas.

About This Book

Don't let pre-calc scare you. When you realize that you already know a whole bunch from Algebra I and Algebra II, you'll see that pre-calculus is really just using that old information in a new way. And even if you're scared, we're here with you, so no need to panic. Before you get ready to start this new adventure, you need to know a few things about this book.

This book isn't a novel. It's not meant to be read in order from beginning to end. You can read any topic at any time, but we've structured it in such a way that it follows the "normal" curriculum. This is hard to do, because most states don't have state standards for what makes pre-calc pre-calc. We looked at a bunch of curriculums, though, and came up with what we think is a good representation of a pre-calc course. Sometimes, we may include a reference to material in another chapter, and we may send you there for more information.

Instead of placing this book on a shelf and never looking at it again or using it as a doorstop (thanks for the advertisement, in either case), we suggest you follow one of two alternatives:

- Look up what you need to know when you need to know it. The index and the table of contents direct you where to look.
- Start at the beginning and read straight through. This way, you may be reminded of an old topic that you had forgotten (anything to get those math wheels churning inside your head). Besides, practice makes perfect, and the problems in this book are a great representation of the problems found in pre-calc textbooks.

Conventions Used in This Book

For consistency and ease of navigation, this book uses the following conventions:

- Math terms are *italicized* when they're introduced or defined in the text.
- Variables are *italicized* to set them apart from letters.
- The symbol for imaginary numbers is a lowercase *i*.

Foolish Assumptions

We don't assume that you love math the way we do, but we do assume that you picked this book up for a reason of your own. Maybe you want a preview of the course before you take it, or perhaps you need a refresher on the topics in the course, or maybe your kid is taking the course and you're trying to help him to be more successful.

Whatever your reason, we assume that you've encountered most of the topics in this book before, because for the most part, they review what you've seen in algebra or geometry.

How This Book Is Organized

This book is divided into five parts dealing with the most commonly taught topics of pre-calc.

Part I: Setting the Foundation: The Nuts and Bolts of Pre-Calculus

First we review basic material from Algebra II. We then cover real numbers and what you'll be asked to do with them. Next up are functions of all kinds (polynomials, rational, exponential, and logarithmic): graphing them and performing operations with them.

Part II: Trig Is the Key: Basic Review, the Unit Circle, and Graphs

The chapters in Part II review trig ratios and word problems for trig. We show you how to build the unit circle, how to solve trig equations, and how to graph trig functions. Some of these topics may be review for you as well; that really depends on how much trig was covered in your Algebra II course.

Part III: Digging into Advanced Trig: Identities, Theorems, and Applications

The chapters in Part III cover basic and advanced identities. We cover the tricky trig proofs in this part. If you're asked to do trig proofs in your pre-calculus course, you definitely want to check out our tips on how to handle them like a pro. We also cover some trig applications that can be solved using the Law of Sines or the Law of Cosines.

Part IV: Poles, Cones, Variables, Sequences, and Finding Your Limits

The chapters in Part IV cover the topics from the remainder of the pre-calculus course. We introduce complex numbers and how to work with them, and we explain conic sections and how to graph them. Because systems of equations tend to get harder in pre-calc, we begin with a review and build up to the tougher topics. Your pre-calc course may focus on only a couple of these topics, so be sure to pay attention to the table of contents here. Next, we move into sequences and series and introduce the binomial theorem, which helps you raise binomials to high powers. Last, we introduce the first topics of a calc course. Sometimes, these are the last topics you'll see in pre-calc, so we want to be sure to go over them.

Part V: The Part of Tens

This book has two handy lists at the end. The first list includes ten tips for using parent graphs: how to recognize them, how to graph them, and how to transform them. The second list covers common mistakes we often see that we'd like to help you avoid.

Icons Used in This Book

Throughout this book you'll see icons in the margins to draw your attention to something important that you need to know.

Pre-calc rules are exactly what they say they are — the rules of pre-calculus. Theorems, laws, and properties all make pre-calc an ironclad course — they must be followed at all times.

You see this icon when we present an example problem whose solution we walk you through step by step. You get a problem and a detailed answer.

Tips are great, especially if you wait tables for a living! These tips are designed to make your life easier, which are the best tips of all!

The Remember icon is used one way: It asks you to remember old material from a previous math course.

Warnings are big red flags that draw your attention to common mistakes that may trip you up.

Where to Go from Here

Pick a starting point in the book and go practice the problems there. If you'd like to review the basics first, start at Chapter 1. If you feel comfy enough with your algebra skills, you may want to skip that chapter and head over to Chapter 2. Most of the topics there are reviews of Algebra II material, but don't skip over something because you think you have it under control. You'll find in pre-calc that the level of difficulty in some of these topics gets turned up a notch or two. Go ahead — dive in and enjoy the world of pre-calculus!

Part I

Setting the Foundation: The Nuts and Bolts of Pre-Calculus

In this part . . .

Pre-calculus is really just another stop on the road to calculus. The chapters in this part begin with a review of the basics: using the order of operations, solving and graphing equations and inequalities, and using the distance and midpoint formulas. Some new material pops up in the form of interval notation, so be sure and check that out. Then we move on to real numbers, including radicals. Everything you ever wanted to know about functions is covered in one of the chapters: graphing and transforming parent graphs, rational functions, and piecewise functions. We also go over performing operations on functions and how to find the inverse. We then move on to solving higher-degree polynomials using techniques like factoring, completing the square, and the quadratic formula. You also find out how to graph these complicated polynomials. Lastly, you discover exponential and logarithmic functions and what you're expected to know about them.

Chapter 1

Beginning at the Very Beginning: Pre-Pre-Calculus

In This Chapter

- Brushing up on the order of operations
- Solving equalities
- Graphing equalities and inequalities
- Finding distance, midpoint, and slope

Pre-calculus is the stepping stone for calculus. It's the final hurdle after all those years of math: pre-algebra, Algebra I, geometry, and Algebra II. Now all you need is pre-calculus to get to that ultimate goal — calculus. And as you may recall from your Algebra II class, you were subjected to much of the same material you saw in algebra and even pre-algebra (just a couple steps up in terms of complexity — but really the same stuff). As the stepping stone, pre-calculus begins with certain concepts that you're expected to understand.

Therefore, we're starting here, at the very beginning, reviewing those concepts. If you feel you're already an expert at everything algebra, feel free to skip past this chapter and get the full swing of pre-calc going. If, however, you need to review, then read on.

If you don't remember some of the concepts we discuss in this chapter, or even in this book, you can pick up another *For Dummies* math book for review. The fundamentals are important. That's why they're called fundamentals. Take the time now to review and save yourself countless hours of frustration in the future!

Reviewing Order of Operations: The Fun in Fundamentals

You can't put on your sock after you put on your shoe, can you? The same concept applies to mathematical operations. There's a specific order to which operation you perform first, second, third, and so on. At this point, it should be second nature, but because the concept is so important (especially when you start doing more complex calculations), a quick review is worth it, starting with everyone's favorite mnemonic device.

Please excuse who? Oh, yeah, you remember this one — my dear Aunt Sally! The old mnemonic still stands, even as you get into more complicated problems. Please Excuse My Dear Aunt Sally is a mnemonic for the acronym PEMDAS, which stands for

- Parentheses (including absolute value, brackets, and radicals)
- Exponents
- Multiplication and Division (from left to right)
- Addition and Subtraction (from left to right)

The order in which you solve algebraic problems is very important. Always work what's in the parentheses first, then move on to the exponents, followed by the multiplication and division (from left to right), and finally, the addition and subtraction (from left to right).

You should also have a good grasp on the properties of equality. If you do, you'll have an easier time simplifying expressions. Here are the properties:

- **Reflexive property: $a = a$.** For example, $4 = 4$.
- **Symmetric property: If $a = b$, then $b = a$.** For example, if $2 + 8 = 10$, then $10 = 2 + 8$.
- **Transitive property: If $a = b$ and $b = c$, then $a = c$.** For example, if $2 + 8 = 10$ and $10 = 5 \cdot 2$, then $2 + 8 = 5 \cdot 2$.
- **Commutative property of addition: $a + b = b + a$.** For example, $3 + 4 = 4 + 3$.
- **Commutative property of multiplication: $a \cdot b = b \cdot a$.** For example, $3 \cdot 4 = 4 \cdot 3$.
- **Associative property of addition: $a + (b + c) = (a + b) + c$.** For example, $3 + (4 + 5) = (3 + 4) + 5$.
- **Associative property of multiplication: $a \cdot (b \cdot c) = (a \cdot b) \cdot c$.** For example, $3 \cdot (4 \cdot 5) = (3 \cdot 4) \cdot 5$.
- **Additive identity: $a + 0 = a$.** For example, $4 + 0 = 4$.
- **Multiplicative identity: $a \cdot 1 = a$.** For example, $-18 \cdot 1 = -18$.
- **Additive inverse property: $a + (-a) = 0$.** For example, $5 + (-5) = 0$.
- **Multiplicative inverse property: $a \cdot \frac{1}{a} = 1$.** For example, $-2 \cdot (-\frac{1}{2}) = 1$.
- **Distributive property: $a(b + c) = a \cdot b + a \cdot c$.** For example, $5(3 + 4) = 5 \cdot 3 + 5 \cdot 4$.
- **Multiplicative property of zero: $a \cdot 0 = 0$.** For example, $4 \cdot 0 = 0$.
- **Zero product property: If $a \cdot b = 0$, then $a = 0$ or $b = 0$.** For example, if $x(2x - 3) = 0$, then $x = 0$ or $2x - 3 = 0$.

Following are a couple examples so you can see the order of operations and the properties of equality in action before diving into some practice questions.

Q. Simplify: $= \dfrac{6^2 - 4\left(3-\sqrt{20+5}\right)^2}{|4-8|}$

A. The answer is 5.

Following the order of operations, simplify everything in parentheses first. (Remember that radicals and absolute value marks act like parentheses, so do operations within them first before simplifying the radicals or taking the absolute value.)

Simplify the parentheses by taking the square root of 25 and the absolute value of –4, like so:

$$\frac{6^2 - 4\left(3-\sqrt{25}\right)^2}{|-4|} = \frac{6^2 - 4(3-5)^2}{4} =$$

$$\frac{6^2 - 4(-2)^2}{4}$$

Now you can deal with the exponents by squaring the 6 and the –2: $\frac{36-4(4)}{4}$.

Note: Although they're not written, parentheses are implied around the terms above and below a fraction bar. In other words, the expression $\frac{36-4(4)}{4}$ can also be written as $\frac{[36-4(4)]}{4}$. Therefore, you must simplify the numerator and denominator before dividing the terms following the order of operations, like this: $\frac{36-4(4)}{4} = \frac{36-16}{4} = \frac{20}{4} = 5$.

Q. Simplify: $\dfrac{\left(\frac{1}{8}+\frac{1}{3}\right)+\frac{3}{8}}{\frac{3}{18}+\frac{1}{9}}$

A. The answer is 3.

Using the associative property and the commutative property of addition, rewrite the expression to make the fractions easier to add.

$$\frac{\left(\frac{1}{8}+\frac{3}{8}\right)+\frac{1}{3}}{\frac{3}{18}+\frac{1}{9}}$$

Add the fractions with common denominators.

$$\frac{\frac{4}{8}+\frac{1}{3}}{\frac{3}{18}+\frac{1}{9}}$$

Then reduce the resulting fraction to get the following:

$$\frac{\frac{1}{2}+\frac{1}{3}}{\frac{3}{18}+\frac{1}{9}}$$

Next, find a common denominator for the fractions in the numerator and denominator.

$$\frac{\frac{3}{6}+\frac{2}{6}}{\frac{3}{18}+\frac{2}{18}}$$

Add them, like so:

$$\frac{\frac{5}{6}}{\frac{5}{18}}$$

Recognizing that this expression is a division problem, $\frac{5}{6}+\frac{5}{18}$, multiply by the inverse and simplify:

$$\frac{5}{6}\cdot\frac{18}{5} = \frac{5\cdot 18}{6\cdot 5} = \frac{\not{5}\cdot 18}{6\cdot \not{5}} = \frac{3}{1} = 3$$

1. Simplify: $\dfrac{3\sqrt{(4-6)^2+[2-(-1)]^2}}{|-3-(-1)|}$

Solve It

2. Simplify: $\dfrac{|-3|-|2|+(-1)}{|-7+2|}$

Solve It

3. Simplify: $(2^3-3^2)^4(-5)$

Solve It

4. Simplify: $\dfrac{|5(1-4)+6|}{3\left(-\frac{1}{6}+\frac{2}{3}\right)-\frac{1}{2}}$

Solve It

Keeping Your Balance While Solving Equalities

Just as simplifying expressions is the basis of pre-algebra, solving for variables is the basis of algebra. Why should you care? Because both are essential to the more complex concepts covered in pre-calculus.

Solving linear equations with the general format of $ax+b=c$, where a, b, and c are constants, is relatively easy using the properties of numbers. The goal, of course, is to isolate the variable, x.

One type of equation you can't forget is absolute value equations. The *absolute value* is defined as the distance from 0. In other words, $|x| = \begin{cases} x, & x \geq 0 \\ -x, & x < 0 \end{cases}$. As such, an absolute value has two possible solutions: one where the quantity inside the absolute value bars is positive and another where it's negative. To solve these equations, you must isolate the absolute value term (find the value of the absolute value term) and then set the quantity inside the absolute value bars to the positive and negative values (see the second example question that follows).

Check out the following examples or skip ahead to the practice questions if you think you're ready to tackle them.

Q. Solve for x: $3(2x - 4) = x - 2(-2x + 3)$

A. $x = 6$

First, using the distributive property, distribute the 3 and the –2 to get $6x - 12 = x + 4x - 6$. Then combine like terms and solve using algebra, like so: $6x - 12 = 5x - 6$; $x - 12 = -6$; $x = 6$.

Q. Solve for x: $|x - 3| + (-16) = -12$

A. $x = 7$ or -1

Isolate the absolute value: $|x - 3| = 4$. Next, set the quantity inside the absolute value bars to the positive solution: $x - 3 = 4$. Then set the quantity inside the absolute value bars to the negative solution: $-(x - 3) = 4$. Solve both equations to find two possible solutions: $x - 3 = 4$, which gives $x = 7$; and $-(x - 3) = 4$, which yields $-x + 3 = 4$ and the solution $x = -1$.

5. Solve: $3 - 6[2 - 4x(x + 3)] = 3x(8x + 12) + 27$

Solve It

6. Solve: $\frac{x}{2} + \frac{x-2}{4} = \frac{x+4}{2}$

Solve It

7. Solve: $|x - 3| + |3x + 2| = 4$

Solve It

8. Solve: $3 - 4(2 - 3x) = 2(6x + 2)$

Solve It

9. Solve: $2|x-3|+12=6$

Solve It

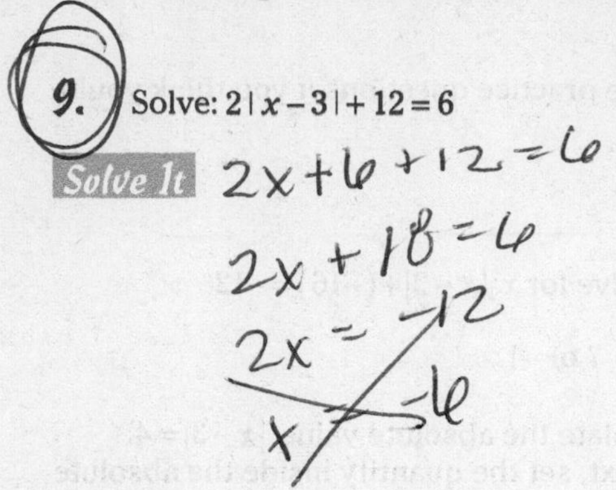

10. Solve: $3(2x+5)+10=2(x+10)+4x+5$

Solve It

When Your Image Really Counts: Graphing Equalities and Inequalities

Graphs are visual representations of mathematical equations. In pre-calculus, you'll be introduced to many new mathematical equations and then be expected to graph them. We give you lots of practice graphing these equations when we cover the more complex equations. In the meantime, it's important to practice the basics: graphing linear equalities and inequalities.

Both graphs exist on the *Cartesian coordinate system,* which is made up of two axes: the horizontal, or *x*-axis, and the vertical, or *y*-axis. Each point on the coordinate plane is called a *Cartesian coordinate pair* and has an *x* coordinate and a *y* coordinate. So the notation for any point on the coordinate plane looks like this: (x, y). A set of these ordered pairs that can be graphed on a coordinate plane is called a *relation.* The *x* values of a relation are its *domain*, and the *y* values are its *range*. For example, the domain of the relation R = {(2, 4), (–5, 3), (1, –2)} is {2, –5, 1}, and the range is {4, 3, –2}.

You can graph a linear equation, whether it's an equality or an inequality, in two ways: by using the plug-and-chug method or by using the slope-intercept form. We review both approaches in the following sections.

Graphing with the plug-and-chug method

To graph the plug-and-chug way, start by picking domain (x) values. Plug them into the equation to solve for the range (y) values. For linear equations, after you plot these points (x, y) on the coordinate plane, you can connect the dots to make a line. The process also works if you choose range values first and then plug in to find the corresponding domain values.

There's also a helpful method for finding *intercepts,* the points that fall on the *x*- or *y*-axes. To find the *x*-intercept $(x, 0)$, plug in 0 for y and solve for x. To find the *y*-intercept $(0, y)$, plug in 0 for x and solve for y. For example, to find the intercepts of the linear equation $2x + 3y = 12$, start by plugging in 0 for y: $2x + 3(0) = 12$. Then, using properties of numbers, solve for x: $2x + 0 = 12$; $2x = 12$; $x = 6$. So the *x*-intercept is $(6, 0)$. For the *y*-intercept, plug in 0 for x and solve for y: $2(0) + 3y = 12$; $0 + 3y = 12$; $3y = 12$; $y = 4$. Therefore, the y-intercept is (0, 4). At this point, you can plot those two points and connect them to graph the line ($2x + 3y = 12$), because, as you learned in geometry, two points make a line. See the resulting graph in Figure 1-1.

As equations become more complex, you can use the plug-and-chug method to get some key pieces of information.

Graphing by using the slope-intercept form

The slope-intercept form of a linear equation gives a great deal of helpful information in a cute little package. The equation $y = mx + b$ immediately gives you the y-intercept (b) that you worked to find in the plug-and-chug method; it also gives you the slope (m). *Slope* is a fraction that gives you the rise over the run. To change equations that aren't written in slope-intercept form, you simply solve for y. For example, if you use the linear equation $2x + 3y = 12$, you start by subtracting $2x$ from each side: $3y = -2x + 12$. Next, you divide all the terms by 3: $y = -(2/3)x + 4$. Now that the equation is in slope-intercept form, you know that the y-intercept is 4, and you can graph this point on the coordinate plane. Then, you can use the slope to plot the second point. From the slope-intercept equation, you know that the slope is $-2/3$. This tells you that the rise is -2 and the run is 3. From the point (0, 4), plot the point 2 down and 3 to the right. In other words, (3, 2). Lastly, connect the two points to graph the line. The resulting graph in Figure 1-2 is identical to Figure 1-1.

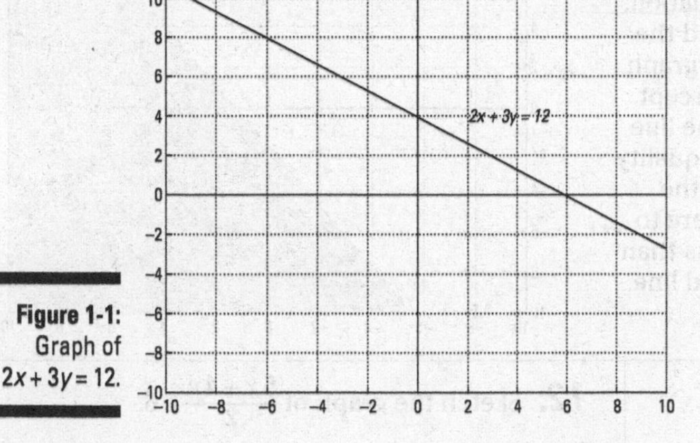

Figure 1-1: Graph of $2x + 3y = 12$.

Figure 1-2: Graph of $y = -(2/3)x + 4$.

Graphing inequalities by either method

Similar to graphing equalities, graphing inequalities begins with plotting two points by either method. However, because *inequalities* are used for comparisons — greater than, less than, or equal to — you have two more questions to answer after finding two points:

- Is the line dashed (< or >) or solid (≤ or ≥)?
- Do you shade under the line ($y <$ or $y \leq$) or above the line ($y >$ or $y \geq$)?

Here's an example of an inequality followed by a few practice questions.

Q. Sketch the graph of the inequality: $3x - 2y > 4$

A. Put the inequality into slope-intercept form by subtracting $3x$ from each side of the equation to get $-2y > -3x + 4$ and dividing each term by –2 to get $y < (3/2)x - 2$. (***Remember:*** When you multiply or divide an inequality by a negative, you need to reverse the inequality.) From the resulting equation, you can find the y-intercept, –2, and the slope, (3/2). Use this information to graph two points by using the slope-intercept form. Next, decide the nature of the line (solid or dashed). Because the inequality is strict, the line is dashed. Graph the dashed line so you can decide where to shade. Because $y < (3/2)x - 2$ is a less-than inequality, shade below the dashed line, as shown in the following figure.

This method works only if the boundary line is first converted to slope-intercept form. An alternative is to graph the boundary line using any method and then use a sample point (such as (0,0)) to determine which half-plane to shade.

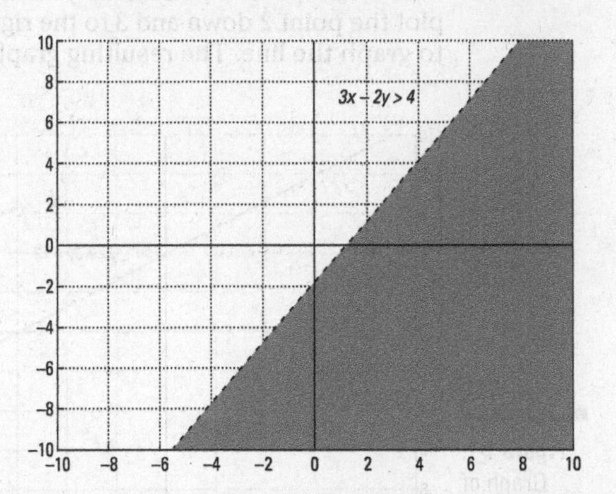

11. Sketch the graph of $\frac{1}{3}(6x + 2y) = 16$.

Solve It

12. Sketch the graph of $\frac{5x+4y}{2} \geq 6$.

Solve It

13. Sketch the graph of $4x + 5y \geq 2(3y + 2x + 4)$.

Solve It

14. Sketch the graph of $x - 3y = 4 - 2y - y$.

Solve It

Using Graphs to Find Distance, Midpoint, and Slope

Graphs are more than just pretty pictures. From a graph, it's possible to determine two points. From these points, you can figure out the distance between them, the midpoint of the segment connecting them, and the slope of the line connecting them. As graphs become more complex in both pre-calculus and calculus, you're asked to find and use all three of these pieces of information. Aren't you lucky?

Finding the distance

Distance refers to how far apart two things are. In this case, you're finding the distance between two points. Knowing how to calculate distance is helpful for when you get to conics (see Chapter 12). To find the distance between two points (x_1, y_1) and (x_2, y_2), use the following formula: $d = \sqrt{(x_1 - x_2)^2 + (y_1 - y_2)^2}$

Calculating the midpoint

The *midpoint* is the middle of a segment. This concept also comes up in conics (see Chapter 12) and is ever so useful for all sorts of other pre-calculus calculations. To find the midpoint, *M*, of the points (x_1, y_1) and (x_2, y_2), you just need to average the *x* and *y* values and express them as an ordered pair, like so: $M = \left(\frac{x_1 + x_2}{2}, \frac{y_1 + y_2}{2}\right)$

Discovering the slope

Slope is a key concept for linear equations, but it also has applications for trigonometric functions and is essential for differential calculus. *Slope* describes the steepness of a line on the coordinate plane (think of a ski slope). Use this formula to find the slope, m, of the line (or segment) connecting the two points (x_1, y_1) and (x_2, y_2): $m = \frac{y_2 - y_1}{x_2 - x_1}$

Note: Positive slopes move up and to the right (+/+) or down and to the left (–/–). Negative slopes move down and to the right (–/+) or up and to the left (+/–). Horizontal lines have a slope of 0, and vertical lines have an undefined slope.

Following is an example question for your reviewing pleasure. Look it over and then try your hand at the practice questions.

Q. Find the distance, slope, and midpoint of $\overline{AB}$.

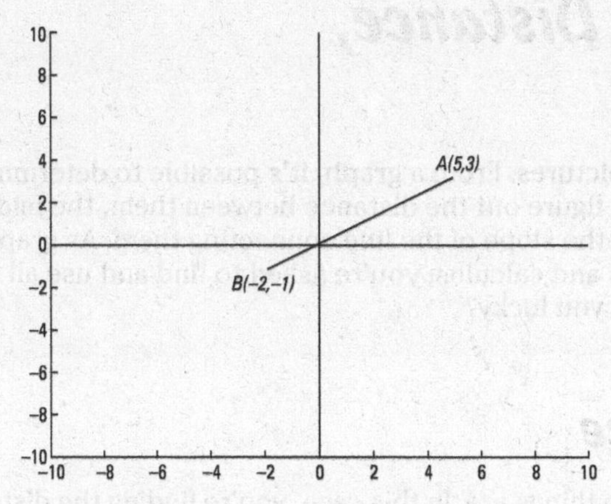

A. The distance is $\sqrt{65}$, the slope is $4/7$, and the midpoint is $m = (3/2, 1)$

Plug the x and y values into the distance formula and, following the order of operations, simplify the terms under the radical (keeping in mind the implied parentheses of the radical itself).

$$d = \sqrt{[5-(-2)]^2 + [3-(-1)]^2} = \sqrt{(5+2)^2 + (3+1)^2} = \sqrt{(7)^2 + (4)^2} = \sqrt{49+16} = \sqrt{65}$$

Because 65 doesn't contain any perfect squares as factors, this is as simple as you can get. To find the midpoint, plug the points into the midpoint equation and simplify using the order of operations.

$$M = \left(\frac{5+(-2)}{2}, \frac{3+(-1)}{2}\right) = \left(\frac{3}{2}, \frac{2}{2}\right) = \left(\frac{3}{2}, 1\right)$$

To find the slope, use the formula, plug in your x and y values, and use the order of operations to simplify.

$$m = \frac{-1-3}{-2-5} = \frac{-4}{-7} = \frac{4}{7}$$

15. Find the length of segment CD, where C is (–2, 4) and D is (3, –1).

Solve It

16. Find the midpoint of segment EF, where E is (3, –5) and F is (7, 5).

Solve It

17. Find the slope of line *GH*, where *G* is (–3, –5) and *H* is (–3, 4).

Solve It

18. Find the perimeter of triangle *CAT*.

Solve It

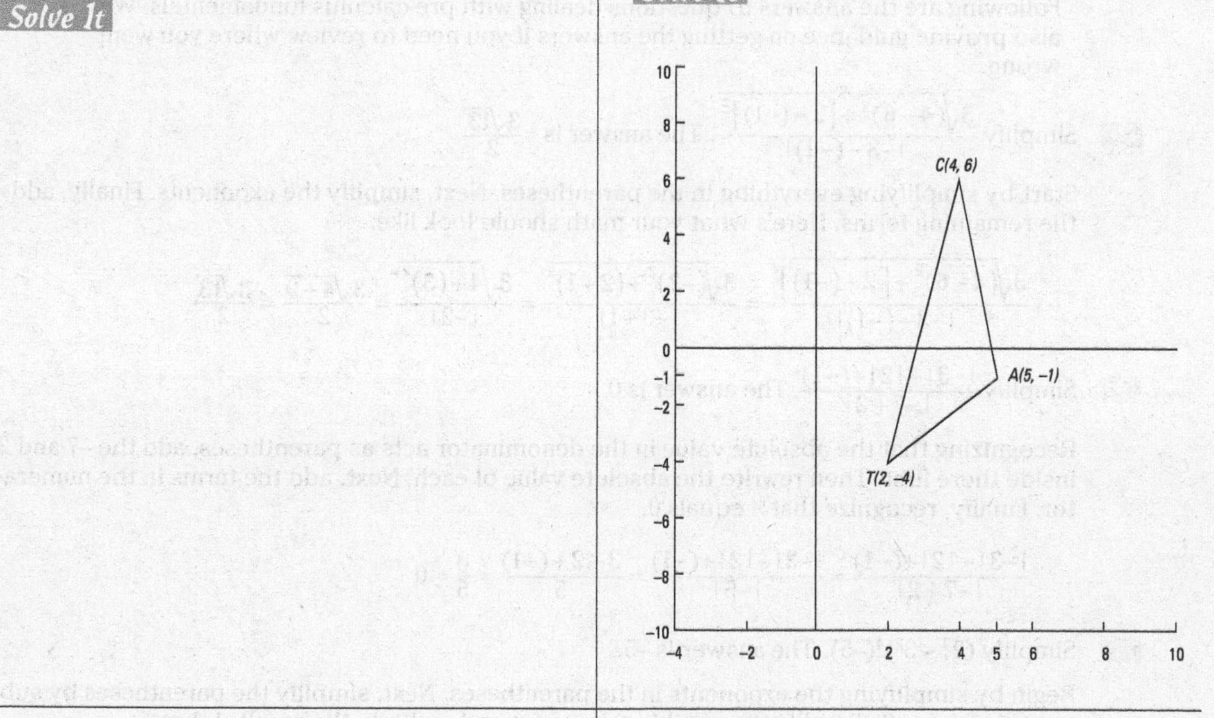

19. Find the center of the rectangle *NEAT*.

Solve It

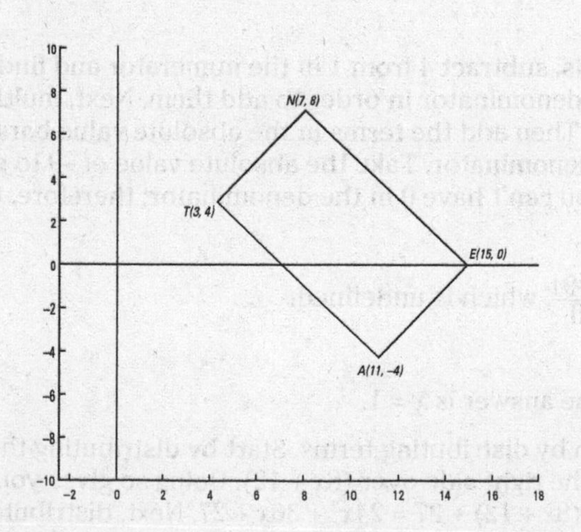

20. Determine whether triangle *DOG* is a right triangle.

Solve It

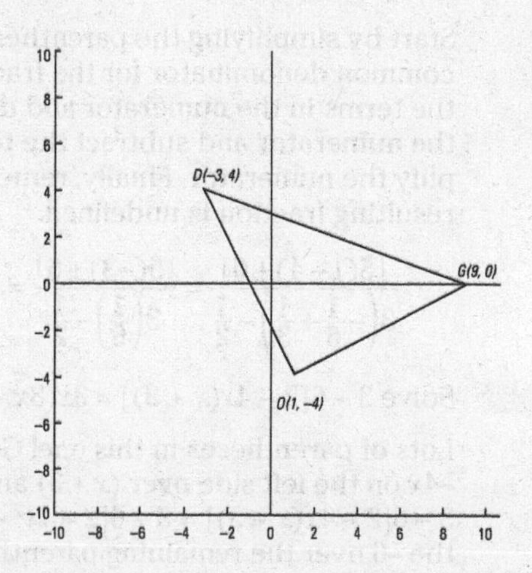

Answers to Problems on Fundamentals

Following are the answers to questions dealing with pre-calculus fundamentals. We also provide guidance on getting the answers if you need to review where you went wrong.

1 Simplify $\frac{3\sqrt{(4-6)^2+[2-(-1)]^2}}{|-3-(-1)|}$. The answer is $=\frac{3\sqrt{13}}{2}$.

Start by simplifying everything in the parentheses. Next, simplify the exponents. Finally, add the remaining terms. Here's what your math should look like:

$$\frac{3\sqrt{(4-6)^2+[2-(-1)]^2}}{|-3-(-1)|}=\frac{3\sqrt{(-2)^2+(2+1)^2}}{|-3+1|}=\frac{3\sqrt{4+(3)^2}}{|-2|}=\frac{3\sqrt{4+9}}{2}=\frac{3\sqrt{13}}{2}$$

2 Simplify $\frac{|-3|-|2|+(-1)}{|-7+2|}$. The answer is 0.

Recognizing that the absolute value in the denominator acts as parentheses, add the –7 and 2 inside there first. Then rewrite the absolute value of each. Next, add the terms in the numerator. Finally, recognize that $\frac{0}{5}$ equals 0.

$$\frac{|-3|-|2|+(-1)}{|-7+2|}=\frac{|-3|-|2|+(-1)}{|-5|}=\frac{3-2+(-1)}{5}=\frac{0}{5}=0$$

3 Simplify $(2^3-3^2)^4(-5)$. The answer is –5.

Begin by simplifying the exponents in the parentheses. Next, simplify the parentheses by subtracting 9 from 8. Simplify the resulting exponent and multiply the result, 1, by –5.

$$(2^3-3^2)^4(-5)=(8-9)^4(-5)=(-1)^4(-5)=1(-5)=-5$$

4 Simplify $\frac{|5(1-4)+6|}{3\left(-\frac{1}{6}+\frac{1}{3}\right)-\frac{1}{2}}$. The answer is undefined.

Start by simplifying the parentheses. To do this, subtract 4 from 1 in the numerator and find a common denominator for the fractions in the denominator in order to add them. Next, multiply the terms in the numerator and denominator. Then add the terms in the absolute value bars in the numerator and subtract the terms in the denominator. Take the absolute value of –9 to simplify the numerator. Finally, remember that you can't have 0 in the denominator; therefore, the resulting fraction is undefined.

$$\frac{|5(1-4)+6|}{3\left(-\frac{1}{6}+\frac{1}{3}\right)-\frac{1}{2}}=\frac{|5(-3)+6|}{3\left(\frac{1}{6}\right)-\frac{1}{2}}=\frac{|-15+6|}{\frac{1}{2}-\frac{1}{2}}=\frac{|-9|}{0}\text{, which is undefined.}$$

5 Solve $3-6[2-4x(x+3)]=3x(8x+12)+27$. The answer is $x=1$.

Lots of parentheses in this one! Get rid of them by distributing terms. Start by distributing the $-4x$ on the left side over $(x+3)$ and the $3x$ on the right side over $(8x+12)$. Doing so gives you $3-6[2-4x(x+3)]=3-6[2-4x^2-12x]$ and $3x(8x+12)+27=24x^2+36x+27$. Next, distribute the –6 over the remaining parentheses on the left side of the equation to get $3-12+24x^2+72x=24x^2+72x-9$. Combine like terms on the left side: $-9+24x^2+72x=24x^2+36x+27$. To isolate x onto one side, subtract $24x^2$ from both sides to get $-9+72x=36x+27$. Subtracting $36x$ from each side gives you $-9+36x=27$. Adding 9 to both sides results in $36x=36$. Finally, dividing both sides by 36 leaves you with your solution: $x=1$.

6 Solve $\frac{x}{2}+\frac{x-2}{4}=\frac{x+4}{2}$. The answer is $x = 10$.

Don't let those fractions intimidate you. Multiply through by the common denominator, 4, to eliminate the fractions altogether. Then solve like normal by combining like terms and isolating x. Here's the math:

$$\frac{x}{2}+\frac{x-2}{4}=\frac{x+4}{2}; 4\left[\frac{x}{2}+\frac{x-2}{4}=\frac{x+4}{2}\right]; 2x+x-2=2x+8; 3x-2=2x+8; 3x=2x+10; x=10$$

7 Solve $|x-3|+|3x+2|=4$. The answer is $x = -3/4, -1/2$.

So you have two absolute value terms? Just relax and remember that absolute value means the distance from 0, so you have to consider all the possibilities to solve this problem. In other words, you have to consider and try four different possibilities: both absolute values are positive, both are negative, the first is positive and the second is negative, and the first is negative and the second is positive.

When you have multiple absolute value terms in a problem, not all the possibilities will work. As you calculate the possibilities, you may create what math people call *extraneous solutions.* These are actually false solutions that don't work in the original equation. You create extraneous solutions when you change the format of an equation. To be sure a solution is real and not extraneous, you need to plug your answer into the original equation to check it. Time to try the possibilities:

- **Positive/positive:** $(x - 3) + (3x + 2) = 4$, $4x - 1 = 4$, $4x = 5$, $x = 5/4$. Plugging this answer back into the original equation, you get $39/4 = 4$. Nope! You have an extraneous solution.
- **Negative/negative:** $-(x - 3) + -(3x + 2) = 4$, $-x + 3 - 3x - 2 = 4$, $-4x + 1 = 4$, $-4x = 3$, $x = -3/4$. Plug that into the original equation and you get $4 = 4$. Voilà! Your first solution.
- **Positive/negative:** $(x - 3) + -(3x + 2) = 4$, $x - 3 - 3x - 2 = 4$, $-2x - 5 = 4$, $-2x = 9$, $x = -9/2$. Put it back into the original equation, and you get $19 = 4$. Nope, again — that's another extraneous solution.
- **Negative/positive:** $-(x - 3) + (3x + 2) = 4$, $-x + 3 + 3x + 2 = 4$, $2x + 5 = 4$, $2x = -1$, $x = -1/2$. Into the original equation it goes, and you get $4 = 4$. Tada! Your second solution.

8 Solve $3 - 4(2 - 3x) = 2(6x + 2)$. The answer is no solution.

Distribute over the parentheses on each side: $3 - 8 + 12x = 12x + 4$. Combine like terms to get $-5 + 12x = 12x + 4$ and subtract $12x$ from each side. The result is $-5 = 4$, which is false. Consequently, there is no solution.

9 Solve $2|x-3|+12=6$. The answer is no solution.

Start by isolating the absolute value: $2|x-3|+12=6; 2|x-3|=-6; |x-3|=-3$. Because an absolute value must be positive or zero, there's no solution to satisfy this equation.

10 Solve $3(2x + 5) + 10 = 2(x + 10) + 4x + 5$. The answer is all real numbers.

Begin by distributing over the parentheses on each side: $3(2x + 5) + 10 = 2(x + 10) + 4x + 5$, $6x + 15 + 10 = 2x + 20 + 4x + 5$. Next, combine like terms on each side: $6x + 25 = 6x + 25$. Subtracting $6x$ from each side gives you $25 = 25$. This is a true statement. Because all these steps are reversible, we see that all real numbers would satisfy this equation.

11 Sketch the graph of $\frac{4}{3}(6x + 2y) = 16$. See the graph for the answer.

Using slope-intercept form, multiply both sides of the equation by the reciprocal of $\frac{4}{3}$, which is $\frac{3}{4}$: $\frac{3}{4} \cdot \frac{4}{3}(6x + 2y) = \frac{3}{4} \cdot 16$. Doing so leaves you with $6x + 2y = 12$. Next, solve for y by subtracting $6x$ from each side and dividing by 2: $2y = -6x + 12$; $y = -3x + 6$. Now, because $y = -3x + 6$ is in slope-intercept form, you can identify the slope (–3) and y-intercept (6). Use these to graph the equation. Start at the y-intercept (0, 6) and move down 3 units and to the right 1 unit. Finally, connect the two points to graph the line.

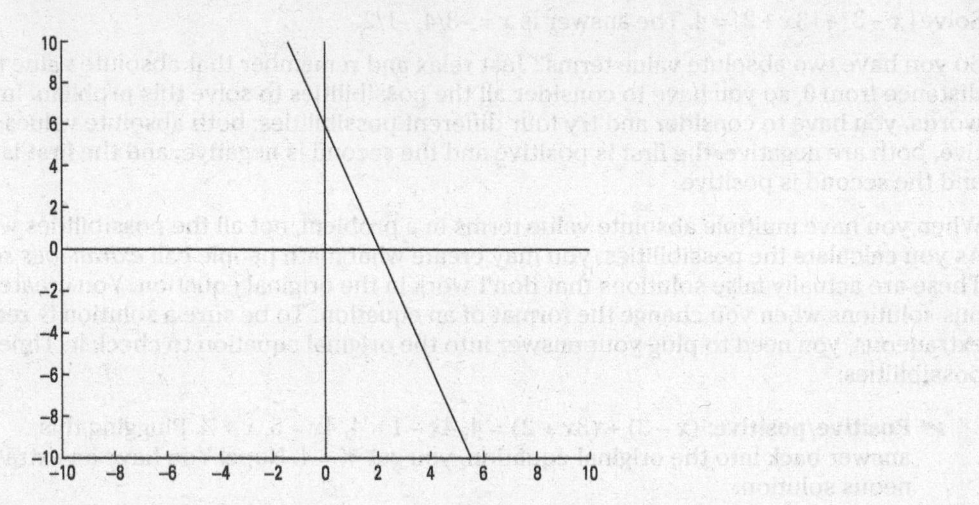

12 Sketch the graph of $\frac{5x + 4y}{2} \geq 6$. See the graph for the answer.

Start by multiplying both sides of the equation by 2: $5x + 4y \geq 12$. Next, isolate y by subtracting $5x$ from each side and dividing by 4, like so: $4y \geq -5x + 12$; $y \geq -(\frac{5}{4})x + 3$. Now that $y \geq -(\frac{5}{4})x + 3$ is in slope-intercept form, you can graph the line $y = -(\frac{5}{4})x + 3$. Because $y \geq -(\frac{5}{4})x + 3$ is a greater-than-or-equal-to inequality, draw a solid line and shade above the line.

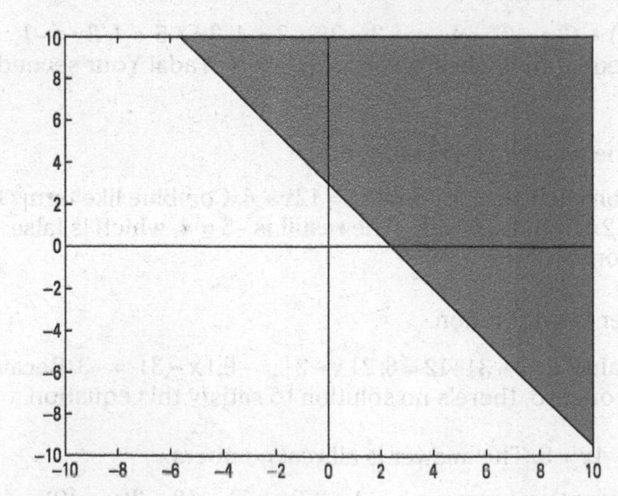

13 Sketch the graph of $4x + 5y \geq 2(3y + 2x + 4)$. See the graph for the answer.

First things first: Get the equation into slope-intercept form by distributing the 2 on the left side. Next, isolate y by subtracting $4x$ from each side, subtracting y from each side, and then dividing by -1 (don't forget to switch your inequality sign, though). Here's what your equation string should look like: $4x + 5y \geq 2(3y + 2x + 4)$; $4x + 5y \geq 6y + 4x + 8$; $5y \geq 6y + 8$; $-y \geq 8$; $y \leq -8$. Because there's no x term, you can think of this as $0x$, which tells you that the slope is 0. Therefore, the resulting line is a horizontal line at -8. Because the inequality is less than or equal to, you shade below the line.

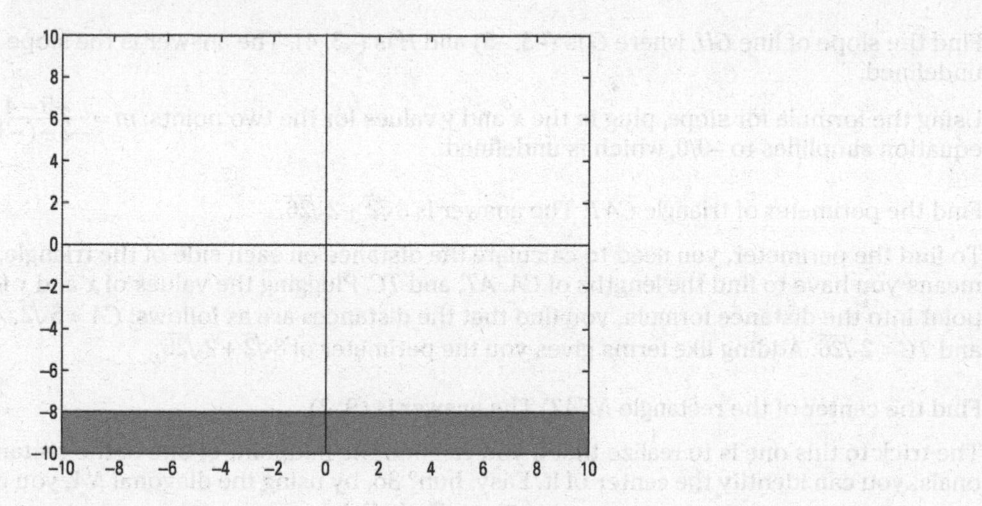

14 Sketch the graph of $x - 3y = 4 - 2y - y$. See the graph for the answer.

Simplify the equation to put it in slope-intercept form. Combine like terms and add $3y$ to each side: $x - 3y = 4 - 2y - y$; $x - 3y = 4 - 3y$; $x = 4$. Here, the resulting line is a vertical line at 4.

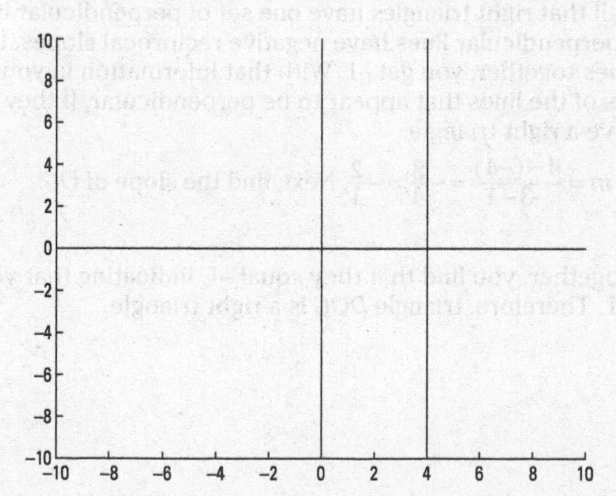

15 Find the length of segment CD, where C is (–2, 4) and D is (3, –1). The answer is $d = 5\sqrt{2}$.

Using the distance formula, plug in the x and y values: $d = \sqrt{(-2-3)^2 + [4-(-1)]^2}$. Then, simplify using the order of operations: $d = \sqrt{(-5)^2 + (5)^2} = \sqrt{25+25} = \sqrt{50} = 5\sqrt{2}$.

16 Find the midpoint of segment EF, where E is (3, –5) and F is (7, 5). The answer is M = (5, 0).

Using the midpoint formula, you get $M = \left(\frac{3+7}{2}, \frac{-5+5}{2}\right)$. Simplify from there to find that $M = (10/2, 0/2)$, M = (5, 0).

17 Find the slope of line GH, where G is (–3, –5) and H is (–3, 4). The answer is the slope is undefined.

Using the formula for slope, plug in the x and y values for the two points: $m = \frac{-5-4}{-3-(-3)}$. This equation simplifies to –9/0, which is undefined.

18 Find the perimeter of triangle CAT. The answer is $8\sqrt{2} + 2\sqrt{26}$.

To find the perimeter, you need to calculate the distance on each side of the triangle, which means you have to find the lengths of CA, AT, and TC. Plugging the values of x and y for each point into the distance formula, you find that the distances are as follows: $CA = 5\sqrt{2}$, $AT = 3\sqrt{2}$, and $TC = 2\sqrt{26}$. Adding like terms gives you the perimeter of $8\sqrt{2} + 2\sqrt{26}$.

19 Find the center of the rectangle $NEAT$. The answer is (9, 2).

The trick to this one is to realize that if you can find the midpoint of one of the rectangle's diagonals, you can identify the center of it. Easy, huh? So, by using the diagonal NA, you can find the midpoint and thus the center: $M = \left(\frac{7+11}{2}, \frac{8+(-4)}{2}\right)$. That simplifies to M = (9, 2).

20 Determine whether triangle DOG is a right triangle. The answer is yes.

This problem forces you to recall that right triangles have one set of perpendicular lines (which form that right angle) and that perpendicular lines have negative reciprocal slopes. In other words, if you multiply their slopes together, you get –1. With that information in your head, all you have to do is find the slopes of the lines that appear to be perpendicular. If they multiply to equal –1, then you know you have a right triangle.

Start by finding the slope of DO: $m = \frac{4-(-4)}{-3-1} = -\frac{8}{4} = -\frac{2}{1}$. Next, find the slope of OG: $m = \frac{-4-0}{1-9} = \frac{-4}{-8} = \frac{1}{2}$.

By multiplying the two slopes together, you find that they equal –1, indicating that you have perpendicular lines: $(-2)(\frac{1}{2}) = -1$. Therefore, triangle DOG is a right triangle.

Chapter 2

Real Numbers Come Clean

In This Chapter

- Finding solutions to equations with inequalities
- Using interval notation to express inequality
- Simplifying radicals and exponents
- Rationalizing the denominator

If fundamentals such as the order of operations (see Chapter 1) are the foundation of your pre-calc house, then the skills you pick up in Algebra I and II are the mortar between your pre-calc bricks. You need this knowledge if you're going to move forward in your pre-calculus studies. Without it, your pre-calc house — also known as your grade — is going to suffer. Never fear, though. We're here to refresh your memory.

In this chapter, we assume you know most of your algebra skills well, so we review only the tougher algebra concepts — as in the ones that give a lot of students trouble if they don't review them. In addition to reviewing inequalities, radicals, and exponents, we also introduce a purely pre-calculus idea: interval notation. (***Note:*** If you feel confident with the other review sections in this chapter, feel free to skip ahead in the book, but make sure you practice some of the interval notation problems before moving on to Chapter 3.)

Solving Inequalities

Inequalities are mathematical sentences that indicate that two expressions aren't equal. They're expressed using the following symbols:

Greater than: $>$

Greater than or equal to: $\geq$

Less than: $<$

Less than or equal to: $\leq$

Solving equations with inequalities is almost exactly the same as solving equations with equalities. There's just one key exception: multiplying and dividing by negative numbers.

When you multiply or divide each side of an inequality by a negative number, you must switch the direction of the inequality symbol. In other words, $<$ becomes $>$ and vice versa.

This is also a good time to put together two key concepts: inequalities and absolute values, or *absolute value inequalities.* With these, you need to remember that absolute values have two possible solutions: one when the quantity in the absolute value bars is positive, and one when it's negative. Therefore, you have to solve for these two possible solutions.

The easiest way to solve for the two possible solutions in absolute value inequalities is to drop the absolute value bars and apply this simple rule:

$|ax \pm b| < c$ becomes $ax \pm b < c$ AND $ax \pm b > -c$

$|ax \pm b| > c$ becomes $ax \pm b > c$ OR $ax \pm b < -c$

Need an easy way to remember this pre-calc rule? Notice the pattern: < is AND, whereas > is OR. Just think: "less thAND" and "greatOR than."

The solutions for these absolute value inequalities can be expressed graphically, as you can see in Figure 2-1. Note that the empty dots aren't included in the solutions.

$|ax \pm b| < c$

$|ax \pm b| > c$

Figure 2-1: Graphical solutions for $|ax \pm b| < c$ and $|ax \pm b| > c$.

One more trick those pesky pre-calculus professors may try and pull on you has to do with absolute value inequalities involving negative numbers. Here are two possible scenarios you may encounter:

- **If the absolute value is less than or equal to a negative number, a solution doesn't exist.** Because an absolute value must be positive or zero, it can never be less than a negative number. For example, $|2x + 3| \leq -5$ doesn't have a solution.
- **If the absolute value is greater than or equal to a negative number, there are infinitely many solutions, and the answer is "all real numbers."** An absolute value indicates a positive number or zero, which is always greater than a negative number. For instance, it doesn't matter which number you plug into the equation $|3x + 5| > -2$, you always get a true statement. Therefore, the solution to this statement is "all real numbers."

Following are a couple example questions, as well as a handful of practice problems, to reacquaint you with the process for solving inequalities.

Q. Solve for x in $5 - 2x > 4$.

A. $x < \frac{1}{2}$

Subtract 5 from each side to get $-2x > -1$. Next, divide both sides by -2 (don't forget to switch that inequality!). You wind up with $x < \frac{1}{2}$.

Q. Solve for x in $|4x + 4| - 3 \geq 9$.

A. $x \geq 2$ or $x \leq -4$

First, you have to isolate the absolute value. To do this, add 3 to both sides. Next, drop the absolute value bars and set up your two inequalities: $4x + 4 \geq 12$ *or* $4x + 4 \leq -12$. Solving each inequality algebraically, you get $x \geq 2$ or $x \leq -4$.

1. Solve for x in $|4 - 2x| > 12$.

Solve It

2. Solve for x in $x^2 - 5x - 20 > 4$.

Solve It

3. Solve for x in $|2x + 16| + 15 > 5$.

Solve It

4. Solve for x in $x^3 - 5x > 4x^2$.

Solve It

Expressing Inequality Solutions in Interval Notation

We're guessing you didn't experience interval notation in algebra. *Interval notation,* although scary sounding, is simply another way of expressing a solution set. Why have another way to write the same thing? Well, this notation is important to know because it's the one most often used in pre-calculus and calculus. And because we know you're incredibly dedicated to becoming a math wizard (uh-huh), you need to know how to cast this spell.

The key to writing a solution in interval notation is to locate the beginning and end of a set of solutions. You can do this by using interval notation or by graphing the solution in order to visualize it. After you locate your key points, you need to decide which of the following types of intervals you're dealing with:

- An **open interval** is always indicated by parentheses in interval notation. For instance, (1, 2). You show an open interval on a graph by using an open dot at the left endpoint, 1, and another open dot at the right endpoint, 2.
- A **closed interval** is always indicated by brackets in interval notation. For example, [0, 3]. To graph a closed interval, use a filled dot at the left endpoint, 0, and another filled dot at the right endpoint, 3.
- A **mixed interval** is indicated by a mix of parentheses and brackets. For instance, both [–4, 3) and (4, 6] are the interval notation for a mixed interval. (Note that [–4, 3) is the interval notation for the solution set $-4 \leq x < 3$, shown in Figure 2-2. Another way to think of this solution set is $x \geq -4$ *and* $x < 3$.) You show a mixed interval on a graph by using either a filled dot at the left endpoint and an open dot at the right endpoint, or an open dot at the left endpoint and a filled dot at the right endpoint.

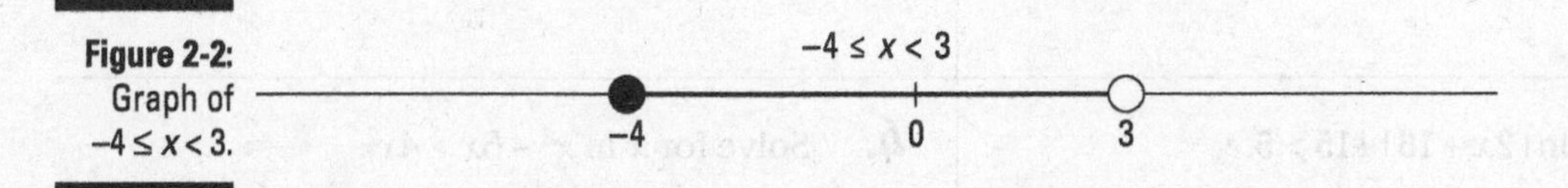

Figure 2-2: Graph of $-4 \leq x < 3$.

To indicate a solution set that includes nonoverlapping sections (also known as *disjoint sets*), you need to state all the intervals of the solution separated by the word *or.* For example, to write the solution set of $x < 2$ or $x \geq 5$ (as shown in Figure 2-3), you need to write both intervals in interval notation: $(-\infty, 2) \cup [5, \infty)$. The symbol between the two sets is the *union* symbol, ∪, and it means that the solution can belong in either interval.

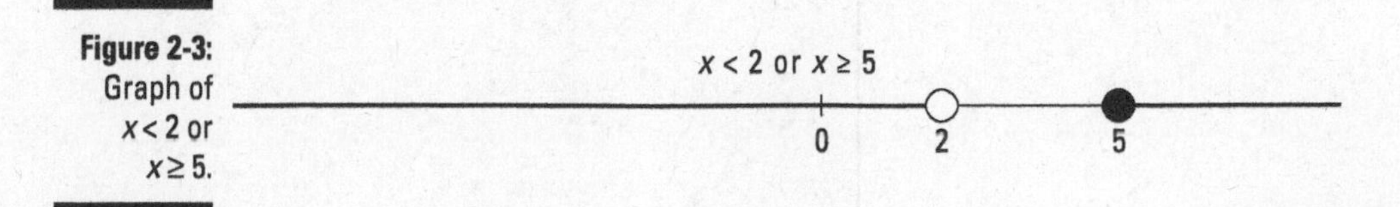

Figure 2-3: Graph of $x < 2$ or $x \geq 5$.

You always use parentheses for ∞ or –∞ because they're not real numbers.

Here's your chance to play around with interval notation to your heart's content.

Q. Write the solution for $5 - 2x > 4$ in interval notation.

A. $(-\infty, \frac{1}{2})$

There's your interval notation. To see how to find the solution set, check out the first example problem in the earlier "Solving Inequalities" section.

Q. Graph the interval set $(-2, 3] \cup (5, \infty)$ on a number line.

A. See the graph.

To create your own graph of this interval set, put your key points on a number line and then place the correct open or filled dots on your key points, depending on whether they're closed, open, or mixed intervals. Finally, thicken the interval by shading it.

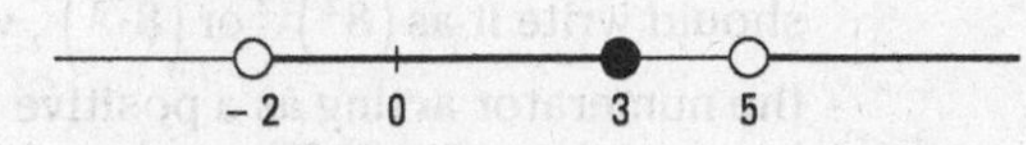

5. Write the solution for $|2x + 16| + 15 > 5$ in interval notation.

Solve It

6. Write the solution of $x^3 - 5x > 4x^2$ in interval notation and graph the solution on a number line.

Solve It

7. Graph the interval set $(-\infty, -7) \cup [-5, 2) \cup (4, \infty)$ on a number line.

Solve It

8. Graph the solution of $|2x - 1| \leq 3$.

Solve It

Don't Panic with Radicals and Exponents — Just Simplify!

Radicals and exponents (also known as *roots* and *powers*) are fundamental algebra concepts. A *radical* signifies the principal root of a number and is indicated by the radical symbol ($\sqrt{\ }$). A *root* of a number is a value that must be multiplied by itself to equal that number. For example, the second root (or *square root*) of 9 is 3 because 3 multiplied by itself is 9 ($3 \cdot 3 = 9$). However ($-3 \cdot -3$) is also equal to 9, but when dealing with the radical symbol, you take +3, the positive and principal root to be the answer rather than –3. For the same reason, the third root (or *cube root*) of 8 is 2 because 2 multiplied by itself two times is 8 ($2 \cdot 2 \cdot 2 = 8$).

When an exponent is a positive integer, it indicates the number of times a number (the base) is multiplied by itself. For example, 2 to the power of 3 is the same as $2^3 = 2 \cdot 2 \cdot 2 = 8$.

When an exponent is a rational number, like in the expression $8^{2/3}$, the exponent, ⅔, isn't equivalent to the number of times 8 is multiplied by itself. To evaluate this expression, you should write it as $\left(8^2\right)^{1/3}$ or $\left(8^{1/3}\right)^2$, which is equivalent to $\sqrt[3]{8^2}$ and $\left(\sqrt[3]{8}\right)^2$, respectively, with the numerator acting as a positive integer exponent and the denominator acting as the index of the radical. (The index of the radical is the superscript you see before the radical sign.) If the index of a radical is odd, then the radical may be negative, but if the index of a radical is even, the radical can't be negative.

Radicals and exponents are closely related to each other. In fact, they're inverse operations. To solve an equation in which the variable is under a radical, simply take the power of both sides. For example, to solve $\sqrt{x} = 4$, you need to square both sides to get $x = 16$. Similarly, you can often solve an equation in which the variable is raised to a power (or has an exponent) by taking the root of both sides. For instance, to solve $x^3 = 27$, you can take the cube root of each side $\sqrt[3]{x} = \sqrt[3]{27}$ to get $x = 3$. You can now use this simple fact to solve equations with radicals and exponents. Who's got the power now, huh?

Sometimes it's easier to solve expressions with radicals and exponents by rewriting them as *rational exponents* — exponents written as fractions. To do this, remember that the numerator of the rational exponent (the top number) is the power, and the denominator (the bottom number) is the index of the radical: $x^{m/n} = \sqrt[n]{x^m} = \left(\sqrt[n]{x}\right)^m$. So, for example, you can rewrite $\sqrt[3]{8^2}$ or $\left(\sqrt[3]{8}\right)^2$ as $8^{2/3}$.

The following example questions show you how to work through equations featuring radicals and exponents before you face the practice problems.

Q. Solve for x in $x^2 - 3x^{3/2} - 4x = 0$.

A. $x = 0, 16$

Don't let this one scare you! Just remember your basic fraction rules and look for a pattern for factoring. Start by factoring out the x. Doing so leaves you with $x\left(x - 3x^{1/2} - 4\right) = 0$. (Remember that $x^{3/2} = x \cdot x^{1/2}$. Recall that to multiply exponentials with the same base, you just add the exponents.) Now, letting y be the square root of x, you can recognize that what's left over in the parentheses is merely a polynomial: $y^2 - 3y - 4 = 0$. Factor this polynomial: $(y - 4)(y + 1) = 0$. When you recognize this, you need only deal with the fractions as the exponents. You can factor the polynomial into $x\left(x^{1/2} - 4\right)\left(x^{1/2} + 1\right) = 0$. Next, set each factor of $x\left(x^{1/2} - 4\right)\left(x^{1/2} + 1\right) = 0$ equal to 0 to find your solutions. They should look something like this:

$$x = 0 \qquad x^{1/2} = 4 \qquad x^{1/2} = -1$$

$$\left(x^{1/2}\right)^2 = 4^2$$

However, $x^{1/2} = -1$ has no solution because the principal square root of a number is always nonnegative. Hence, $x = 0, 16$. (If you need a refresher of exponential rules, flip to Chapter 5 for a quick review. To remind yourself how to solve quadratics, check out Chapter 4.)

Q. Solve for x in $\sqrt{2x-1} + 4 = x + 2$.

A. $x = 5$

Start this one by subtracting 4 from both sides of the equation to isolate the radical: $\sqrt{2x-1} = x - 2$. Next, square each side to get rid of the square root. So $\left(\sqrt{2x-1}\right)^2 = (x-2)^2$ becomes $2x - 1 = (x - 2)(x - 2)$. ***Note:*** A common mistake is to forget that $(x - 2)^2$ is $(x - 2)(x - 2)$ and not $x^2 + 4$, nor $x^2 - 4$; don't fall for this.

Multiply out the right-hand side of the equation by using FOIL — First, Outer, Inner, Last. That leaves you with $2x - 1 = x^2 - 4x + 4$. Now you can bring all the terms to the right-hand side, as in $0 = x^2 - 6x + 5$, and then factor to get $0 = (x - 5)(x - 1)$. Setting each factor equal to 0, you get two possible solutions: $x = 5$ or $x = 1$. Plug both solutions back into the original equation to check for extraneous roots (refer to Chapter 1 for more on these). When you plug in your solutions, here's what you find:

$$\begin{array}{rr} \sqrt{2(5)-1} + 4 = 5 + 2 & \sqrt{2(1)-1} + 4 = 1 + 2 \\ \sqrt{10-1} + 4 = 7 & \sqrt{2-1} + 4 = 3 \\ \sqrt{9} + 4 = 7 & \sqrt{1} + 4 = 3 \\ 3 + 4 = 7 & 1 + 4 = 3 \\ 7 = 7 & 5 = 3 \end{array}$$

Because these steps are reversible, we see that $x = 1$ is an extraneous root, and $x = 5$ is the solution!

9. Simplify $27^{4/3}$

Solve It

10. Solve for x in $x^{5/3} - 6x = x^{4/3}$

Solve It

11. Solve for x in $\sqrt{x-3}-5=0$

Solve It

12. Solve for x in $x^{8/9}=16x^{2/9}$

Solve It

13. Solve for x in $\sqrt{x-7}-\sqrt{2x-7}=-2$

Solve It

14. Solve for x in $x^{2/3}+7x^{1/3}+10=0$

Solve It

Getting Out of a Sticky Situation, or Rationalizing

Ever find yourself justifying why you deserve a day off after a pre-calc test? It may sound irrational, but rationalizing can come in handy sometimes. To simplify a radical expression, you often need to rationalize your denominators. In this section, we review and practice *rationalizing the denominator.*

First up are *monomials,* which are equations with one term in the denominator. With monomials, you're dealing with an expression and not an equation, so you need to remember equivalent fractions.

Keep in mind that a monomial is an expression, *not* an equation. You can't simply square the term to find a solution because you can't counterbalance that action. Instead, you need to multiply the numerator and denominator by the same term (which is the same as multiplying by 1). For example, if you need to rationalize the expression $\frac{3}{\sqrt{2}}$, you can multiply the expression by $\frac{\sqrt{2}}{\sqrt{2}}$, which equals 1. You then get $\frac{3\sqrt{2}}{2}$.

The same idea works for other radicals, but it requires a little more thinking. For example, if you need to rationalize the expression $\frac{2}{\sqrt[3]{5}}$, you need to multiply the numerator and denominator by $\sqrt[3]{5}$ to the second power, or by $\sqrt[3]{5^2}$, because raising a cube root to the third power cancels the root. After multiplying, you get $\frac{2\sqrt[3]{25}}{5}$.

To rationalize expressions with binomials in the denominator, you must multiply both the numerator and denominator by the conjugate. A *conjugate* is a fancy name for the binomial that gives you the difference of two squares when multiplied by the first binomial. It's found by changing the sign of the second term of the binomial. For example, the conjugate of $x + y$ is $x - y$ because when you multiply the two conjugates $(x + y)(x - y)$, you get $x^2 - y^2$, or the difference of two squares. So to rationalize a denominator with a binomial, multiply the numerator and denominator by the conjugate and then simplify. For example, to simplify $\frac{3}{2-\sqrt{3}}$, multiply the numerator and denominator by $2+\sqrt{3}$. The steps look like this:

$$\frac{3\left(2+\sqrt{3}\right)}{\left(2-\sqrt{3}\right)\left(2+\sqrt{3}\right)}=\frac{6+3\sqrt{3}}{4-3}=6+3\sqrt{3}$$

To minimize the amount of work you need to do for rationalizing denominators, rewrite your denominator in factored form (if it's not like that already) so you can identify the base numbers you're dealing with.

Q. Simplify $\frac{12}{\sqrt[3]{9}}$

A. $4\sqrt[3]{3}$

First things first: Rewrite the denominator to get $\frac{12}{\sqrt[3]{3^2}}$. Then multiply the numerator and denominator by $\sqrt[3]{3}$ to arrive at $\frac{12}{\sqrt[3]{3^2}} \cdot \frac{\sqrt[3]{3}}{\sqrt[3]{3}}$. Simplifying gives you $\frac{12\sqrt[3]{3}}{\sqrt[3]{3^3}} = \frac{12\sqrt[3]{3}}{3} = 4\sqrt[3]{3}$.

Q. Simplify $\frac{2+\sqrt{5}}{3-\sqrt{5}}$

A. $\frac{11+5\sqrt{5}}{4}$

Multiply the numerator and denominator by the conjugate of the denominator: $\frac{\left(2+\sqrt{5}\right)}{\left(3-\sqrt{5}\right)} \cdot \frac{\left(3+\sqrt{5}\right)}{\left(3+\sqrt{5}\right)}$.

Note: A common mistake is to simply distribute the second term to your conjugate in the numerator, but you need to remember that you're multiplying two binomials together: $\frac{\left(2+\sqrt{5}\right)}{\left(3-\sqrt{5}\right)} \cdot \frac{\left(3+\sqrt{5}\right)}{\left(3+\sqrt{5}\right)}$, not $\frac{2+\sqrt{5}\left(3+\sqrt{5}\right)}{\left(3-\sqrt{5}\right)\left(3+\sqrt{5}\right)}$.

If you correctly multiply out the expression by using FOIL, you should get: $\frac{6+2\sqrt{5}+3\sqrt{5}+5}{9+3\sqrt{5}-3\sqrt{5}-5}$. When you combine like terms, the final answer is $\frac{11+5\sqrt{5}}{4}$.

15. Simplify $\sqrt{\frac{3}{2x+4}}$

Solve It

16. Simplify $\frac{\sqrt{6}+\sqrt{8}}{\sqrt{10}-\sqrt{2}}$

Solve It

17. Simplify $\frac{3\sqrt[5]{2}}{2\sqrt[5]{18}}$

Solve It

18. Simplify $\frac{8}{4^{2/3}}$

Solve It

Answers to Problems on Real Numbers

This section contains the answers for the practice problems presented in this chapter. We suggest you read the following explanations if your answers don't match up with ours (or if you just want a refresher on solving a particular type of problem).

1 Solve for x in $|4 - 2x| > 12$. The answer is $x > 8$ or $x < -4$.

Start by dropping the absolute value bars and setting up your two inequalities: $4 - 2x > 12$ or $4 - 2x < -12$. Solve algebraically (be careful when you divide by that negative, though): $-2x > 8$ or $-2x < -16$; $x < -4$ or $x > 8$.

2 Solve for x in $x^2 - 5x - 20 > 4$. The answer is $x > 8$ or $x < -3$.

First, you need to recognize that you're dealing with a quadratic and recall that to solve a quadratic, you need to isolate it (for a refresher on quadratics, see Chapter 4). Subtract 4 from both sides of the equation: $x^2 - 5x - 24 > 0$. Next, factor your quadratic: $(x - 8)(x + 3) > 0$. Setting each factor to 0 gives you your key points: 8 and –3. If you put these on a number line, you can see that you have three possible solutions: less than –3, between –3 and 8, and greater than 8. All you have to do is plug in numbers in each interval to see whether you get a positive or negative number. Because you're looking for a solution that's greater than 0, you need a positive result when you multiply your factors, $(x - 8)$ and $(x + 3)$, together. In other words, you want both of your factors to be positive or both of them to be negative. Looking at the number line, you see that your solutions are $x > 8$ or $x < -3$. See? Problems like this one aren't so tough when you take the time to really think about what you're looking at.

3 Solve for x in $|2x + 16| + 15 > 5$. The answer is "all real numbers."

Subtracting 15 from both sides in order to isolate the absolute value gives you $|2x + 16| > -10$. Because you know that absolute values are positive or zero and therefore greater than any negative, you also know that the solution is "all real numbers."

4 Solve for x in $x^3 - 5x > 4x^2$. The answer is $-1 < x < 0$ or $x > 5$.

To solve this problem, gather all of your variables to one side of the equation by subtracting $4x^2$ from both sides: $x^3 - 4x^2 - 5x > 0$. Next, factor out x from each term: $x(x^2 - 4x - 5) > 0$. Then factor the quadratic: $x(x - 5)(x + 1) > 0$. Set your factors equal to 0 so you can find your key points. When you have them, put these points on a number line and plug in test numbers from each possible section to determine whether the factor would be positive or negative. Then, given that you're looking for a positive solution, think about the possibilities: (+)(+)(+) = (+), (+)(+)(–) = (–), (–)(+)(–) = (+), (–)(–)(–) = (–). Therefore, your solution is $-1 < x < 0$ or $x > 5$.

5 Write the solution for $|2x + 16| + 15 > 5$ in interval notation. The answer is $(-\infty, \infty)$.

The solution is all real numbers, and you write that in interval notation by writing it as infinity to negative infinity.

6 Write the solution of $x^3 - 5x > 4x^2$ in interval notation and graph the solution on a number line. The answer is $(-1, 0) \cup (5, \infty)$.

The graph looks like this:

$x^3 - 5x > 4x^2$

–1 0 5

7 **Graph the interval set $(-\infty, -7) \cup [-5, 2) \cup (4, \infty)$ on a number line. See the graph.**

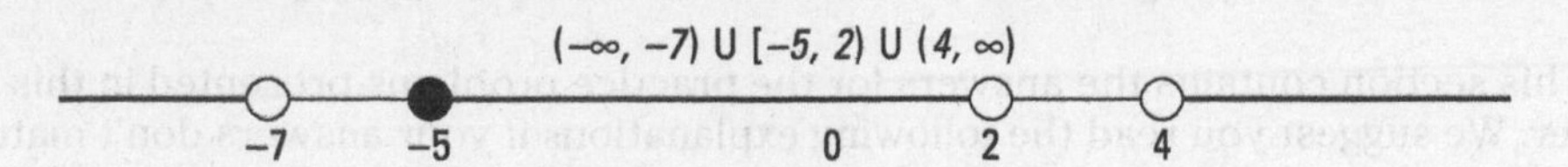

8 **Graph the solution of $|2x - 1| \leq 3$. See the graph.**

Start by dropping the absolute value sign and setting up your two inequalities: $2x - 1 \leq 3$ and $2x - 1 \geq -3$. Then solve both inequalities to find your solution: $2x \leq 4$ and $2x \geq -2$; $x \leq 2$ and $x \geq -1$. You can rewrite these solutions as $-1 \leq x \leq 2$, which you can then graph.

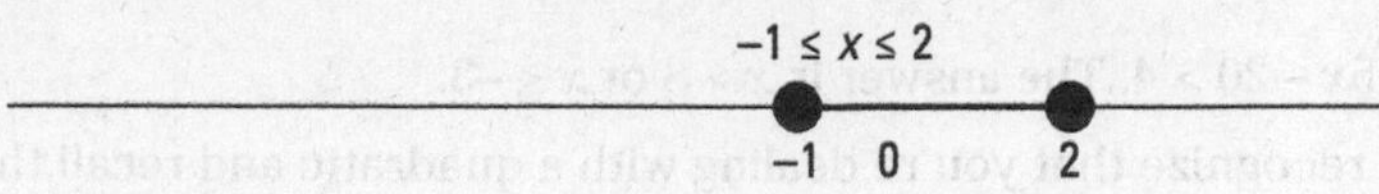

9 **Simplify $27^{4/3}$. The answer is 81.**

If you take a minute to look at this problem, you realize that you can think of it in two ways: $\sqrt[3]{(27)^4}$ and $\left(\sqrt[3]{27}\right)^4$. Either way gives you the correct answer, but one is easier to deal with than the other.

Starting with $\sqrt[3]{(27)^4}$, the order of operations tells you to take the 27 to the power of 4 first. Doing so gives you $\sqrt[3]{531{,}441}$ — and that's just plain gross.

Solving this problem is much easier when you choose to deal with it written like this: $\left(\sqrt[3]{27}\right)^4$. In this case, the order of operations tells you to take the cube root of 27, which is 3, and then take 3 to the fourth power, which is 81. Ah . . . much better.

10 **Solve for x in $x^{5/3} - 6x = x^{4/3}$. The answer is $x = 0, -8, 27$.**

Your first task is to bring all the terms to one side in descending order: $x^{5/3} - x^{4/3} - 6x = 0$. Next, factor out an x from each term: $x\left(x^{2/3} - x^{1/3} - 6\right) = 0$. The resulting expression is similar to $y^3(y^2 - y - 6)$, which factors into $y^3(y + 2)(y - 3)$. Similarly, you can factor $x\left(x^{2/3} - x^{1/3} - 6\right) = 0$ into $x\left(x^{1/3} + 2\right)\left(x^{1/3} - 3\right) = 0$, and then set each factor equal to 0 and solve. Here's the solution set:

$$x\left(x^{1/3} + 2\right)\left(x^{1/3} - 3\right) = 0$$

$$x^{1/3} + 2 = 0 \qquad x^{1/3} - 3 = 0$$

$$x^{1/3} = -2 \qquad x^{1/3} = 3$$

$$\left(x^{1/3}\right)^3 = (-2)^3 \qquad \left(x^{1/3}\right)^3 = (3)^3$$

$$x = 0 \qquad x = -8 \qquad x = 27$$

11 **Solve for x in $\sqrt{x - 3} - 5 = 0$. The answer is $x = 28$.**

Isolate the radical by adding 5 to each side of the equation to get $\sqrt{x - 3} = 5$. Then square both sides to get rid of the square root, like so: $\left(\sqrt{x - 3}\right)^2 = 5^2$. This equation simplifies to $x - 3 = 25$, which is the same as $x = 28$. It is easy to see that $x = 28$ is indeed a solution.

12 Solve for x in $x^{8/9} = 16x^{2/9}$. The answer is $x = 0, -64, 64$.

You know the drill: Bring all the terms to one side in descending order, like this: $x^{8/9} - 16x^{2/9} = 0$. Next, factor out $x^{2/9}$ from each term to get $x^{2/9}\left(x^{6/9} - 16\right) = 0$ and simplify the exponent in the parentheses to get $x^{2/9}\left(x^{2/3} - 16\right) = 0$. You can see that the expression in parentheses has the same form as $y^2 - 16$, which factors into $(y + 4)(y - 4)$. Similarly, you can factor $x^{2/9}\left(x^{2/3} - 16\right) = 0$ into $x^{2/9}\left(x^{1/3} + 4\right)\left(x^{1/3} - 4\right) = 0$.

Last but not least, set each factor equal to 0 so you can find your solution set:

$$x^{2/9} = 0 \qquad x^{1/3} + 4 = 0 \qquad x^{1/3} - 4 = 0$$

$$\left(x^{2/9}\right)^{9/2} = (0)^{9/2} \qquad x^{1/3} = -4 \qquad x^{1/3} = 4$$

$$x = 0 \qquad \left(x^{1/3}\right)^3 = (-4)^3 \qquad \left(x^{1/3}\right)^3 = (4)^3$$

$$x = -64 \qquad x = 64$$

13 Solve for x in $\sqrt{x-7} - \sqrt{2x-7} = -2$. The answer is $x = 8, 16$.

Begin by isolating one of the radicals: $\sqrt{x-7} = -2 + \sqrt{2x-7}$. Then, square both sides to get rid of that radical, $\left(\sqrt{x-7}\right)^2 = \left(-2 + \sqrt{2x-7}\right)^2$, and make sure you multiply your binomials correctly: $x - 7 = \left(-2 + \sqrt{2x-7}\right)\left(-2 + \sqrt{2x-7}\right)$. Multiplying the terms on the right side of the equation gives you $x - 7 = 4 - 4\sqrt{2x-7} + 2x - 7$. Next, isolate the remaining radical by using basic algebra: $4\sqrt{2x-7} = x + 4$.

Then, you can square both sides again to remove the remaining radical: $\left(4\sqrt{2x-7}\right)^2 = (x+4)^2$. Using algebra, multiply the two binomials and combine like terms: $16(2x - 7) = (x + 4)(x + 4)$; $32x - 112 = x^2 + 8x + 16$; $0 = x^2 - 24x + 128$. This quadratic factors into $0 = (x - 8)(x - 16)$. When you set both factors equal to 0, you get two possible solutions: $x = 8$ and $x = 16$. Plug both back into the original equation to find that both solutions work.

14 Solve for x in $x^{2/3} + 7x^{1/3} + 10 = 0$. The answer is $x = -8, -125$.

Start by recognizing that this trinomial is similar to $y^2 + 7y + 10 = 0$, which factors to $(y + 5)(y + 2) = 0$.

Similarly, $x^{2/3} + 7x^{1/3} + 10 = 0$ factors into $\left(x^{1/3} + 5\right)\left(x^{1/3} + 2\right) = 0$. By setting each one equal to 0, you can easily solve for the solutions. Just take each side to the power of 3. In other words, $x^{1/3} = -5$ becomes $\left(x^{1/3}\right)^3 = (-5)^3$, so $x = -125$. Similarly, $x^{1/3} = -2$ becomes $\left(x^{1/3}\right)^3 = (-2)^3$, so $x = -8$.

15 Simplify $\sqrt{\frac{3}{2x+4}}$. The answer is $\frac{\sqrt{6x+12}}{2x+4}$.

First, you need to separate the fraction into two radicals: one in the numerator and one in the denominator: $\frac{\sqrt{3}}{\sqrt{2x+4}}$. Next, multiply the numerator and denominator by the square root in the denominator: $\frac{\sqrt{3}}{\sqrt{2x+4}} \cdot \frac{\sqrt{2x+4}}{\sqrt{2x+4}}$. Did we trick you here? This one doesn't require the use of a conjugate because there isn't another term added to the radical. Just simplify the numerator by multiplying the radicals: $\frac{\sqrt{3(2x+4)}}{2x+4} = \frac{\sqrt{6x+12}}{2x+4}$.

16 **Simplify $\frac{\sqrt{6}+\sqrt{8}}{\sqrt{10}-\sqrt{2}}$. The answer is $\frac{\sqrt{15}+\sqrt{3}+2\sqrt{5}+2}{4}$.**

Multiply the numerator and denominator by the conjugate of the denominator to get $\frac{\left(\sqrt{6}+\sqrt{8}\right)}{\left(\sqrt{10}-\sqrt{2}\right)}\cdot\frac{\left(\sqrt{10}+\sqrt{2}\right)}{\left(\sqrt{10}+\sqrt{2}\right)}$ and then use FOIL to multiply the binomials in the numerator and denominator; the result should be $\frac{\sqrt{60}+\sqrt{12}+\sqrt{80}+\sqrt{16}}{10-2}$. Simplify each radical: $\frac{2\sqrt{15}+2\sqrt{3}+4\sqrt{5}+4}{8}$. Finally, because each term in the numerator and denominator is divisible by 2, divide both terms by 2: $\frac{\sqrt{15}+\sqrt{3}+2\sqrt{5}+2}{4}$.

17 **Simplify $\frac{3\sqrt[5]{2}}{2\sqrt[5]{18}}$. The answer is $\frac{\sqrt[5]{3^3}}{2}$.**

Begin by factoring the denominator: $\frac{3\sqrt[5]{2}}{2\sqrt[5]{2\cdot 3^2}}=\frac{3\sqrt[5]{2}}{2\sqrt[5]{2}\cdot\sqrt[5]{3^2}}$. Notice the $\sqrt[5]{2}$ in both the numerator and denominator? Cancel them so you have one less term to worry about. Next, multiply the numerator and denominator of $\frac{3}{2\sqrt[5]{3^2}}$ by $\sqrt[5]{3^3}$ to eliminate the radical in the denominator: $\frac{3}{2\sqrt[5]{3^3}}\cdot\frac{\sqrt[5]{3^3}}{\sqrt[5]{3^3}}$. Multiply through to get $\frac{3\sqrt[5]{3^3}}{2\sqrt[5]{3^5}}=\frac{3\sqrt[5]{3^3}}{2\cdot 3}$ and then cancel the 3 from the numerator and denominator to arrive at your final answer, which should be $\frac{\sqrt[5]{3^3}}{2}$.

18 **Simplify $\frac{8}{4^{2/3}}$. The answer is $2\sqrt[3]{4}$.**

Change the fractional exponent into a radical to get $\frac{8}{\sqrt[3]{4^2}}$ and then multiply the numerator and denominator by one more cube root of 4: $\frac{8}{\sqrt[3]{4^2}}\cdot\frac{\sqrt[3]{4}}{\sqrt[3]{4}}$. Multiply to get $\frac{8\sqrt[3]{4}}{4}$ and simplify the resulting fraction.

Here's your answer: $2\sqrt[3]{4}$.

Chapter 3

Controlling Functions by Knowing Their Function

In This Chapter

- Determining whether a function is even or odd
- Introducing parent functions and revealing how to graph them
- Graphing rational functions and piece-wise functions
- Performing operations on functions and finding their domains and ranges
- Working with the inverses of functions

By this point in your math studies, you're familiar with the *coordinate plane* (it's what you get when two number lines meet at a 90-degree angle). You know that the horizontal axis is called the x-axis and that the vertical one is called the y-axis. You also know that each point, or *ordered pair*, on the plane is named (x, y). But did you know that a set of ordered pairs is called a *relation?* The *domain* of the relation is the set of all the x values, and the *range* is the set of all the y values. For convenience, the domain variable always precedes the range variable in order and alphabetically in this book.

A *function* is a relation where every x in the domain pairs with one (and only one) y in the range. Think of a function as a computer. Domain is input, and range is output. You can't put input in a computer and get out different outputs (if you do, your computer's broken). A common symbol for a function is $f(x)$, and it translates into "f of x." We explore the concept of functions, as well as some of their properties, in this chapter.

Using Both Faces of the Coin: Even and Odd

If you've ever taken an art class, you've probably heard the term *symmetry.* It means that what you're seeing is balanced, with equal or similar parts on both sides of the object, be it a painting, sculpture, or photograph. A graph can be symmetrical as well. The graph of a relation can have any of the following three basic types of symmetry:

- ***Y*-axis symmetry:** Each point on the left side of the y-axis is mirrored by a point on the right side (and vice versa).
- ***X*-axis symmetry:** Each point above the x-axis is mirrored by a point below it (and vice versa).

- **Origin symmetry:** A graph has this type of symmetry if it's unchanged when reflected across both the x-axis and y-axis. In other words, if you turn a graph with origin symmetry upside down and then flip it left to right, it looks exactly the same.

In pre-calculus, functions take the idea of symmetry and use different terms to describe the same idea.

- **Even function:** This is a function whose graph is symmetrical with respect to the y-axis. Basically, each input x and the opposite input $-x$ give the same y value, which means $f(x) = f(-x)$.
- **Odd function:** This is any function whose graph is symmetrical with respect to the origin. In plain English, each x value gives a y value, and its opposite $-x$ gives the opposite $-y$, which means that $f(-x) = -f(x)$.

Q. Determine whether $f(x) = x^4 - x^2$ is even, odd, or neither.

A. This function is even.

Replace x with $-x$ in the equation and see what happens: $f(-x) = -x^4 - (-x)^2$. A negative number to an even power is a positive number, so $f(-x) = x^4 - x^2$. Because you get the same exact function as the original one, this function is even.

1. Is $f(x) = x^3 - 1$ even, odd, or neither?

Solve It

2. Does the given graph appear to be even, odd, or neither?

Solve It

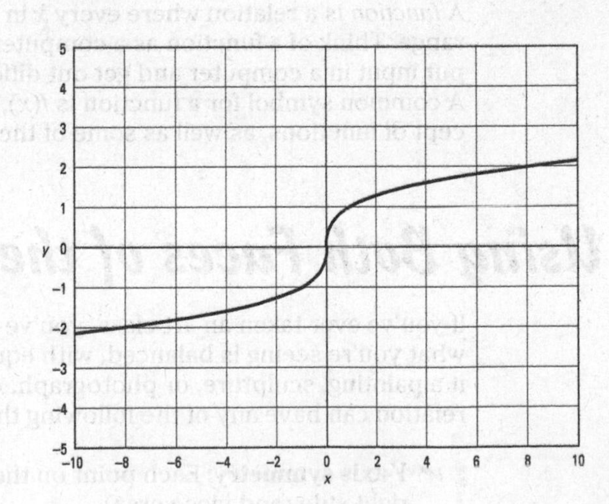

3. Sketch half the graph of $f(x) = \sqrt{x^2 - 4}$ and use symmetry to complete the graph.

Solve It

4. Sketch half the graph of $f(x) = 4x^3$ and use symmetry to complete the graph.

Solve It

Leaving the Nest: Transforming Parent Graphs

You see certain functions over and over again in pre-calc, and sometimes you're even asked to graph them. You can use the plug-and-chug method to graph functions, but the more complicated the function, the longer it's going to take you to graph it using this method. That's why we recommend you memorize a few basic graphs of functions, called *parent graphs*.

Common parent graphs include quadratic functions, square roots, absolute values, cubics, and cube roots. Moving these basic graphs around the coordinate plane is known as *transforming the function* and is easier than using the plug-and-chug method. Several types of function transformations exist:

- Horizontal transformations
- Vertical transformations
- Reflections
- Horizontal translations
- Vertical translations

In the sections that follow, we take a look at each parent function and then show you how to transform it. ***Note:*** Even though in most sections we take a look at only one function when discussing the transformations, the rules we present apply to *all* functions in the same way. So if we talk about a quadratic function in the section on vertical transformations, that's not the only function that has vertical transformations — they all do.

Quadratic functions

Quadratic functions are defined by second-degree *polynomials* (any expression with more than one term in it; see Chapter 4), and the highest exponent on any one variable is 2. The parent quadratic function is $f(x) = x^2$, and its graph is known as a *parabola* (a type of conic section; head to Chapter 12 for the scoop on these). Furthermore, $f(x) = x^2$ is an even function, so the graph of the parent quadratic function is symmetric with respect to the y-axis.

Begin the graph at the *vertex,* which is also known as the *origin;* here the vertex is the point (0, 0). To get to another point on the right portion of the graph, move right 1 and up 1. From there, another point is over 1, up 3 (= $f(2) - f(1)$); continuing from there, another point is over 1, up 5 (= $f(3) - f(2)$). For this particular function, in order to plot points on the right portion of the graph, you can always move right 1 and up to the next odd number. Then you can use symmetry to complete the graph. Check out Figure 3-1 for the graph of the parent quadratic function.

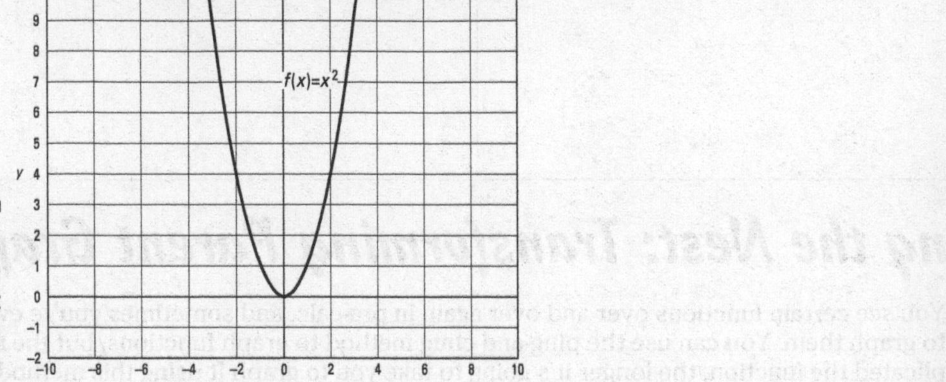

Figure 3-1: Graphing the parent quadratic function.

Square root functions

Not surprisingly, *square root functions* feature square roots. The parent square root function is $g(x) = \sqrt{x}$. As you can see in Figure 3-2, this graph looks like half a parabola that's turned on its side. Because the square root of a negative number isn't a real number, you can't have x values that are negative, which is why the parent graph of the square root function doesn't cross into the left side of the coordinate plane. (For more on the square roots of negative numbers, flip to Chapter 11.) The graph begins at the origin (0, 0) and then moves up 1, right 1; up 1, right 3; up 1, right 5; and so on.

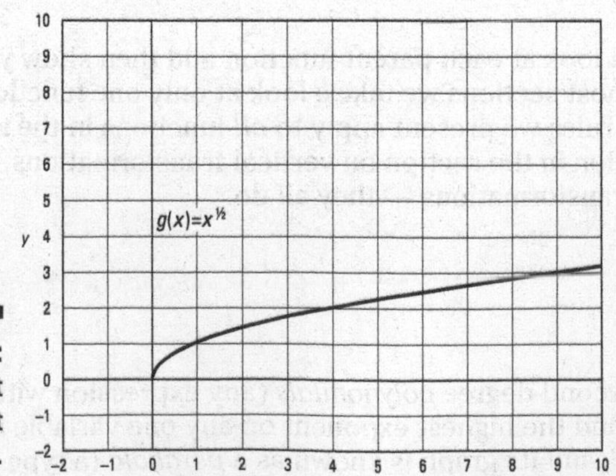

Figure 3-2: The parent square root function.

Absolute value functions

The parent *absolute value function* is $h(x) = |x|$. You should recognize the absolute value bars and know that this function always gives the distance from the origin, so it always gives a nonnegative output. Furthermore, it is an even function, so the graph of the parent absolute value function is symmetric with respect to the *y*-axis.

To graph this function, begin at the vertex (0, 0) and move right 1 and up 1. From there, still working on the right-hand side of the coordinate plane, move over 1 and up 1, over 1 and up 1. Continue this pattern to complete the right side of the graph; use symmetry to complete the left half of the graph (see Figure 3-3).

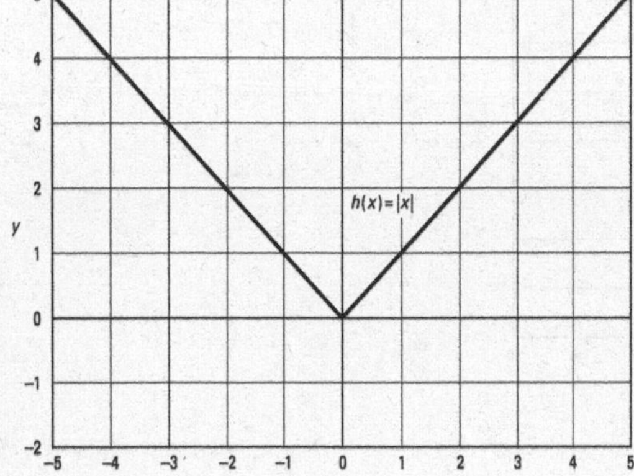

Figure 3-3: The parent absolute value function.

Cubic functions

Cubic functions are defined by third-degree polynomials, so the highest exponent on any one variable is 3. The parent cubic function is $p(x) = x^3$, which is an odd function. If you turn its graph upside down and flip from left to right (see Figure 3-4), it looks exactly the same. If you start the graph at the origin (0, 0), a point on the graph to the left of (0, 0) is over 1 and down 1, and a point on the graph to the right of (0, 0) is over 1 and up 1.

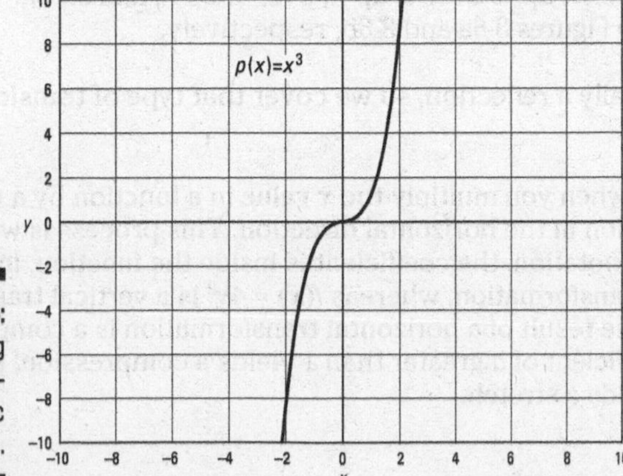

Figure 3-4: Graphing the parent cubic function.

Cube root functions

Cube root functions are related to cubic functions similar to the way that quadratic and square root functions are related. The parent cube root function is $r(x) = \sqrt[3]{x}$. Like the cubic function (see the preceding section), from the origin, a point to the left on the graph is over 1 and down 1, and a point to the right on the graph is over 1 and up 1. However, the graph of the parent cube root function is longer than it is tall. Take a look at Figure 3-5 to see for yourself.

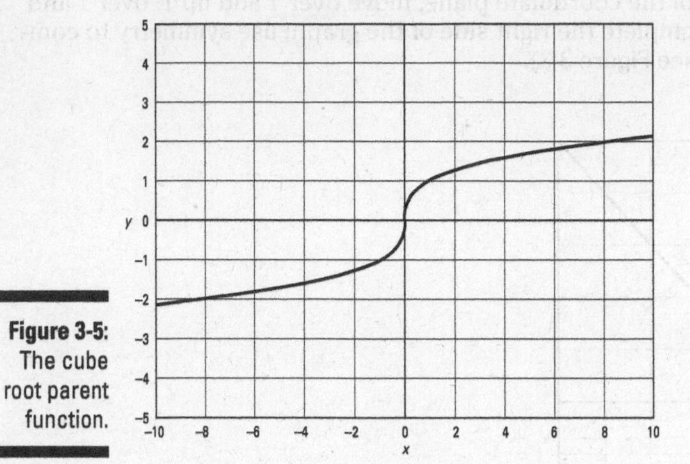

Figure 3-5: The cube root parent function.

Transformations

Pre-calculus features two kinds of transformations: horizontal and vertical.

You create a *vertical transformation* when you multiply a function by a constant because doing so changes the graph of the function in the vertical direction. This process is written as $af(x)$. Think of the result as a vertical stretch or a vertical shrink. A coefficient of a greater than 1 yields a stretch, and a coefficient of a between 0 and 1 yields a shrink. For example, $f(x) = 2x^2$ multiplies each y-value by 2. So from the vertex, instead of moving over 1 and up 1, you now move over 1 and up 2. From there, instead of moving over 1 and up 3, you now move over 1 and up 6. You get the idea. For another example, if $g(x) = \frac{x^2}{4}$, you move in this manner: From the vertex, go over 1, up $\frac{1}{4}$; over 1, up $\frac{3}{4}$; over 1, up $\frac{5}{4}$; and so on. The graphs of $f(x)$ and $g(x)$ are shown in Figures 3-6a and 3-6b, respectively.

Note: A negative coefficient is actually a reflection, so we cover that type of transformation in the later "Reflections" section.

A *horizontal transformation* occurs when you multiply the x value in a function by a constant that changes the graph of the function in the horizontal direction. This process is written as $f(a \cdot x)$. As you can see from this notation, the coefficient is inside the function. In other words, $f(x) = (4x)^2$ is a horizontal transformation, whereas $f(x) = 4x^2$ is a vertical transformation (see the preceding section). The result of a horizontal transformation is a compression or a stretch along the x-axis. A coefficient of a greater than 1 yields a compression, and a coefficient of a between 0 and 1 yields a stretch.

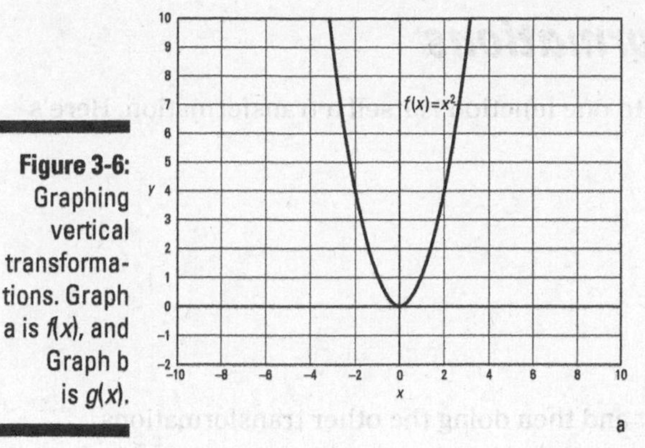

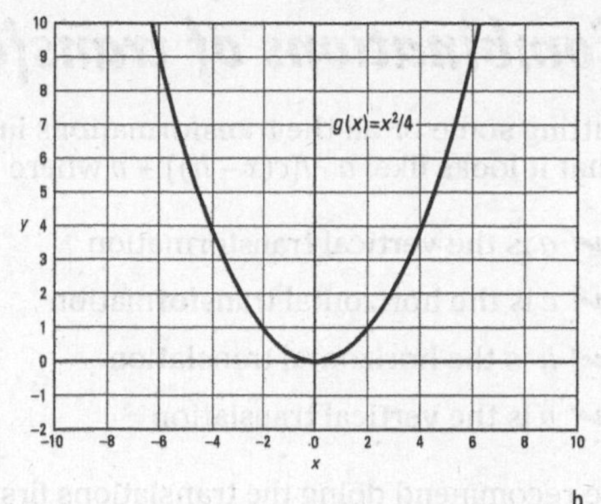

Figure 3-6: Graphing vertical transformations. Graph a is $f(x)$, and Graph b is $g(x)$.

To sketch the graph of a horizontal transformation of a parent function, set the expression inside the function equal to 1 and solve for x. So if $g(x)=|3x|$, then set $3x = 1$ and solve for $x = \frac{1}{3}$. From the origin, you move over ⅓, up 1; over ⅓, up 1; and over ⅓, up 1.

Translations

An action that moves a graph horizontally or vertically on the coordinate plane is called a *translation.* The result of a translation is that every point on the parent graph is moved right, left, up, or down by the same amount. In the next sections, we take a closer look at each kind of translation, or *shift.*

Vertical shifts

Adding a number to or subtracting a number from a given function is a *vertical shift* and is written as $f(x) + v$, where v is the vertical shift. For example, $p(x) = x^3 - 1$ moves the parent cubic function down by 1, and $p(x) = x^3 + 4$ moves the parent cubic function up by 4.

Horizontal shifts

Adding a number to or subtracting a number from the function's independent variable (input) is a *horizontal shift.* This type of shift is always written in the form $f(x - h)$. For example, $h(x)=\sqrt{(x-2)}$ moves the parent square root function to the right by 2, and $h(x)=\sqrt{(x+3)}$ moves the parent square root function to the left by 3.

Reflections

Reflections take the parent function and reflect it over a horizontal or vertical line. When the vertical transformation coefficient is negative, the function is flipped upside down over the x-axis. For example, $f(x)= -5x^2$ affects the vertical transformation by a factor of 5 and turns the graph upside down at the same time. If the horizontal transformation is negative, the function is flipped backward over the y-axis. So, for example, $h(x)=\sqrt{(-x)}$ turns the function to the left rather than to the right.

Combinations of transformations

Putting some or all the transformations into one function is itself a transformation. Here's what it looks like: $a \cdot f[c(x-h)] + v$ where

- a is the vertical transformation
- c is the horizontal transformation
- h is the horizontal translation
- v is the vertical translation

We recommend doing the translations first and then doing the other transformations.

Following are two examples of how to transform the parent graph of a function, followed by six practice questions. You should be prepared to face these puppies on a test if you try solving all these problems.

Q. Graph the function $f(x) = (x-3)^2$ by transforming the parent graph.

A. See the graph. This transformation is done in one step. Because the constant is subtracted inside the quadratic function, you recognize it as a horizontal shift to the right by 3. Take the parent quadratic function and move each point to the right by 3, as shown in the graph.

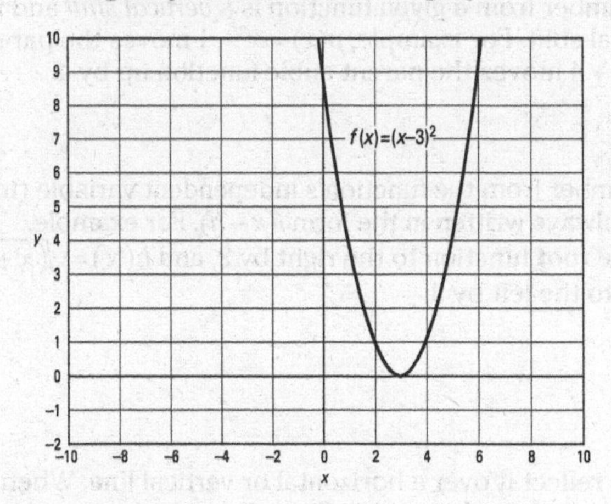

Q. Sketch the graph of $g(x) = \sqrt{3-x} + 1$ by transforming the parent function.

A. See the graph. This one takes some work before you can begin graphing it. You first have to rewrite it in the proper form before you can recognize the various transformations to the parent square root function. Rewrite the stuff inside the square root first so it's in the right order: $g(x) = \sqrt{-x+3} + 1$. Next, factor out the leading coefficient to get the horizontal transformation: $g(x) = \sqrt{-1(x-3)} + 1$. Looks like the graph is flipped horizontally. Notice that factoring out the coefficient affects the horizontal translation — it's to the right by 3, and the vertical translation is up 1. These transformations can be seen in the final graph.

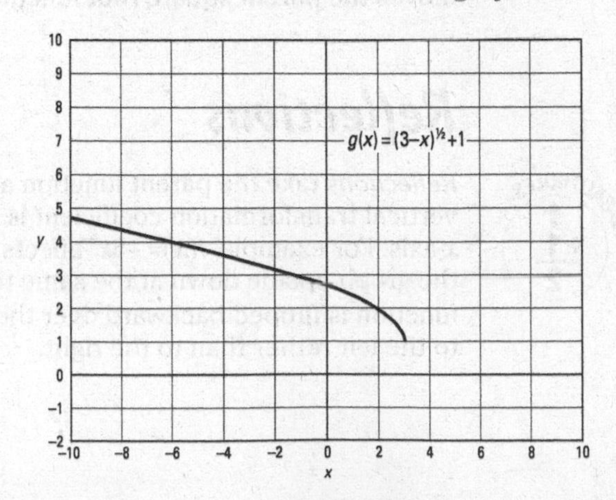

5. Graph the function $a(x) = -2(x-1)^2$.

Solve It

6. Graph the function $b(x) = |x+4| - 1$.

Solve It

7. Graph the function $c(x) = \sqrt{x+3}$.

Solve It

8. Graph the function $f(x) = -x^2 - 6x$.

Solve It

Given the graph of the function g(x) in the following figure, sketch the graph of the functions in Questions 9 and 10.

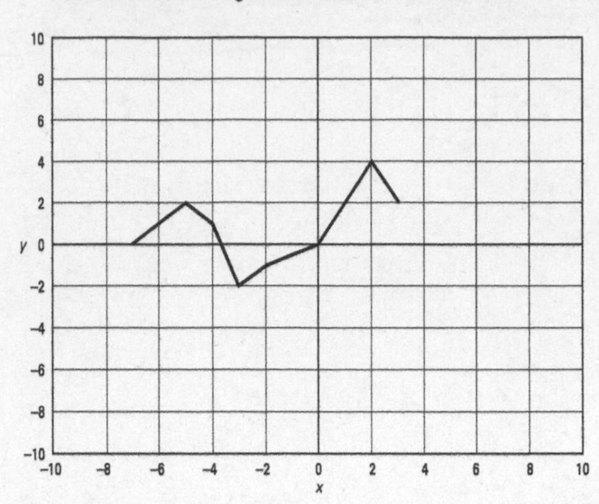

9. $g(x) - 3$

Solve It

10. $2g(x - 1)$

Solve It

Graphing Rational Functions

A *rational function,* such as $f(x) = \frac{(x+1)}{(x-3)^2}$, is the quotient of two polynomials, excluding division by zero (we cover polynomials in Chapter 4). By this point in your math studies, you know that when the denominator of a fraction is 0, the result is undefined. The same is true for rational functions. Because the denominator polynomial (usually) has a variable, it's possible that certain values of the variable — like $x = 3$ for $f(x) = \frac{(x+1)}{(x-3)^2}$ — will make the denominator 0. When a rational function has values that make it undefined, the graph of that function may feature a *vertical asymptote,* which is a vertical line that the graph never crosses.

To find the vertical asymptote, if there is one, first simplify the rational function to ensure that numerator and denominator don't have any common factors other than constants (no value will simultaneously make the numerator and denominator polynomials zero); then set the denominator equal to 0 and solve it. These solutions (roots) identify locations for vertical asymptotes.

Some rational functions also have a *horizontal asymptote,* which is a horizontal line that a graph approaches. To find the horizontal asymptote, take a look at both the numerator's degree and the denominator's degree. (If you don't recall how to find the degree of a polynomial, head to Chapter 4.) Here are the three possibilities for horizontal asymptotes:

- **The degree of the denominator is greater:** The bottom of the fraction is getting bigger faster, and the fraction goes to 0 as x gets larger. Your horizontal asymptote is the x-axis, or $y = 0$.
- **The degree of both is the same:** The top and bottom of the fraction are moving at the same rate. The quotient of the leading coefficients gives you the horizontal asymptote.
- **The degree of the numerator exceeds the degree of the denominator:** The top of the fraction is getting bigger faster. In short, as x gets larger, so does y; therefore, there's no horizontal asymptote.

Work through these example questions before trying your hand at graphing rational functions.

Q. Graph the function $f(x) = \dfrac{3x-1}{4-x}$

A. See the graph. First, find the vertical asymptote (if there is one) by setting the denominator equal to 0 and solving. If $4 - x = 0$, then $x = 4$ and the numerator is not zero when $x = 4$. Draw a coordinate plane and add in a dotted vertical line at $x = 4$ to mark your vertical asymptote. Now, look at the numerator and the denominator; the degree on each is one. Divide the leading coefficients to find the horizontal asymptote. In this case, the numerator's leading coefficient is 3, and the denominator's leading coefficient is −1. This means your horizontal asymptote is $y = 3/{-1} = -3$.

Now that you have both asymptotes, use them to help you get the graph. The vertical asymptote divides your domain into two intervals: $(-\infty, 4)$ and $(4, \infty)$. Pick a couple of x-values on each interval and plug them into the function to determine whether the graph lives above or below the horizontal asymptote. For example, if $x = -5$, then $y = -1.77$. And if $x = 0$, then $y = -0.25$. If you graph those two points, you see that they're both above the horizontal asymptote. Keep checking points such as x and y intercepts and looking at x values close to the vertical asymptote until you have a good idea of what the graph looks like.

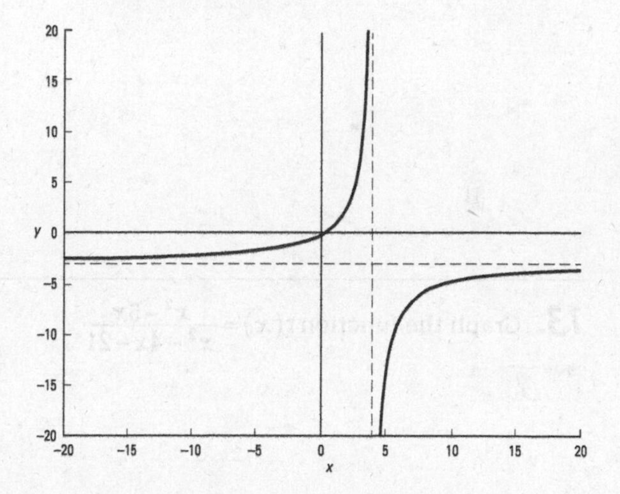

Q. Graph the function $g(x)=\frac{2x-6}{x^2+4}$

A. See the graph. When you try to find the vertical asymptote, you notice that $x^2 + 4 = 0$ doesn't have a solution because $x^2 = -4$ has no solution (in real numbers anyway). Also, because the denominator has a bigger degree, the horizontal asymptote is the x-axis, or $y = 0$. However, by setting the numerator equal to 0, you do get a solution: $2x - 6 = 0$; $2x = 6$; $x = 3$. This means the graph crosses the x-axis at $x = 3$. Because there's no vertical asymptote, use $x = 3$ to give you the intervals to look at to get the graph.

On the first interval $(-\infty, 3)$, y is negative, and the whole graph is below the horizontal asymptote. On the next interval $(3, \infty)$, y happens to be positive, and the function is above the horizontal asymptote. If you pick x values bigger than 3 that keep getting bigger, you see y increase slowly and then decrease again and get closer and closer to 0. This gives you the graph of this function.

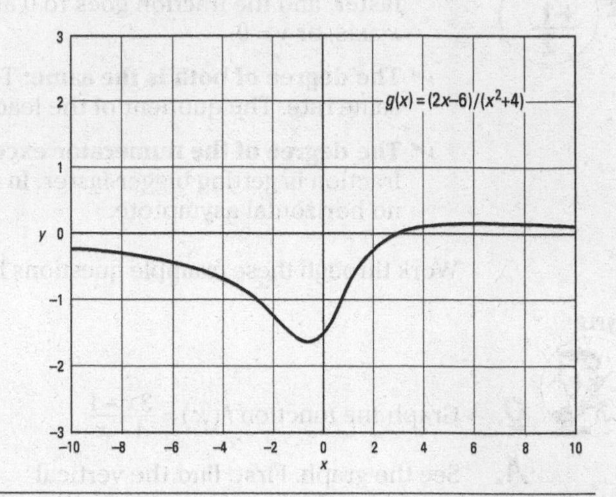

11. Graph the function $q(x)=\frac{-2}{(x-4)(x+5)}$

Solve It

12. Graph the function $r(x)=\frac{x+3}{x^2-x-6}$

Solve It

13. Graph the function $t(x)=\frac{x^2-5x}{x^2-4x-21}$

Solve It

14. Graph the function $u(x)=\frac{x^2-10x-24}{x+1}$

Solve It

Piecing Together Piece-Wise Functions

A *piece-wise function* is broken into pieces. In other words, it actually contains several functions, each of which is defined on a restricted interval. The output depends on what the input is. The graphs of these functions may look like they've been broken into pieces. Because of this broken quality, a piece-wise function that jumps is called *discontinuous*.

Q. Graph $f(x)=\begin{cases} x^2+2 & \text{if } x\leq 1 \\ 3x-1 & \text{if } x>1 \end{cases}$

A. See the graph. This function has been broken into two pieces: When $x \leq 1$, the function follows the graph of the quadratic function, and when $x > 1$, the function follows the graph of the linear function. (Notice the hole in this second piece of the graph to indicate that the point isn't actually there.)

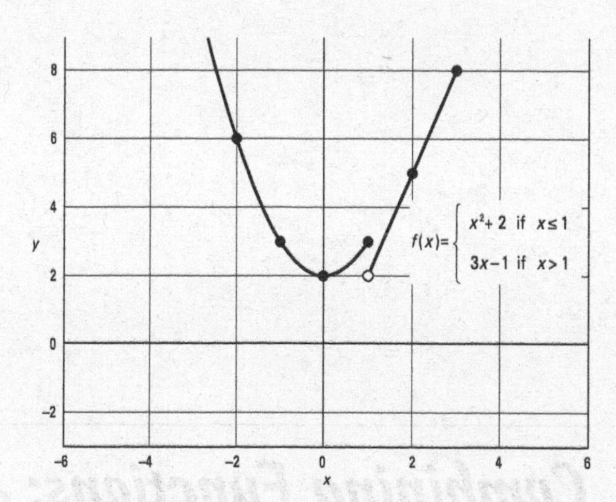

15. Graph $g(x)=\begin{cases} \sqrt{x+3} & \text{if } x\leq -1 \\ (x+3)^2 & \text{if } x>-1 \end{cases}$

Solve It

16. Graph $h(x)=\begin{cases} \frac{1}{2}x-4 & \text{if } x\leq -2 \\ 3x+3 & \text{if } -2<x<2 \\ 4-x & \text{if } x\geq 2 \end{cases}$

Solve It

17. Graph $m(x)=\begin{cases} x^3+2 & \text{if} \quad x<0 \\ x^2+2 & \text{if} \quad 0\le x<2 \\ x+2 & \text{if} \quad x\ge 2 \end{cases}$

Solve It

18. Graph $n(x)=\begin{cases} |x-1| & \text{if} \quad x<-3 \\ -3 & \text{if} \quad x=-3 \\ |x|-1 & \text{if} \quad x>-3 \end{cases}$

Solve It

Combining Functions: A Guaranteed Successful Outcome

You've come to know (and, heck, maybe even love) the four basic operations in math: addition, subtraction, multiplication, and division. Well, in pre-calculus, you take functions and add, subtract, multiply, and divide them in order to create a brand-new function — a process that's sometimes called *combining functions.*

If you're asked to graph a combined function without using a graphing calculator, you have to plug and chug your way through it by picking plenty of *x*-values to make sure you get an accurate representation of the graph. You may also be asked to find one specific value for a combined function — you get an *x* value and you just plug it in and see what happens.

For all the example and practice questions that follow in this section, you use three functions:

- $f(x) = x^2 - 6x + 2$
- $g(x) = 2x^2 - 5x$
- $h(x)=\sqrt{3x+2}$

Q. Find $(f-g)(x)$.

A. $(f-g)(x) = -x^2 - x + 2$

Because these two functions are both polynomials, solving this problem is really about collecting like terms and subtracting them. (Just be sure to watch your negative signs!) Here's the math: $(f-g)(x) = (x^2 - 6x + 2) - (2x^2 - 5x) = x^2 - 6x + 2 - 2x^2 + 5x = -x^2 - x + 2$.

19. Find $(f + h)(x)$.

Solve It

20. Find $(f \cdot g)(x)$.

Solve It

21. Find $(\frac{f}{g})(x)$. Does this new function have any undefined values?

Solve It

22. Find $(g + h)(x)$.

Solve It

Evaluating Composition of Functions

When you place one function inside another one, you're creating a *composition of functions.* For instance, if you have two functions, $f(x)$ and $g(x)$, then the composition $f(g(x))$ takes g and puts it in place of x in the formula for $f(x)$. This is also written as $(f \circ g)(x)$, and it's basically read right to left; the g function goes into the f function.

For convenience, we use the following same three functions we present in the preceding section to illustrate how to evaluate the functions that result from composition of functions:

- $f(x) = x^2 - 6x + 2$
- $g(x) = 2x^2 - 5x$
- $h(x) = \sqrt{3x+2}$

Eager to start working with compositions of functions? Skip straight to the practice problems and then head to the end of this chapter to check your answers.

Q. Find $f(h(x))$.

A. $3x+4-6\sqrt{3x+2}$

Start by substituting the entire h function for every x in the f function: $\left(\sqrt{3x+2}\right)^2 - 6\sqrt{3x+2} + 2$. A square root and a square cancel each other out: $3x+2-6\sqrt{3x+2}+2$. Simplify by combining any like terms: $3x+4-6\sqrt{3x+2}$.

23. Find $(f \circ g)(x)$.

Solve It

24. Find $(g \circ f)(x)$.

Solve It

25. Find $h(f(x))$.

Solve It

26. Find $(f \circ f)(x)$.

Solve It

27. Find $f(g(-1))$.

Solve It

28. Find $g(h(3))$.

Solve It

Working Together: Domain and Range

When you're combining and composing functions (see the two preceding sections) and you want to find out what's happening to the domain and range of the new function, you must keep in mind that in a function, the domain is the input, usually x, and the range is the output, usually y. The truth is that the domain of the given function depends on the operation being performed and the original functions. It's possible that something changed, and it's also possible that nothing did. Typically in pre-calc, you're asked to find the domain of a combined function and not the range simply because determining the range is often more complicated.

Pre-calc teachers and textbooks talk a lot about two functions whose domains aren't all real numbers:

- **Rational functions:** The denominator of any fraction can't be 0, so it's possible that some rational functions are undefined because of this fact. Set the denominator equal to 0 and solve to find the restrictions on your domain.
- **Square root functions (or any even root):** The *radicand* (what's under the root sign) can't ever be negative, a fact that affects domain. To find out how, set the radicand greater than or equal to 0 and solve. The solution to this inequality is your domain.

Undefined values are also called *excluded values,* so be on the lookout for your textbook to use that terminology as well. When you're asked to find the domain of a combined function, take your time. We can't give you one rule that works all the time for finding a combined function's domain. You just have to take a look at both of the original functions and ask yourself whether their domains have any restrictions. If they do, these restrictions often carry through and may be added to those additional restrictions arising from the combined function.

In the following example and problems, you're asked to use those same three functions you've been using for the last two sections to determine domains of some functions resulting from various combinations or compositions of those three functions:

$f(x) = x^2 - 6x + 2$ $\quad g(x) = 2x^2 - 5x$ $\quad h(x) = \sqrt{3x+2}$

Q. Find the domain of $f(h(x))$.

A. The domain is all numbers greater than or equal to $-\frac{2}{3}$.

Take a look at the original two functions first. Because $f(x)$ is a polynomial, there are no restrictions on the domain. However, $h(x)$ is a square root function, which means the radicand has to be positive: $3x + 2 \geq 0$; $3x \geq -2$; $x \geq -\frac{2}{3}$. The new, combined function must honor this domain as well.

29. Find the domain of $(f \circ g)(x)$.

Solve It

30. Find the domain of $h(f(x))$.

Solve It

31. Find the domain of $(f + h)(x)$.

Solve It

32. Find the domain of $(f/g)(x)$.

Solve It

Unlocking the Inverse of a Function: Turning It Inside Out

An *inverse function* undoes what a function does. You've seen inverse operations before: Addition undoes subtraction, and division undoes multiplication. It shouldn't surprise you, then, that functions have inverses. If $f(x)$ is the original function, then $f^{-1}(x)$ is the symbol for the inverse. This notation is used only for the inverse function and is *never* meant to represent $\frac{1}{f(x)}$.

In the course of your pre-calculus studies, you'll be asked to do three main things with inverses:

- Given a function, graph its inverse.
- Find the inverse of a given function.
- Show that two functions are inverses of each other.

In all three cases, it's all about input and output. If (a, b) is a point in the original function, then (b, a) is a point in the inverse function. Domain and range swap places from a function to its inverse. So if you're asked to graph the inverse function, graph the original and then swap all x and y values in each point to graph the inverse. To find the inverse of a given function, literally take x and y, or $f(x)$ and switch them. After the swap, change the name to the symbol for an inverse function, $f^{-1}(x)$, and solve for the inverse. Lastly, to show that two functions, $f(x)$ and $g(x)$, are inverses of each other, place one inside the other by using composition of functions, $f(g(x))$, and simplify to show that you get x. Then do it the other way around with $g(f(x))$ to make sure it works both ways.

In case that seems a bit confusing, here are a couple example questions so you can see what we mean. We suggest you review them before moving on to the practice questions that follow.

Q. Find the inverse of $f(x) = 5x - 4$.

A. $f^{-1}(x) = \frac{x+4}{5}$.

First, switch x and $f(x)$: $x = 5f(x) - 4$. Name the new function by its correct name, the inverse function: $x = 5f^{-1}(x) - 4$. Now solve for the inverse: $x + 4 = 5f^{-1}(x)$; $\frac{x+4}{5} = f^{-1}(x)$.

Q. Determine whether $f(x) = 3x - 1$ and $g(x) = \frac{x+1}{3}$ are inverses of each other.

A. These two functions are inverses.

First, find $(f \circ g)(x)$: $3\left(\frac{x+1}{3}\right) - 1$. Then simplify this expression: $x + 1 - 1 = x$. That's what it's supposed to be, so move on to the next one: $(g \circ f)(x)$: $\frac{3x-1+1}{3} = \frac{3x}{3} = x$. That one works, too, so these two functions are inverses of each other.

33. Graph the inverse of $g(x)=\sqrt{x-2}$.

Solve It

34. Find the inverse of $k(x)=\frac{3x}{x-1}$.

Solve It

35. Determine whether $f(x) = x^3 - 1$ and $g(x)=\sqrt[3]{x}+1$ are inverses of each other.

Solve It

36. Determine whether $f(x)=\frac{1-x}{2}$ and $g(x) = 1 - 2x$ are inverses of each other.

Solve It

Answers to Problems on Functions

Following are the answers to problems dealing with functions. We also provide guidance on getting the answers if you need to review where you went wrong.

1 Is $f(x) = x^3 - 1$ even, odd, or neither? The answer is neither.

Find $f(-x) = (-x)^3 - 1 = -x^3 - 1$. This isn't the same function as the original one, so the answer isn't even. It's also not the exact opposite of the original, so the answer isn't odd either. Therefore, this function is neither odd nor even.

2 Determine whether the given graph is even, odd, or neither. The answer is odd.

If you look at the graph upside down, it looks exactly the same — that means it's odd.

3 Sketch half of the graph of $f(x) = \sqrt{x^2 - 4}$ and use symmetry to complete the graph. See the graph for the answer.

If you find $f(-x)$ first, you discover that the function doesn't change at all, which means you have an even function. Plug and chug some negative values for x to find that the positive values for each corresponding x are the same. For example, $f(-2) = 0$, so you know that $f(2)$ is also 0. $f(-3) = \sqrt{5}$, and so does $f(3)$. $f(-5) = \sqrt{21}$, and so does $f(5)$. Knowing these points gives you the graph.

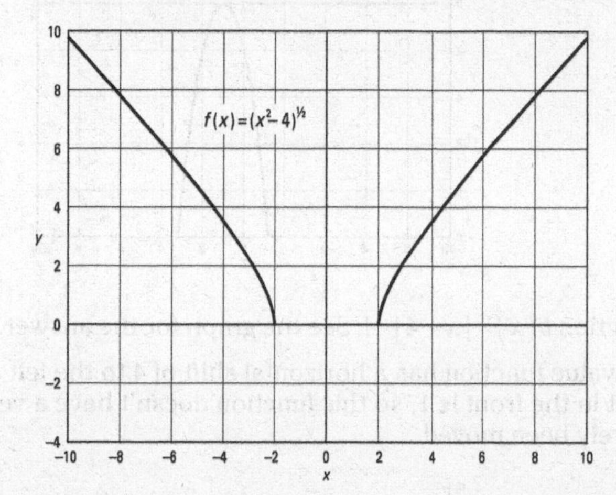

4 Sketch half the graph of $f(x) = 4x^3$ and use symmetry to complete the graph. See the graph for the answer.

Well, $f(-x)$ is $-4x^3$, which is the exact opposite of the original function. Looks like you have an odd graph. Each x gives you a value that's $f(x)$, and each opposite $-x$ gives the opposite $-f(x)$. Plug and chug some values to get the graph: $f(-3) = -108$, so $f(3) = 108$. $f(2) = 32$, so $f(-2) = -32$. $f(-1) = -4$, so $f(1) = 4$. Put these and as many other points as you want on the graph.

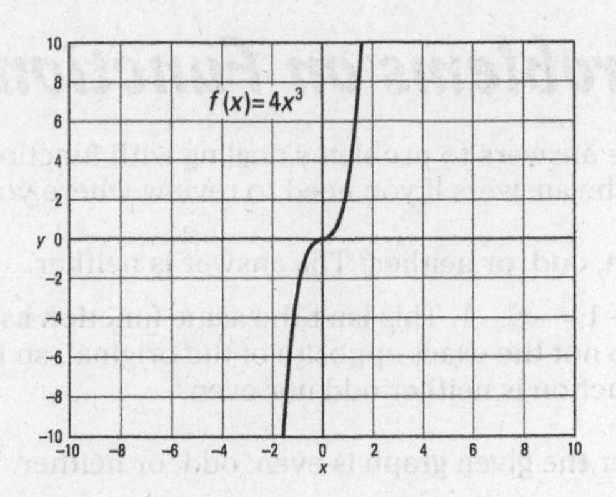

5 Graph the function $a(x) = -2(x - 1)^2$. See the graph for the answer.

This function takes the parent quadratic graph and moves it to the right by 1. The vertical stretch is 2, making each point twice as tall. The negative sign is a reflection, turning the graph upside down. Put all of these pieces together to get the graph.

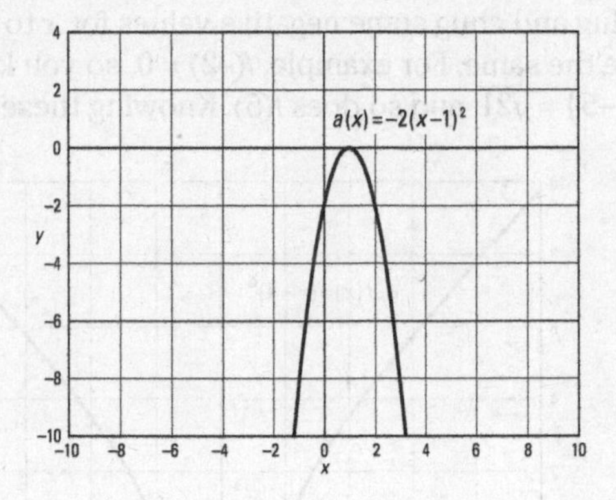

6 Graph the function $b(x) = |x + 4| - 1$. See the graph for the answer.

This absolute value function has a horizontal shift of 4 to the left and a vertical shift of 1 down. The coefficient in the front is 1, so this function doesn't have a vertical stretch or shrink — the graph has merely been moved.

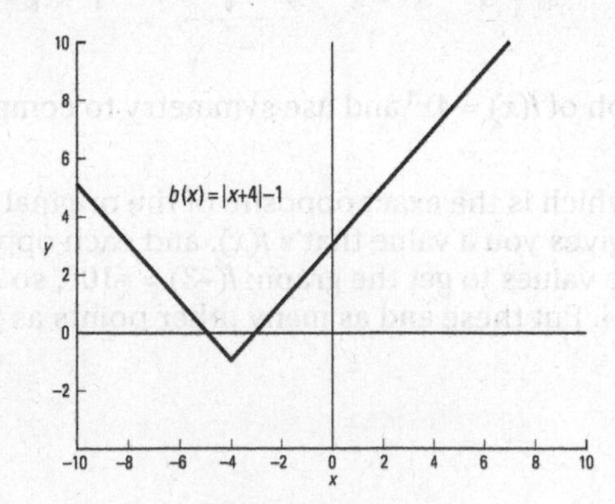

7 Graph the function $c(x)=\sqrt{x+3}$. See the graph for the answer.

This square root function is shifted horizontally to the left by 3. (Don't forget that those horizontal shifts are always the opposite of what they appear to be.)

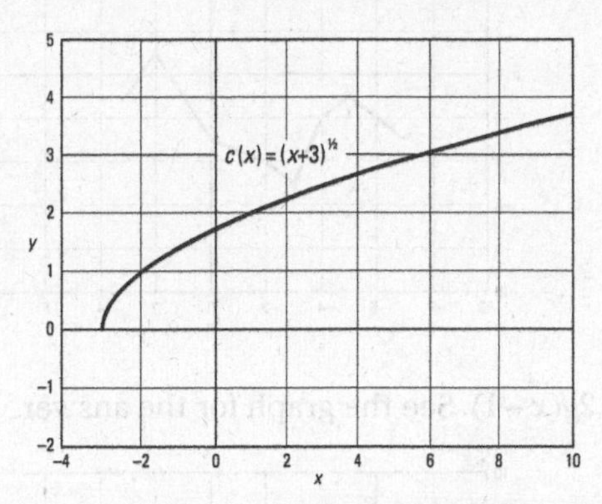

8 Graph the function $f(x) = -x^2 - 6x$. See the graph for the answer.

This problem actually brings up the topic of conic sections, which we cover in depth in Chapter 12. This function creates a parabola, and in order to get this parabola into its graphable form, you have to follow a procedure known as *completing the square* (see Chapters 4 and 12 for more information on this procedure). We do it for you here and tell you that the function will become $f(x) = -1(x + 3)^2 + 9$. We include it here because a few textbooks (though not many) teach completing the square early so that you can graph these types of problems. If your teacher is mean enough to include one of these *without* teaching you how to complete the square, you have to plug and chug this type of problem in order to graph it — pick x-values to find the corresponding y-values. Just be sure that your final graph is a parabola, like ours.

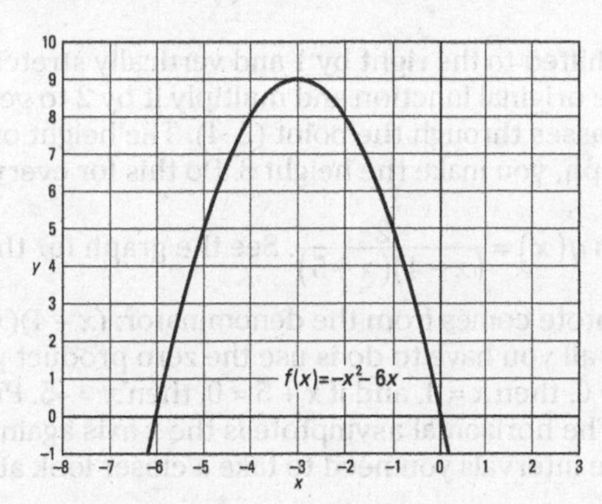

9 Graph the function $g(x) - 3$. See the graph for the answer.

To solve this problem, take every single point on the given $g(x)$ function and shift each one down by 3. If you do, you'll end up with a graph that looks like ours, which is really just a graph that we made up. Because it's a function, it still follows all the rules of transforming functions.

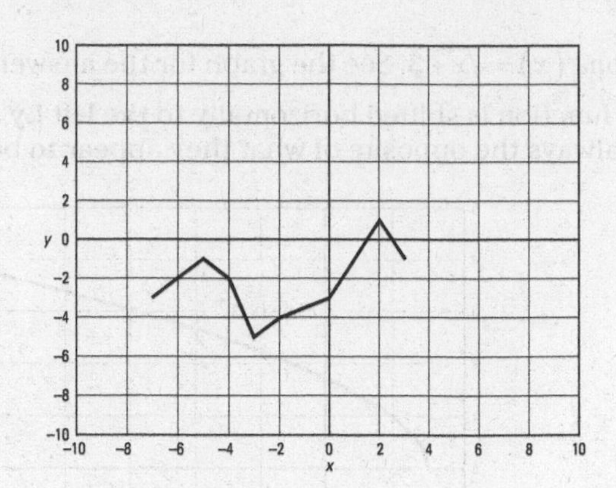

10 Graph the function $2g(x-1)$. See the graph for the answer.

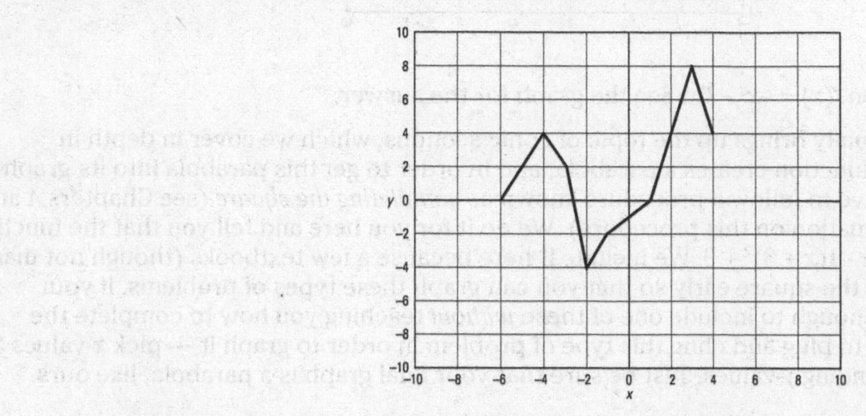

This time, g(x) is shifted to the right by 1 and vertically stretched by a factor of 2. Take the height of each point in the original function and multiply it by 2 to get the new height. For example, the original function passes through the point (2, 4). The height of this point is 4, so when you double that in the new graph, you make the height 8. Do this for every single point to get the graph.

11 Graph the function $q(x)=\frac{-2}{(x-4)(x+5)}$. See the graph for the answer.

The vertical asymptote comes from the denominator: $(x-4)(x+5)=0$. This equation is already neatly factored, so all you have to do is use the zero product property, set each factor equal to 0, and solve. If $x-4=0$, then $x=4$, and if $x+5=0$, then $x=-5$. Put both of these on the graph as vertical asymptotes. The horizontal asymptote is the x-axis again because the denominator has the greater degree. The intervals you need to take a closer look at are $(-\infty,-5)$, $(-5,-4)$, and $(4,\infty)$.

Pick a couple x-values from each interval to get an idea of what the graph is doing. For example, when $x=-7$, $y=-0.09$, and when $x=-6$, $y=-0.2$. Both of these points are below the horizontal asymptote. When $x=-4$, $y=0.25$; when $x=-1$, $y=0.1$; when $x=1$, $y=0.11$; and when $x=3$, $y=0.25$. These points are all above the horizontal asymptote. On the final interval, when $x=5$, $y=-0.2$ and when $x=6$, $y=-0.09$, these points are both below the horizontal asymptote. Put all the pieces together in the final graph.

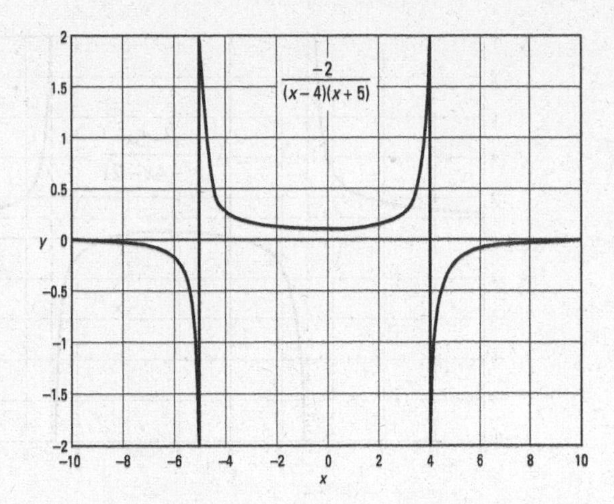

12 Graph the function $r(x)=\dfrac{x+3}{x^2-x-6}$. See the graph for the answer.

First up are vertical asymptotes. Set $x^2 - x - 6 = 0$ and factor to get $(x - 3)(x + 2) = 0$. Set each factor equal to 0 and solve. If $x - 3 = 0$, then $x = 3$, and if $x + 2 = 0$, then $x = -2$. Add these two vertical asymptotes to your graph. Next up is the horizontal asymptote. Because the denominator has the greater degree, the horizontal asymptote is the x-axis again. Notice, however, that now that the variable is in the numerator as well, there may be an x-intercept. Set the numerator equal to 0 and solve. $x + 3 = 0$ tells you that $x = -3$ is an intercept. The graph crosses the x-axis. Use this fact to set up the intervals: $(-\infty, -3)$, $(-3, -2)$, $(-2, 3)$, and $(3, \infty)$. Each interval is, respectively, below, above, below, and above the horizontal asymptote. The graph looks a little weird, but then, which of these problems doesn't look weird?

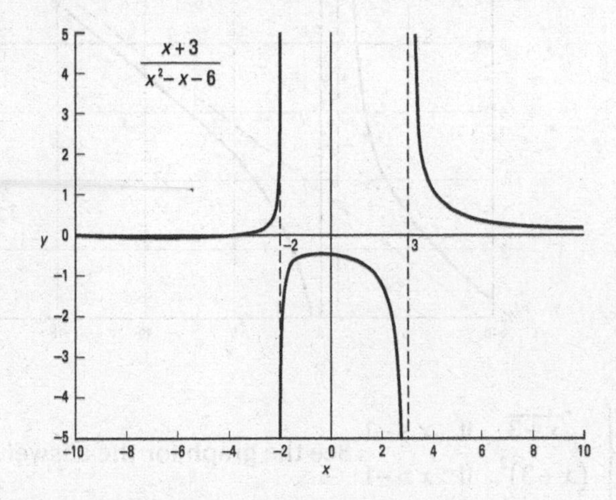

13 Graph the function $t(x)=\dfrac{x^2-5x}{x^2-4x-21}$. See the graph for the answer.

Find the vertical asymptotes for this one by factoring the denominator. If $x^2 - 4x - 21 = 0$, then $(x - 7)(x + 3) = 0$. This gives you two solutions: $x = 7$ and $x = -3$. The degrees are the same, so the horizontal asymptote is $y = 1$. Put the asymptotes onto the graph and then pick x-values and x-intercepts to get the graph.

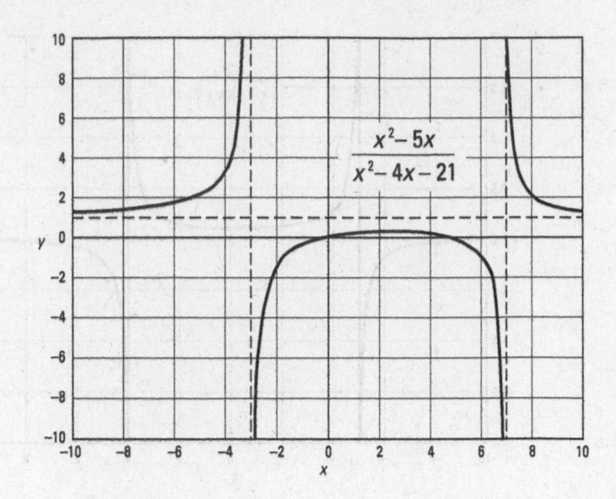

14 Graph the function $u(x)=\frac{x^2-10x-24}{x+1}$. See the graph for the answer.

This is the toughest of all the problems here because there's no horizontal asymptote — the numerator has the greater degree. Use long division to find the quotient $x-11$; graph this as an equation, $y=x-11$, with a dotted line to mark your oblique asymptote. Next, add the vertical asymptote by solving the equation $x+1=0$ to get $x=-1$. Finally, plug and chug some values on each interval to get the graph.

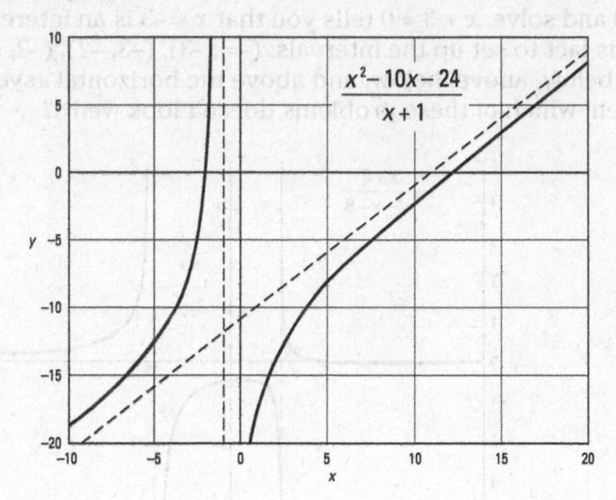

15 Graph $g(x)=\begin{cases}\sqrt{x+3} & \text{if } x\le -1\\ (x+3)^2 & \text{if } x>-1\end{cases}$. See the graph for the answer.

Take a look at each interval of the domain to determine the graph's shape. For this piece-wise function, the top piece is only defined when $x\le -1$. Consequently, this part of the graph looks like a square root graph shifted 3 to the left. The bottom piece is defined when $x>-1$, which makes this part of the graph a parabola that's shifted 3 to the left.

If it helps you to lightly sketch the whole graph and then erase the part you don't need, we highly recommend doing that.

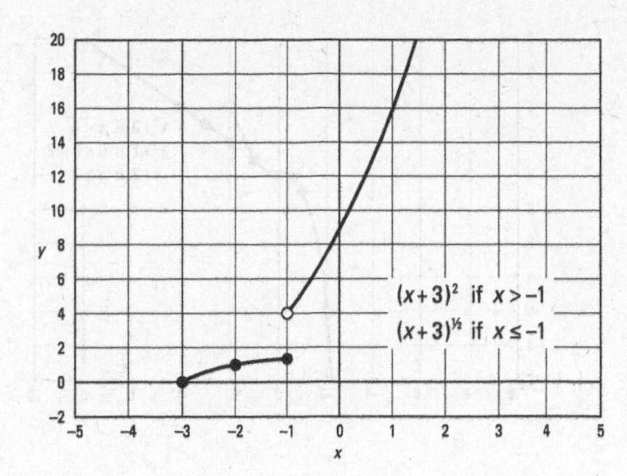

16 Graph $h(x)=\begin{cases}\frac{1}{2}x-4 & \text{if} \quad x\le-2\\ 3x+3 & \text{if} \quad -2<x<2\\ 4-x & \text{if} \quad x\ge2\end{cases}$. See the graph for the answer.

The first piece is a linear function that's defined only when $x\le-2$. The second piece is also a linear function, defined between –2 and 2. The third piece is yet another linear function, defined when $x\ge2$.

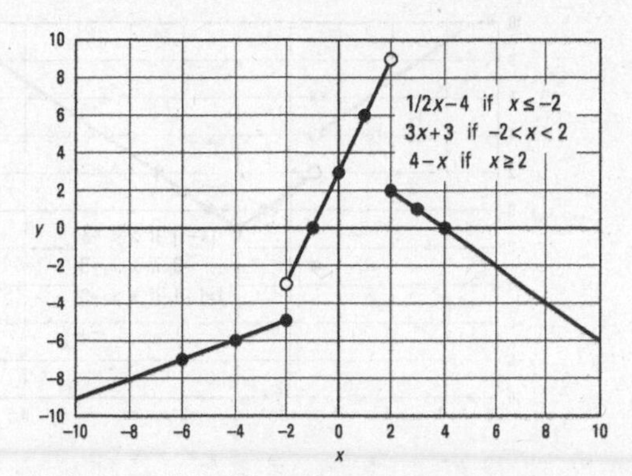

17 Graph $m(x)=\begin{cases}x^3+2 & \text{if} \quad x<0\\ x^2+2 & \text{if} \quad 0\le x<2\\ x+2 & \text{if} \quad x\ge2\end{cases}$. See the graph for the answer.

The first piece is a cubic function that's shifted up by 2 — its right endpoint should be open. However, when you graph the second piece, you see that it's a parabola that's shifted up by 2. Its left endpoint overlaps the right endpoint of the first piece, filling the hole that was there, and the graph carries on until $x = 2$, where it gets broken again. The third piece follows the linear function to the right of $x = 2$.

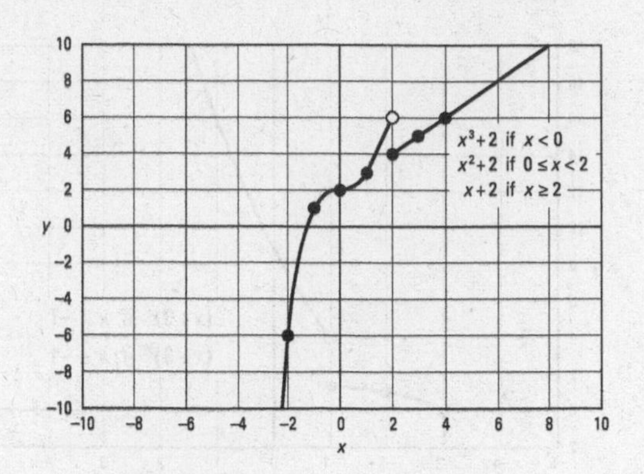

18 Graph $n(x)=\begin{cases} |x-1| & \text{if} \quad x<-3 \\ -3 & \text{if} \quad x=-3 \\ |x|-1 & \text{if} \quad x>-3 \end{cases}$. See the graph for the answer.

This piece-wise function is different because the middle piece is defined at just one point. When $x = 3$, $y = -3$. That's it. The first piece follows the absolute value graph that has been shifted to the right by 1. The third piece is also an absolute value graph, but it has been shifted down 1.

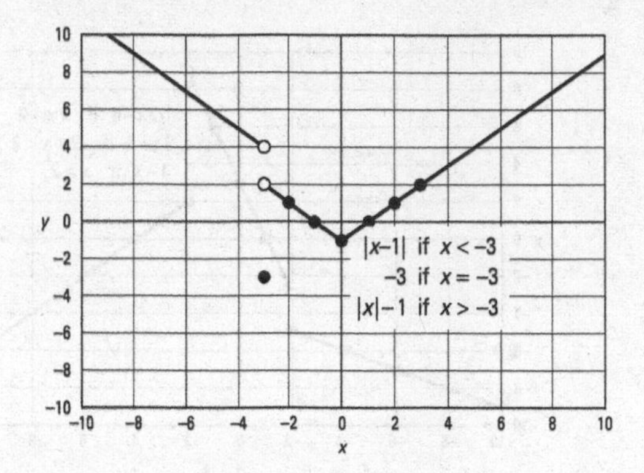

19 Find $(f + h)(x)$. The answer is $x^2-6x+2+\sqrt{3x+2}$.

Take the f function and add the h function to it. Because one is a polynomial and the other is a square root, there are no pesky like terms. The answer is therefore $x^2-6x+2+\sqrt{3x+2}$.

20 Find $(f \cdot g)(x)$. The answer is $2x^4 - 17x^3 + 34x^2 - 10x$.

Start off by writing out what you've been asked to find — the product of f and g: $(f \cdot g)(x) = (x^2 - 6x + 2)(2x^2 - 5x)$. Distribute each term of the left polynomial to each term of the right polynomial to get $2x^4 - 12x^3 + 4x^2 - 5x^3 + 30x^2 - 10x$. Next, combine the like terms: $2x^4 - 17x^3 + 34x^2 - 10x$.

21 Find $(h/g)(x)$. Does this new function have any undefined values? The answer is $\frac{\sqrt{3x+2}}{2x^2-5x}$, and yes, there are undefined values — specifically, $x = 0$ and $x = 5/2$.

You're being asked to find the quotient of h and g, with h on the top and g on the bottom. Because they're different types of functions, you can't simplify them, which means your answer is, simply, $\frac{\sqrt{3x+2}}{2x^2-5x}$.

Because the denominator now has a polynomial, there may be undefined values. Set the denominator equal to 0 to start: $2x^2 - 5x = 0$. Solve by factoring out the greatest common factor: $x(2x - 5)$. This has two solutions, $x = 0$ and $x = \frac{5}{2}$. These are the undefined values.

22 Find $(g + h)(x)$. The answer is $2x^2 - 5x + \sqrt{3x+2}$.

No like terms exist because you're adding another polynomial and a square root. That means the answer is $2x^2 - 5x + \sqrt{3x+2}$.

23 Find $(f \circ g)(x)$. The answer is $4x^4 - 20x^3 + 13x^2 + 30x + 2$.

Take the g function and start plugging into f everywhere it says x: $(2x^2 - 5x)^2 - 6(2x^2 - 5x) + 2$. Multiply everything out first: $4x^4 - 20x^3 + 25x^2 - 12x^2 + 30x + 2$. Then combine like terms to get this answer: $(f \circ g)(x) = 4x^4 - 20x^3 + 13x^2 + 30x + 2$.

24 Find $(g \circ f)(x)$. The answer is $2x^4 - 24x^3 + 75x^2 - 18x - 2$.

This time, place f into g where it says x: $2(x^2 - 6x + 2)^2 - 5(x^2 - 6x + 2)$. Square the polynomial on the left first by multiplying $x^2 - 6x + 2$ by itself and distributing each term by each term. The squared polynomial is quite long: $2(x^4 - 6x^3 + 2x^2 - 6x^3 + 36x^2 - 12x + 2x^2 - 12x + 4) - 5(x^2 - 6x + 2)$. Combine like terms: $2(x^4 - 12x^3 + 40x^2 - 24x + 4) - 5(x^2 - 6x + 2)$. Distribute the coefficients next: $2x^4 - 24x^3 + 80x^2 - 48x + 8 - 5x^2 + 30x - 10$. Combine the like terms to arrive at the final answer: $(g \circ f)(x) = 2x^4 - 24x^3 + 75x^2 - 18x - 2$.

25 Find $h(f(x))$. The answer is $\sqrt{3x^2 - 18x + 8}$.

Substitute f for x in the h function: $\sqrt{3(x^2 - 6x + 2) + 2}$. Distribute that 3 inside the root to get $\sqrt{3x^2 - 18x + 6 + 2}$ and combine those like terms to get $\sqrt{3x^2 - 18x + 8}$.

26 Find $(f \circ f)(x)$. The answer is $x^4 - 12x^3 + 34x^2 + 12x - 6$.

This time, place f into itself everywhere it says x: $(x^2 - 6x + 2)^2 - 6(x^2 - 6x + 2) + 2$. Because we go through the process of squaring that polynomial in Question 24, we won't do it again here. When you multiply everything out, you get $x^4 - 12x^3 + 40x^2 - 24x + 4 - 6x^2 + 36x - 12 + 2$. Combine like terms to end up with this answer: $(f \circ f)(x) = x^4 - 12x^3 + 34x^2 + 12x - 6$.

27 Find $f(g(-1))$. The answer is 9.

If you've already tackled Question 23, you've already found that $f(g(x))$ is $4x^4 - 20x^3 + 13x^2 + 30x + 2$. Now you can substitute -1 for x: $4(-1)^4 - 20(-1)^3 + 13(-1)^2 + 30(-1) + 2$. Simplify by dealing with all the exponents first: $4(1) - 20(-1) + 13(1) + 30(-1) + 2$. Simplify further by multiplying to get $4 + 20 + 13 - 30 + 2$. Add and subtract to (finally!) end up with 9.

28 Find $g(h(3))$. The answer is $22 - 5\sqrt{11}$.

Remember that composition of a function is read from right to left. That means this question is asking you to plug 3 into h and then plug that answer into g. Start with $h(3) = \sqrt{3(3)+2} = \sqrt{9+2} = \sqrt{11}$. Now plug this value into g and find $g(\sqrt{11})$:

$$2(\sqrt{11})^2 - 5\sqrt{11} = 2(11) - 5\sqrt{11} = 22 - 5\sqrt{11}.$$

29 Find the domain of $(f \circ g)(x)$. The domain is all real numbers.

Question 23 has you find the composition $(f \circ g)(x)$. What you wind up with is a longish polynomial, specifically $(f \circ g)(x) = 4x^4 - 20x^3 + 13x^2 + 30x + 2$. Because it's a polynomial, there's nothing weird about the domain. It's all real numbers.

30 Find the domain of $h(f(x))$. The domain is $x \leq \frac{18-\sqrt{228}}{6}$ or $x \geq \frac{18+\sqrt{228}}{6}$.

In Question 25, you find that $h(f(x)) = \sqrt{3x^2 - 18x + 8}$, which has a quadratic expression under the square root. The quantity therefore needs to be nonnegative. Find where $3x^2 - 18x + 8$ is nonnegative by setting it greater than or equal to zero and solving the resulting quadratic inequality by using the quadratic formula. Doing so gives you two key values: $x \leq \frac{18-\sqrt{228}}{6}$, which is approximately 0.48, and $x \geq \frac{18+\sqrt{228}}{6}$, which is approximately 5.52. Using the approximations as stand-ins, look at the intervals determined by them: $x < 0.48$, $0.48 < x < 5.52$, and $x > 5.52$. If you plug test values from each interval into the inequality, you discover which intervals work and which don't. In this case, for $3x^2 - 18x + 8$ to be nonnegative, x has to be less than or equal to 0.48 or greater than or equal to 0.52. Thus, the domain is $x \leq \frac{18-\sqrt{228}}{6}$ and $x \geq \frac{18+\sqrt{228}}{6}$.

31 Find the domain of $(f + h)(x)$. The domain is $x \geq -\frac{2}{3}$.

$(f+h)(x) = x^2 - 6x + 2 + \sqrt{3x+2}$. This adds a square root to a polynomial, so the new combined function must follow all the rules that the square root function did by itself. In this case, $3x + 2$ is nonnegative, yielding the answer.

32 Find the domain of $(\frac{h}{g})(x)$. The domain is $x \geq -\frac{2}{3}$, except $x = 0$ and $x = \frac{5}{2}$.

The square root in the numerator restricts the domain to $(h/g)(x) = \frac{\sqrt{3x+2}}{2x^2 - 5x}$. The polynomial in the denominator has undefined values: $x = 0$ and $x = \frac{5}{2}$. These are both in the restricted domain, so they become part of the answer. You express it as one neat sentence: "The domain is $x \geq -\frac{2}{3}$, except $x = 0$ and $x = \frac{5}{2}$."

33 Graph the inverse of $g(x) = \sqrt{(x-2)}$. See the graph for the answer.

Start off by graphing the square root function shifted to the right by 2. Points on this graph include (2, 0), (3, 1), and (6, 2). Recall that the inverse is the reflection of the original function across the line $y = x$. Flip them to get (0, 2), (1, 3), and (2, 6) — all points on the inverse function graph.

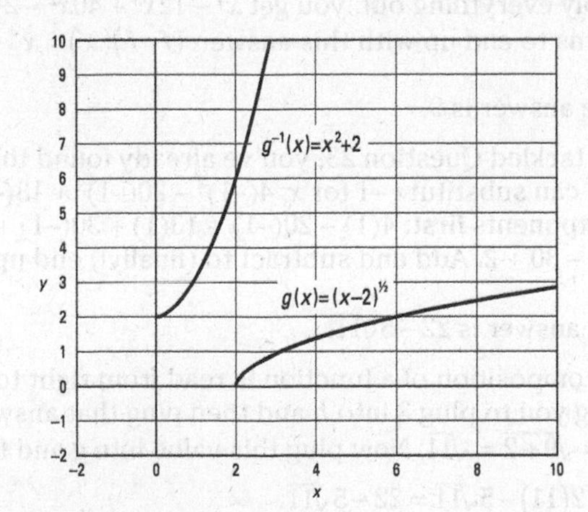

34 Find the inverse of $k(x) = \frac{3x}{x-1}$. The answer is $k^{-1}(x) = \frac{x}{x-3}$.

Switch x and $k(x)$ and name the new inverse by its real name: $x = \frac{3k^{-1}(x)}{k^{-1}(x)-1}$. Notice the inverse is in the numerator and the denominator. The only way you can solve for it is to get rid of the fraction first by multiplying both sides by the denominator and getting $x(k^{-1}(x) - 1) = 3k^{-1}(x)$.

Distribute the x and get $x(k^{-1}(x) - 1) = 3k^{-1}(x)$. Get everything with the inverse in it to one side and everything else to the other side: $xk^{-1}(x) - 3k^{-1}(x) = x$. Factor out the greatest common factor $k^{-1}(x)$ on the left: $k^{-1}(x)(x - 3) = x$. Now divide both sides by $x - 3$ to solve for the inverse: $k^{-1}(x) = \frac{x}{x-3}$.

35 Determine whether $f(x) = x^3 - 1$ and $g(x) = \sqrt[3]{x} + 1$ are inverses of each other. The answer is that they're not inverses.

First, find the composition: $(f \circ g)(x) = \left(\sqrt[3]{x} + 1\right)^3 - 1$. This doesn't simplify to get x, so you can stop. They're not inverses.

36 Determine whether $f(x) = \frac{1-x}{2}$ and $g(x) = 1 - 2x$ are inverses of each other. The answer is yes, they are inverses.

$(f \circ g)(x) = \frac{1-(1-2x)}{2} = \frac{1-1+2x}{2} = \frac{2x}{2} = x$. One down, one to go.

$(g \circ f)(x) = 1 - 2\left(\frac{1-x}{2}\right) = 1 - (1 - x) = 1 - 1 + x = x$. Okay, you checked both, and it's official: They're inverses.

Chapter 4

Searching Roots to Get the Degree

In This Chapter

- Solving quadratic equations by factoring
- Exploring methods to solve quadratic equations that don't factor
- Finding the roots of a polynomial
- Using roots and the leading coefficient test to graph polynomials

A *polynomial* is an algebraic expression of one or more terms involving a sum of powers of a variable multiplied by coefficients. The highest exponent on any term in a polynomial is its *degree*. In this chapter, we review solving polynomial equations to find the solutions, which are also called *roots* or *zeros*. We start with a review of solving *quadratic equations* — polynomials where the highest exponent is two. Then we move into equations with higher degrees and show you how to solve them. We also take a look at using roots to factor polynomials and how to graph polynomials.

Reasoning Through It: Factoring a Factorable Polynomial

Before getting started on the nitty-gritty, here's some vocabulary you should know to be successful in this chapter (and hereafter):

- **Standard form:** What most textbooks use to write a quadratic equation: $ax^2 + bx + c = 0$
- **Quadratic term:** The term with the second degree: ax^2
- **Linear term:** The term with the linear degree: bx
- **Constant:** The term with zero degree: c
- **Leading coefficient:** The number multiplying the term with the highest degree: a

In math, the process of breaking down a polynomial into the product of two or more polynomials with a smaller degree is called *factoring*. In general, factoring is often easy for quadratic equations and is the first thing you want to try when asked to solve second-degree polynomial equations. Some types of factoring (like the difference of cubes or grouping — more on those later in this section) may work on higher-degree polynomials, and you should always check to see whether this is the case first. When presented with a polynomial equation and asked to solve it, you should always try the following methods of factoring, in order:

- **Finding the greatest common factor:** The *greatest common factor,* or GCF, is the biggest expression that will divide into all the other terms. It's a little like doing the distributive property backwards.

  Look at all those factors to see what they share in common (that's your GCF), factor the GCF out from every term by putting it in front of a set of parentheses, and leave the factors that aren't the GCF inside the parentheses.

- **Working with a binomial polynomial:** If the polynomial has two terms, check to see whether it's a difference of squares or the sum or difference of cubes.
  - **Difference of squares:** $a^2 - b^2$ always factors to $(a - b)(a + b)$
  - **Difference of cubes:** $a^3 - b^3$ always factors to $(a - b)(a^2 + ab + b^2)$
  - **Sum of cubes:** $a^3 + b^3$ always factors to $(a + b)(a^2 - ab + b^2)$

- **Working with a trinomial polynomial:** Some instructors teach the "guess-and-check method," where you keep trying different pairs of binomials until you happen to stumble onto the right one. This isn't fun by any means, and you could try all day and never figure it out (maybe the polynomial is prime and won't factor).

  We recommend using the *British method* (also known as the *FOIL method backwards*) instead. Follow these steps to use this method:

  1. **Multiply the quadratic term and the constant term.**

     You do this only in your head (or somewhere else on your paper) and you only do it to proceed to the next step.

  2. **Write down all the factors of the result of Step 1, in pairs.**

     Again, you do this for you only and also to make sure that you list every possibility — that's why it's not guess-and-check. If you list them all and none of them work (see Step 3), you know your trinomial is prime.

  3. **Find the pair from the list in Step 2 that produces the linear term when added.**

     Only one, if any, of them will work. If none of them do, it's prime.

  4. **Break up the linear term into two terms — the winning pair from Step 3.**

     You've now created a polynomial with four terms. Proceed to the next type of factoring — a polynomial with more than three terms.

- **Working with a polynomial that has more than three terms:** Try grouping the polynomial into two sets of two. Find the GCF for each set and factor it out. Find the GCF of the two remaining expressions and factor it out. You end up with two binomials, exactly what you're looking for!

Once you factor the polynomial, you can use the zero product property to solve it by setting each factor equal to 0 and solving.

Q. Solve the equation $3x^2 + x - 2 = 0$.

A. $x = \frac{2}{3}$, $x = -1$. **Multiply the quadratic term and the constant term: $(3x^2)(-2) = -6x^2$. Write down all factors of this, in pairs: $-x$ and $6x$, x and $-6x$, $-2x$ and $3x$, and $2x$ and $-3x$. The pair that adds up to the linear term is $-2x$ and $3x$. Split the middle term into two using this pair: $3x^2 - 2x + 3x - 2 = 0$. Now that you have four terms instead of three, use grouping to factor it: $x(3x - 2) + 1(3x - 2) = 0$. Notice that the second set of two terms only has a GCF of one, but you still factor it out. Now there's a GCF again — both sets of terms share $(3x - 2)$, so that can factor out to the front: $(3x - 2)(x + 1) = 0$. Finally, use the zero product property to solve the equation. If $3x - 2 = 0$, then $x = \frac{2}{3}$; and if $x + 1 = 0$, then $x = -1$.**

Q. Solve the equation $3x^3 - 3x = 0$.

A. $x = 0, 1$, **and** -1. **Always check for the GCF first and factor it out: $3x(x^2 - 1) = 0$. Now recognize the "leftovers" as a difference of squares that factors again: $3x(x - 1)(x + 1) = 0$. Set each factor equal to 0 and solve:**

$3x = 0, x = 0$

$x - 1 = 0, x = 1$

$x + 1 = 0, x = -1$

1. Solve the equation $2y^2 + 5y = 12$.

Solve It

2. Solve the equation $16m^2 - 8m + 1 = 0$.

Solve It

3. Solve the equation $x^3 + x^2 = 9x + 9$.

Solve It

4. Solve the equation $\frac{1}{6}x^2 + \frac{2}{3}x = 2$.

Solve It

Solving a Quadratic Polynomial Equation

What happens when a quadratic equation doesn't factor? You're done, right? Well, not quite. You have two more methods you can use: the quadratic formula, and completing the square. Your teacher may require you to use both, so we include both here. When you graph quadratics, as you do in Chapters 3 and 12, the easiest method is to complete the square and then use the rules of transforming a parent function to get the graph.

Completing the square

Here are the steps for completing the square:

1. Make sure the quadratic is written in standard form: $ax^2 + bx + c = 0$
2. Add (or subtract) the constant term from both sides: $ax^2 + bx = -c$
3. Factor out the leading coefficient from the quadratic term and the linear term: $a\left(x^2 + \frac{b}{a}x\right) = -c$
4. Divide the new linear coefficient by two: $b/a \div 2 = b/2a$; square this: $\left(\frac{b}{2a}\right)^2 = \frac{b^2}{4a^2}$; and add this inside the parentheses: $a\left(x^2 + \frac{b}{a} + \frac{b^2}{4a^2}\right)$
5. Keep the equation balanced by multiplying the leading coefficient by the term you just added in Step 4: $a \cdot \frac{b^2}{4a^2} = \frac{b^2}{4a}$, and adding it to the other side: $a\left(x^2 + \frac{b}{a}x + \frac{b^2}{4a^2}\right) = -c + \frac{b^2}{4a}$
6. Divide the leading coefficient from both sides: $x^2 + \frac{b}{a}x + \frac{b^2}{4a^2} = -\frac{c}{a} + \frac{b^2}{4a^2}$
7. Factor the trinomial on the left side of the equation: $\left(x + \frac{b}{2a}\right)^2 = -\frac{c}{a} + \frac{b^2}{4a^2}$
8. Take the square root of both sides: $x + \frac{b}{2a} = \pm\sqrt{-\frac{c}{a} + \frac{b^2}{4a^2}}$
9. Solve for x: $x = \frac{-b}{2a} \pm \sqrt{-\frac{c}{a} + \frac{b^2}{4a^2}}$

Quadratic formula

Of course, those of you who know the quadratic formula should vaguely recognize the preceding steps — they're the derivation of the quadratic formula. All you have to do is find the common denominator of both the fractions inside the square root, add them together, and watch the square root simplify. Ultimately, you end up with the quadratic formula:

$$x = \frac{-b \pm \sqrt{b^2 - 4ac}}{2a}$$

Q. Solve the equation $5x^2 - 12x - 2 = 0$.

A. $x = \frac{6 \pm \sqrt{46}}{5}$. **This equation doesn't factor, so you use the quadratic formula to solve it. $a = 5$, $b = -12$, $c = -2$. Plug these values into the quadratic formula:**

$$\frac{-(-12) \pm \sqrt{(-12)^2 - 4(5)(-2)}}{2(5)}$$

Now simplify it: $\frac{12 \pm \sqrt{144 + 40}}{10} = \frac{12 \pm \sqrt{184}}{10}$. **Don't forget to check your square roots and simplify them as well:** $\frac{12 \pm 2\sqrt{46}}{10}$. **Finally, 2 goes into every coefficient and constant in the answer, so it simplifies even further to** $\frac{6 \pm \sqrt{46}}{5}$.

5. Solve $x^2 - 10 = 2x$.

Solve It

6. Solve $7x^2 - x + 2 = 0$.

Solve It

7. Solve $x^2 - 4x - 7 = 0$ by completing the square.

Solve It

8. Solve $-2.31x^2 - 4.2x + 6.7 = 0$.

Solve It

Getting to the Summit by Solving High-Order Polynomials

The greater the degree of your given polynomial, the harder it is to solve the equation by factoring. You should always still try that first because you never know . . . it may actually work! When factoring fails, however, you begin anew with a longer and more complicated process for finding the roots. We walk you through each step, one by one.

Always begin by finding the degree of the polynomial because it gives you some very important information about your graph. The degree of the polynomial tells you the maximum number of roots — it's that easy. A fourth-degree polynomial will have up to, but no more than, four roots.

Determining positive and negative roots: Descartes' Rule of Signs

When you have a polynomial with real coefficients, you can use *Descartes' Rule of Signs* to determine the possible number of positive and negative real roots. All you have to do is count! First, make sure that the polynomial $f(x)$ has real coefficients and is written in descending order, from highest to lowest degree. Look at the sign of each term and count how many times the sign changes from positive to negative and vice versa as you view the polynomial from left to right. The number of sign changes represents the maximum number of positive real roots. The rule also says that the possible number of positive real roots decreases by 2 over and over again until you end up with 1 or 0 (more on this in the next section). This gives you the list of the possible number of real positive roots.

Descartes also figured out that if you take a look at $f(-x)$ and count again, you discover the maximum number of negative real roots. Remember that negative numbers raised to even powers are positive, and negative numbers raised to odd powers are negative. This means that $f(-x)$ changes from $f(x)$ only on the odd degrees. Each odd exponent becomes the opposite of what it was in $f(x)$. Count the number of times the sign changes in this function and subtract 2 over and over until you end up at 1 or 0, and you end up with a list of the possible number of real negative roots.

Counting on imaginary roots

Complex roots happen in a quadratic equation that has real coefficients when the *discriminant* (the part of the quadratic formula under the root sign: $b^2 - 4ac$) is negative. The $\pm$ in the quadratic formula also tells you that there are two of these roots, always in pairs. This is why you subtract by 2 in the preceding section; you have to account for the fact that roots may be in pairs of complex numbers. In fact, the pairs will always be *complex conjugates* of each other — if one root is $a + bi$, for example, the other one is $a - bi$.

The *Fundamental Theorem of Algebra* says that every polynomial of degree one or greater has at least one complex root. Chapter 11 explains complex numbers in depth, but for now all you need to know is that complex numbers can be roots of polynomial equations. Kinda cool, huh?

You know the total number of possible roots, and if you also know the list of possible positive and negative roots, you can use all that information to determine how many possible non-real complex roots a polynomial has. Count up every possible combination of positive and negative roots. The remaining number of roots in each situation represents the number of possible non-real complex roots.

Getting the rational roots

When all the coefficients of a polynomial are integers, the *Rational Root Theorem* helps you narrow down the possibilities for the roots even further. Right now, if you've gone through all the steps, you know only the total number of possible roots, how many might be positive or negative real, and how many might be complex. That still leaves an infinite number of possibilities for the values of the roots! The Rational Root Theorem helps you because it finds the possible roots that are *rational* (those that can be written as fractions). The problem with the theorem? Not all roots are rational. Keep in mind that some (or all) of the roots may be irrational or complex numbers.

To use the Rational Root Theorem, take all the factors of the constant term and divide by all the factors of the leading coefficient; both positive and negative divisors must be included. This produces a list of fractions that are all possibilities for roots. You *could* try plugging each one of these possibilities into the original function in the hopes of finding a root (remember, they're also called zeros because the value of the function will be 0). However, this process is long and tedious because you're dealing with the original function each time. If there are 50 roots, you're not helping yourself by plugging and chugging. Instead, move on to the next step.

Finding roots through synthetic division

Armed with the list of fractions we ask you to generate in the preceding section, pick one fraction and try to find its roots through the process of synthetic division, as we describe in this section. If this tactic works, the quotient is a *depressed polynomial.* No, it's not sad — but its degree will be less than the one you started with. You use this quotient to find the next one, each time lessening the degree, which narrows down the roots you have to find. At some point, your polynomial will end up as a quadratic equation, which you can solve using factoring or the quadratic formula. Now that's clever! If the root you try works, you should always try it again to see if it's a root with *multiplicity* — that is, a root that's used more than once.

Here are the steps to use for synthetic division:

1. **Make sure the polynomial is written in descending order. If any degrees are missing, fill in the gaps with zeros.**
2. **Write the number of the root you're testing outside the synthetic division sign. Write the coefficients of the polynomial in descending order and include any zeros from Step 1 inside the synthetic division sign.**
3. **Drop the first coefficient down.**
4. **Multiply the root on the outside and this coefficient. Write this product above the synthetic division line.**
5. **Add the next coefficient and the product from Step 4. This answer goes below the line.**
6. **Multiply the root on the outside and the answer from Step 5.**
7. **Repeat over and over again until you use all the coefficients.**

This process is easier to see with an example, like the one that follows. Just know that when you do synthetic division, you end up with a list of roots that actually work in the polynomial.

Q. Find the roots of the equation $x^3 + x^2 - 5x + 3 = 0$.

A. $x = 1$ (double root), $x = -3$. We go through the whole process described in this section for this example question.

The number of roots: First, this equation is third degree, so it may have up to three different roots.

Descartes' Rule of Signs: Next, by looking at $f(x) = x^3 + x^2 - 5x + 3$, you notice that the sign changes twice (between the second and third terms and the third and fourth terms). This means there could be two or zero positive real roots. Next, look at $f(-x) = -x^3 + x^2 + 5x + 3$ and notice the sign changes only once, giving you only one negative real root.

Complex roots: So if two roots are positive and one is negative, that leaves none left over that are complex. But if none are positive and one is negative, that leaves two complex roots.

Rational Root Theorem: Take all the factors of 3 (the constant term) and divide by all the factors of 1 (the leading coefficient) to determine the possible rational roots: $\pm\frac{1}{1}$, $\pm\frac{3}{1}$. Reduce the fractions and discard any duplicates to get the final list: ±1, ±3.

Synthetic division: Pick a root, any root, and use synthetic division to test and see whether it actually is a root. Because we know the answers (we *did* write the question), we have you start with $x = 1$:

1	1	1	−5	3
		1	2	−3
	1	2	−3	0

The last column on the right is the remainder; because it's 0, you know you have one root: $x = 1$. Also notice that the other numbers are the coefficients of the depressed polynomial you're now working with: $x^2 + 2x - 3 = 0$. Because this is a quadratic, we recommend shifting gears and factoring it to $(x + 3)(x - 1) = 0$ to be able to use the zero product property to solve and get $x = -3$ and $x = 1$ (again — making it a double root!).

Q. Solve the equation $x^3 + 8x^2 + 22x + 20 = 0$.

A. $x = -2$, $x = -3 + i$, $x = -3 - i$. This equation is also a third degree, so it will have a maximum of three roots. Looking at $f(x) = x^3 + 8x^2 + 22x + 20$ reveals that none of them are positive. Looking at $f(-x) = -x^3 + 8x^2 - 22x + 20$ reveals that either one or three of them are negative. If none are positive and three are negative, there can't be any complex roots. However, if none are positive and only one is negative, two of them have to be complex. The Rational Root Theorem generates this list of fractions (and we're only looking at the negatives because we know there aren't any positive roots): $-\frac{1}{1}$, $-\frac{2}{1}$, $-\frac{4}{1}$, $-\frac{5}{1}$, $-\frac{10}{1}$, and $-\frac{20}{1}$. These all reduce, respectively, to -1, -2, -4, -5, -10, and -20. Start off with $x = -2$ to discover one of your roots:

−2	1	8	22	20
		−2	−12	−20
	1	6	10	0

The reduced polynomial you're now working with is $x^2 + 6x + 10$. This quadratic doesn't factor, so you use the quadratic formula to find that the last two roots are indeed complex: $x = -3 \pm i$.

9. Solve the equation $2x^3 + 3x^2 - 18x + 8 = 0$.

Solve It

10. Solve the equation $12x^4 + 13x^3 - 20x^2 + 4x = 0$.

Solve It

11. Solve the equation $x^3 + 7x^2 + 13x + 4 = 0$.

Solve It

12. Find the roots of the polynomial $x^4 + 10x^3 + 38x^2 + 66x + 45$.

Strike That! Reverse It! Using Roots to Find an Equation

The *factor theorem* says that if you know the root of a polynomial, then you also know a factor of the polynomial. These two go back and forth, one to the other — roots and linear factors are interchangeable. Your textbook may ask you to factor a polynomial with a degree higher than two, and you may find that it just won't factor using any of the techniques we describe in the earlier sections. In this case, you must find the roots and use them to find the factors.

If $x = c$ is a root, then $x - c$ is a factor and vice versa. It always works, and that's something you can count on. Nice, huh?

Q. Use the roots of $x^3 + x^2 - 5x + 3 = 0$ to factor the equation.

A. $(x - 1)^2(x + 3) = 0$. This is the question from the first example in the last section. You found that the roots are $x = 1$ (double root) and $x = -3$. Using the factor theorem, if $x = 1$ is a root, then $x - 1$ is a factor (twice); and if $x = -3$ is a root, then $x - (-3)$, or $x + 3$, is a factor. This means that $x^3 + x^2 - 5x + 3 = 0$ factors to $(x - 1)^2(x + 3) = 0$.

13. Find the lowest order polynomial with leading coefficient as 1 that has –3, –2, 4, and 6 as its roots.

Solve It

14. Find the lowest order polynomial with leading coefficient as 1 that has 2, $4 + 3i$, and $4 - 3i$ as its roots.

Solve It

15. Factor the polynomial $6x^4 - 7x^3 - 18x^2 + 13x + 6$.

Solve It

16. Factor the polynomial $x^4 + 10x^3 + 38x^2 + 66x + 45$.

Solve It

Graphing Polynomials

Once you have a list of the roots of your polynomial, you've done the hardest part of graphing the polynomial. Remember that real roots or zeros are x-intercepts — you now know where the graph crosses the x-axis. Follow these steps to get to the graph:

1. **Mark the x-intercepts on your graph.**
2. **Find the y-intercept by letting $x = 0$. The shortcut? It will always be the constant term.**
3. **Use the leading coefficient test to determine which of the four possible ways the ends of your graph will point:**
   - If the degree of the polynomial is even and the leading coefficient is positive, both ends of the graph will point up.
   - If the degree of the polynomial is even and the leading coefficient is negative, both ends of the graph will point down.
   - If the degree of the polynomial is odd and the leading coefficient is positive, the left side of the graph will point down and the right side will point up.
   - If the degree of the polynomial is odd and the leading coefficient is negative, the left side of the graph will point up and the right side will point down.
4. **Figure out what happens in between the x-intercepts by picking any x-value on each interval and plugging it into the function to determine whether it's positive (and, therefore, above the x-axis) or negative (below the x-axis).**
5. **Plot the graph by using all the information you've determined.**

Q. Graph the equation $f(x) = x^3 + x^2 - 5x + 3$.

A. See the graph. This is the first example from the section on solving high-order polynomials again. You found that the roots are $x = 1$ (double root) and $x = -3$. The y-intercept is the constant $y = 3$. The leading coefficient test tells you the graph starts by pointing down and ends by pointing up. The double root at $x = 1$ makes the graph "bounce" and not cross there.

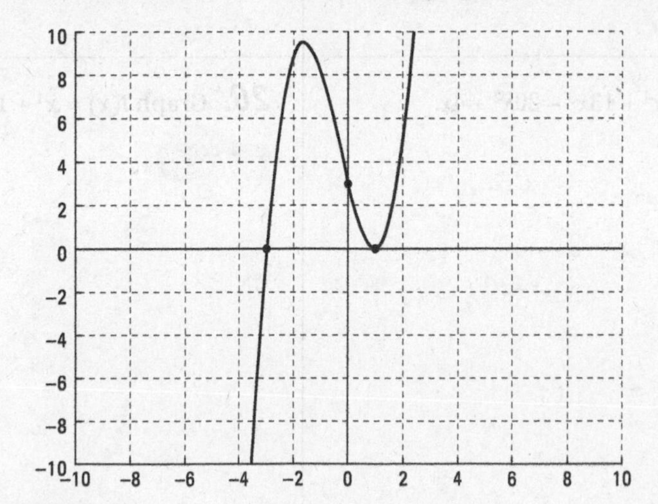

Q. Graph the equation $f(x) = x^3 + 8x^2 + 22x + 20$.

A. **See the graph. This is the second example from the section on solving high- order polynomials. You found one real root of $x = -2$, as well as the complex conjugates $x = -3 \pm i$.**

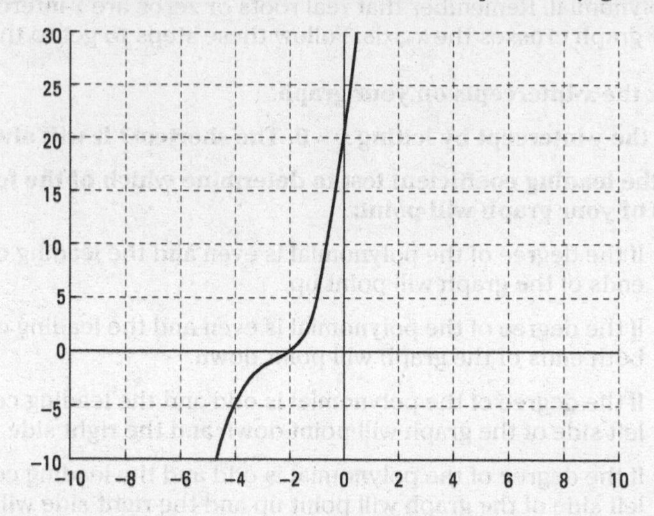

17. Graph $f(x) = x^4 + 2x^3 - 13x^2 - 14x + 24$.

Solve It

18. Graph $f(x) = 6x^4 - 7x^3 - 18x^2 + 13x + 6$.

Solve It

19. Graph $f(x) = 12x^4 + 13x^3 - 20x^2 + 4x$.

Solve It

20. Graph $f(x) = x^4 + 10x^3 + 38x^2 + 66x + 45$.

Solve It

Answers to Problems on Roots and Degrees

This section contains the answers for the practice questions presented in this chapter. We suggest you read the following explanations if your answers don't match up with ours (or if you just want a refresher on solving a particular type of problem).

1 Solve the equation $2y^2 + 5y = 12$. The answer is $y = \frac{3}{2}, -4$.

Begin with any quadratic equation by getting 0 on one side of the equation. In this case, subtract 12 from both sides: $2y^2 + 5y - 12 = 0$. Now, begin factoring by multiplying the leading term ($2y^2$) and the constant term (–12) to get $-24y^2$. List all the factors of this, in pairs: $-y$ and $24y$, y and $-24y$, $-2y$ and $12y$, $2y$ and $-12y$, $-3y$ and $8y$, $3y$ and $-8y$, $-4y$ and $6y$, and $4y$ and $-6y$. The correct pair that adds up to the linear term is $-3y$ and $8y$. Split up the trinomial into a polynomial using this magic pair: $2y^2 - 3y + 8y - 12 = 0$. Factor by grouping: $y(2y - 3) + 4(2y - 3) = 0$. Look at both terms and notice that each contains $(2y - 3)$ — that's a greatest common factor! Factor out the GCF: $(2y - 3)(y + 4) = 0$. Use the zero product property: $2y - 3 = 0$, $2y = 3$, $y = \frac{3}{2}$; and $y + 4 = 0$, $y = -4$.

2 Solve the equation $16m^2 - 8m + 1 = 0$. The answer is $m = \frac{1}{4}$.

Why is there only one answer? Oh right, it's a double root, probably. Factor it to find out: $(16m^2)(1) = 16m^2$, with the factors being m and $16m$, $-m$ and $-16m$, $2m$ and $8m$, $-2m$ and $-8m$, $4m$ and $4m$, and $-4m$ and $-4m$. The winning pair is the last one. Now, create the polynomial $16m^2 - 4m - 4m + 1 = 0$ and group it to get $4m(4m - 1) - 1(4m - 1) = 0$. Next, factor out the GCF: $(4m - 1)(4m - 1) = 0$. Notice that both factors are the same. Your answer is the same root twice! $4m - 1 = 0$, $4m = 1$, $m = \frac{1}{4}$.

3 Solve the equation $x^3 + x^2 = 9x + 9$. The answer is $x = -3, -1$, and 3.

You need to get 0 on one side first: $x^3 + x^2 - 9x - 9 = 0$ will do. If you group the polynomial into two sets of two, you get a greatest common factor: $x^2(x + 1) - 9(x + 1) = 0$. This also has a GCF in it: $(x + 1)(x^2 - 9) = 0$. Notice that the right factor is a difference of squares and will factor again: $(x + 1)(x - 3)(x + 3) = 0$. Set each factor equal to 0 and solve: $x + 1 = 0$, $x = -1$; $x - 3 = 0$, $x = 3$; $x + 3 = 0$, $x = -3$.

4 Solve the equation $\frac{1}{6}x^2 + \frac{2}{3}x = 2$. The answer is $x = 2$ and $x = -6$.

We decided to make things different and get 0 on one side first. We're kidding, of course! You always have to get 0 on one side to solve polynomials that are second degree or higher:

$$\frac{1}{6}x^2 + \frac{2}{3}x - 2 = 0$$

Next, we multiply every term by the least common multiple of 6 to get rid of those pesky fractions. This gives you the polynomial $x^2 + 4x - 12 = 0$. This factors to $(x + 6)(x - 2) = 0$. The zero product property gets you to the two solutions: $x = -6$ and $x = 2$.

5 Solve $x^2 - 10 = 2x$. The answer is $1 \pm \sqrt{11}$.

Get 0 on one side first: $x^2 - 2x - 10 = 0$. This equation doesn't factor, so use the quadratic formula to solve:

$$x = \frac{-(-2) \pm \sqrt{(-2)^2 - 4(1)(-10)}}{2(1)} = \frac{2 \pm \sqrt{4 + 40}}{2} = \frac{2 \pm \sqrt{44}}{2} = \frac{2 \pm 2\sqrt{11}}{2} = 1 \pm \sqrt{11}$$

6 Solve $7x^2 - x + 2 = 0$. The answer is "no real solution."

This equation also doesn't factor, so use the quadratic formula to solve:

$$x = \frac{-(-1) \pm \sqrt{(-1)^2 - 4(7)(2)}}{2(7)} = \frac{1 \pm \sqrt{1-56}}{14} = \frac{1 \pm \sqrt{-55}}{14}$$

The negative sign under the square root tells you that the solution is not a real number — no real solution exists. The solution involves complex numbers, which are presented in Chapter 11.

7 Solve $x^2 - 4x - 7 = 0$ by completing the square. The answer is $2 \pm \sqrt{11}$.

This time you're asked to complete the square. Make sure you always follow your teacher's directions. Begin by adding the 7 to both sides: $x^2 - 4x = 7$. Now factor out the leading coefficient: $1(x^2 - 4x) = 7$. Take half of –4 and square it, and add that inside the parentheses to get $1(x^2 - 4x + 4)$. You need to keep the equation balanced by multiplying the coefficient and the new term you just added inside the parentheses and adding that to the opposite side. Because the coefficient is 1, that's not that hard: $1(x^2 - 4x + 4) = 7 + 4$. Now factor the trinomial: $1(x-2)^2 = 11$, and divide the leading coefficient: $(x-2)^2 = 11$. Take the square root of both sides: $x - 2 = \pm\sqrt{11}$. Add the 2: $x = 2 \pm \sqrt{11}$.

8 Solve $-2.31x^2 - 4.2x + 6.7 = 0$. The solutions are approximately –2.84 and 1.02.

Those ugly decimals should make you reach immediately for a calculator and plug away at the quadratic formula.

$$x = \frac{-(-4.2) \pm \sqrt{(-4.2)^2 - 4(-2.31)(6.7)}}{2(-2.31)} = \frac{4.2 \pm \sqrt{17.64 + 61.908}}{-4.62} = \frac{4.2 \pm \sqrt{79.548}}{-4.62} = \frac{4.2 \pm 8.919}{-4.62}$$

Take your time through these types of problems. Simplify to get the final two answers: –2.84 and 1.02.

9 Solve the equation $2x^3 + 3x^2 - 18x + 8 = 0$. The zeros are $x = -4$, ½, and 2.

This third-degree equation has, at most, three real roots. The two changes in sign in f(x) show two or zero positive roots, and the one change in sign in f($-x$) shows one negative root. The list of possible rational zeros is: ±1, ±½, ±2, ±4, ±8.

Start off by testing $x = 2$.

$$\begin{array}{r|rrrr} 2 & 2 & 3 & -18 & 8 \\ & & 4 & 14 & -8 \\ \hline & 2 & 7 & -4 & 0 \end{array}$$

The depressed polynomial is $2x^2 + 7x - 4$, which factors to $(2x-1)(x+4)$, which tells you that the other two roots are $x = ½$ and –4.

10 Solve the equation $12x^4 + 13x^3 - 20x^2 + 4x = 0$. The roots are $x = 0$, ¼, –2, and ⅔.

Factor out the GCF in all the terms first: $x(12x^3 + 13x^2 - 20x + 4) = 0$. The first factor gives you one solution immediately: $x = 0$. Now concentrate on the leftover polynomial inside the parentheses and solve: $12x^3 + 13x^2 - 20x + 4 = 0$. This third-degree polynomial has at most three real roots, two or zero of which are positive and one of which is negative. The list of possibilities is: ±1, ±½, ±⅓, ±¼, ±⅙, ±1/12, ±2, ±⅔, ±4, and ±⅓.

Start off by testing $x = -2$.

$$\begin{array}{r|rrrr} -2 & 12 & 13 & -20 & 4 \\ & & -24 & 22 & -4 \\ \hline & 12 & -11 & 2 & 0 \end{array}$$

The depressed polynomial this time is $12x^2 - 11x + 2$. This factors to $(4x - 1)(3x - 2)$, which gets you to the last two roots: $x = \frac{1}{4}$ and $x = \frac{2}{3}$.

11 Solve the equation $x^3 + 7x^2 + 13x + 4 = 0$. The answers are $x = -4$ and $\frac{-3 \pm \sqrt{5}}{2}$.

This cubic polynomial has a maximum of three real roots. None of them are positive and three or one of them is negative. The list of possibilities this time (ignoring all the positives) is: –1, –2, and –4.

Start off by testing –4.

$$\begin{array}{r|rrrr} -4 & 1 & 7 & 13 & 4 \\ & & -4 & -12 & -4 \\ \hline & 1 & 3 & 1 & 0 \end{array}$$

The depressed polynomial $x^2 + 3x + 1$ doesn't factor, but the quadratic formula reveals that the last two solutions are $x = \frac{-3 \pm \sqrt{5}}{2}$.

12 Find the roots of the equation $x^4 + 10x^3 + 38x^2 + 66x + 45$. The roots are $x = -3$ (double root) and $x = -2 \pm i$.

This fourth-degree polynomial has no positive roots and 4, 2, or 0 negative roots. The list of possibilities to pick from is: –1, –3, –5, –9, 15, –45.

Start off with $x = -3$.

$$\begin{array}{r|rrrrr} -3 & 1 & 10 & 38 & 66 & 45 \\ & & -3 & -21 & -51 & -45 \\ \hline & 1 & 7 & 17 & 15 & 0 \end{array}$$

This time, when you test it again, it works.

$$\begin{array}{r|rrrr} -3 & 1 & 7 & 17 & 15 \\ & & -3 & -12 & -15 \\ \hline & 1 & 4 & 5 & 0 \end{array}$$

You're left with the depressed polynomial $x^2 + 4x + 5$, which doesn't factor, but you can use the quadratic formula to find that the last two roots are complex: $x = -2 \pm i$.

13 Find the lowest order polynomial with leading coefficient as 1 that has –3, –2, 4, and 6 as its roots. The answer is $x^4 - 5x^3 - 20x^2 + 60x + 144$.

Use the factor theorem to help you figure this one out. If $x = -3$, then $x + 3$ is one of the factors. Similarly, if $x = -2$, then $x + 2$ is a factor; if $x = 4$, then $x - 4$ is a factor; and if $x = 6$, then $x - 6$ is a factor. If you take all the factors and multiply them, you get $(x + 3)(x + 2)(x - 4)(x - 6)$. FOIL the first two binomials to get $x^2 + 5x + 6$ and the second two binomials to get $x^2 - 10x + 24$. Multiply your way through those two polynomials:

$$\begin{array}{r} x^2 + \ \ 5x + \ \ 6 \\ x^2 - \ 10x + \ 24 \\ \hline 24x^2 + 120x + 144 \\ -10x^3 - 50x^2 - \ \ 60x \qquad\qquad \\ x^4 + \ 5x^3 + \ \ 6x^2 \qquad\qquad\qquad\qquad \\ \hline x^4 - \ 5x^3 - 20x^2 + \ 60x + 144 \end{array}$$

You end up with the polynomial $x^4 - 5x^3 - 20x^2 + 60x + 144$.

14 **Find the lowest order polynomial with leading coefficient as 1 that has 2, $4 + 3i$, and $4 - 3i$ as its roots. The answer is $x^3 - 10x^2 + 41x - 50$.**

This time the factors are $x - 2$, $x - 4 - 3i$, and $x - 4 + 3i$. In cases like these, it's easier to multiply the complex numbers first. When you do that, you end up with the trinomial $x^2 - 8x + 25$. Now multiply that by the binomial to end up with the polynomial: $x^3 - 10x^2 + 41x - 50$.

15 **Factor the polynomial $6x^4 - 7x^3 - 18x^2 + 13x + 6 = 0$. The answer is $(x - 2)(x - 1)(3x + 1)(2x + 3)$.**

You're still using the factor theorem, but this time you have to find the roots first. The roots are $x = 2, 1, -1/3$, and $-3/2$. This means that $x - 2$, $x - 1$, $x + 1/3$, and $x + 3/2$ are your factors. You can get rid of those fractions by multiplying each term of the factor by the LCD. In other words, multiply $x + 1/3$ by 3 and $x + 3/2$ by 2. This finally gives you $(x - 2)(x - 1)(3x + 1)(2x + 3)$.

16 **Factor the polynomial $x^4 + 10x^3 + 38x^2 + 66x + 45$. The answer is $(x^2 + 4x + 5)(x + 3)^2$.**

This is the same equation that appears in Question 12. It has two non-real roots: $x = -2 \pm i$ and $x = -3$, a double root. This means your factors are $(x + 2 + i)(x + 2 - i)(x + 3)(x + 3)$. You multiply out the two non-real roots to come up with a polynomial factor and get $(x^2 + 4x + 5)(x + 3)^2$.

17 **Graph $f(x) = x^4 + 2x^3 - 13x^2 - 14x + 24$. See the following graph for the answer.**

The *x*-intercepts are: $x = -4, -2, 1$, and 3. Mark those on the graph first. Then find the *y*-intercept: $y = 24$. The leading coefficient test tells you that both ends of this graph point up. Here's the graph:

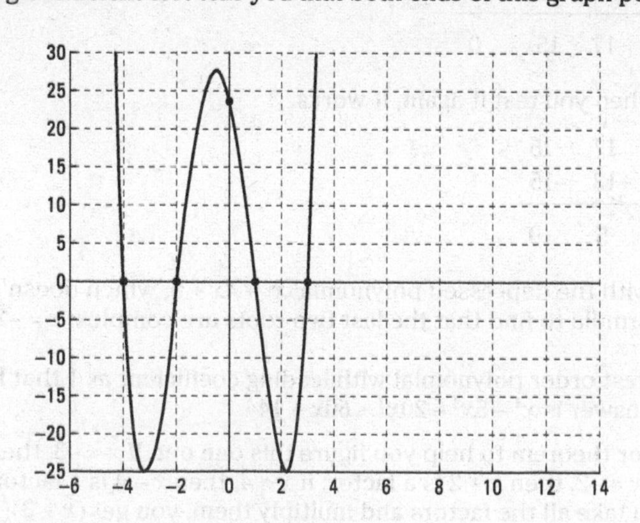

18 Graph $f(x) = 6x^4 - 7x^3 - 18x^2 + 13x + 6$. See the following graph for the answer.

You found the roots for this polynomial in Question 15: $x = -\frac{3}{2}, -\frac{1}{3}, 1$, and 2. Mark those on the graph first. Then find that $y = 6$ is the y-intercept. The leading coefficient test tells you that both ends of this graph point up. Here's the graph:

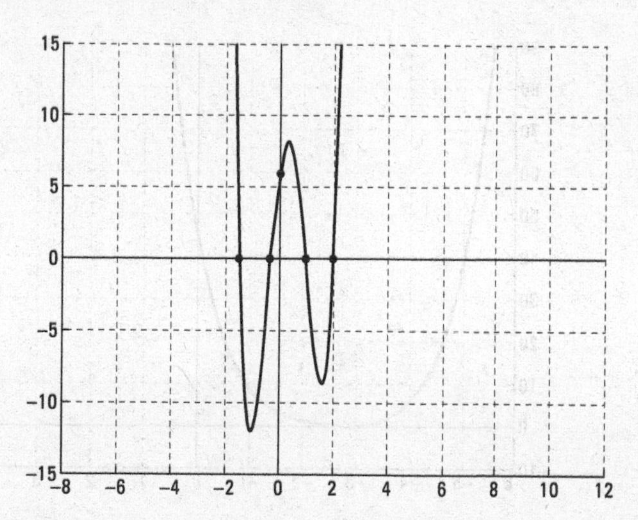

19 Graph $f(x) = 12x^4 + 13x^3 - 20x^2 + 4x$. See the following graph for the answer.

This is the same equation that appears in Question 10, where you found that the solutions are $x = -2, 0, \frac{1}{4}$, and $\frac{2}{3}$. This polynomial has no constant, so $y = 0$ is the y-intercept. This graph crosses at the origin. The leading coefficient test tells you that both ends of the graph point up. Here's the graph:

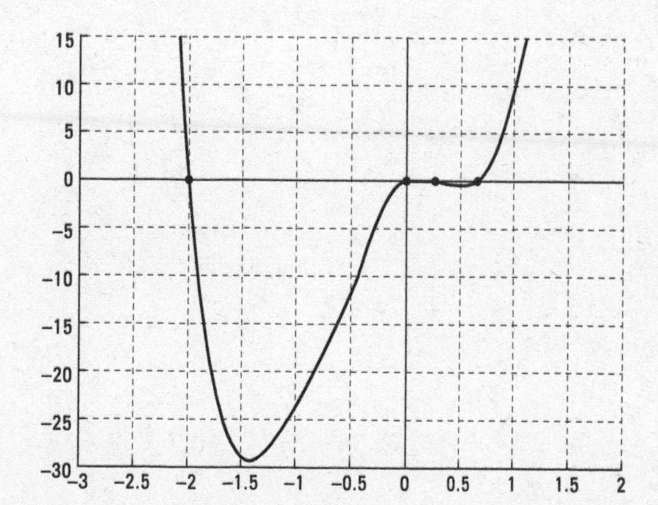

20 Graph $f(x) = x^4 + 10x^3 + 38x^2 + 66x + 45$. See the following graph for the answer.

Question 12 has roots of $x = -3$ (as a double root); the other roots are imaginary. The graph will bounce at this point. The *y*-intercept is 45. The leading coefficient test tells you that both ends of the graph point up. Here's the graph:

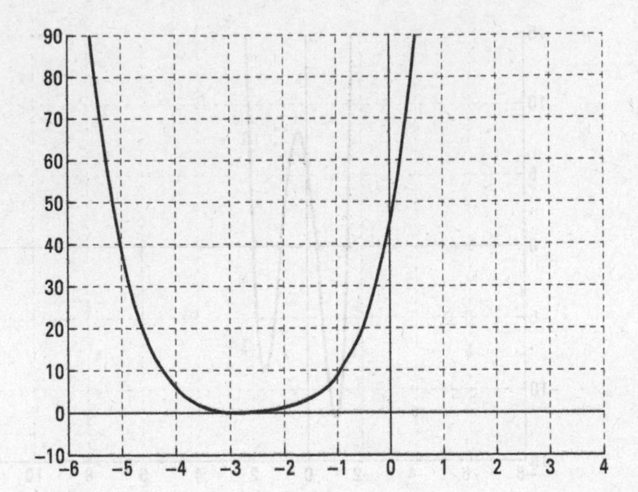

Chapter 5

Exponential and Logarithmic Functions

In This Chapter

- Figuring out exponential functions
- Looking at logarithmic functions
- Using exponents and logs to solve equations
- Working with exponential word problems

Exponential growth is simply the idea that something gets bigger and bigger (or, in the case of exponential decay, smaller and smaller) very fast. Exponential and logarithmic functions can be used to describe growth or decay. They have many practical applications, such as determining population growth, measuring the acidity of a substance, and calculating compound interests and loan payments. In addition, they're central to many concepts in calculus (a good reason to master them in pre-calculus!).

In this chapter, you practice solving equations, simplifying expressions, and graphing exponential and logarithmic functions. In addition, you practice manipulating functions to solve equations and practically applying the concepts to word problems.

Getting Bigger (Or Smaller) All the Time — Solving Exponential Functions

Exponential functions are functions in which the variable is in the exponent. When the base of the exponent is greater than 1, the function gets really big really fast, and when it's less than 1, it gets really small really fast.

The exponent indicates the power of the expression. The base can be any positive constant except 1, including a special constant that mathematicians and scientists define as *e*. This irrational constant, *e*, has a value of approximately 2.7183, and it's extremely useful in exponential expressions (and in logarithms, but we're getting ahead of ourselves).

Solving exponential equations requires that you recall the basic rules of exponents:

$$c^a \cdot c^b = c^{a+b}$$

$$\left(\frac{c}{d}\right)^a = \frac{c^a}{d^a}$$

$$\frac{c^a}{c^b} = c^{a-b}$$

$$c^{-a} = \frac{1}{c^a}$$

$$\left(c^a\right)^b = c^{ab}$$

$$(c \cdot d)^a = c^a \cdot d^a$$

$$c^0 = 1$$

If $c^a = c^b$ then $a = b$, provided c is not 0 or 1.

When graphing exponential equations, it's important to recall the tricks for transforming graphs (see Chapter 3 for a refresher).

Q. Solve for x in $8^{4x+12} = 16^{2x+5}$.

A. $x = -4$. First, in order to utilize the rules of exponents, it's helpful if both expressions have the same base. So, knowing that $8 = 2^3$ and $16 = 2^4$, by factoring and rewriting using exponents, you can rewrite both sides of the equation with a base of 2: $2^{3(4x+12)} = 2^{4(2x+5)}$. Now that your bases are the same, you can set your exponents equal to each other (using properties of exponents): $3(4x + 12) = 4(2x + 5)$. Next, you can simplify using the distributive property of equality: $12x + 36 = 8x + 20$. Finally, you can solve algebraically: $4x + 36 = 20$; $4x = -16$; $x = -4$.

Q. Sketch the graphs of (A) $y = 2^x$, (B) $y = 2^x + 1$, (C) $y = 2^{x+3}$, (D) $y = 2^{-x}$, and (E) $y = -2^x$, all on the same set of axes.

A. Graphs B–E are all transformations of the first graph, Graph A (see Chapter 3 for a review of transformations of graphs). By adding 1 to Graph A, the result is Graph B, a shift up of 1 unit. By adding 3 to the exponent of Graph A, the result is Graph C, shifted 3 units to the left. Graph D is the result of making the exponent negative, which results in a reflection over the y-axis, and Graph E, created by negating the entire function, results in the reflection of the graph over the x-axis. See the following resulting graphs.

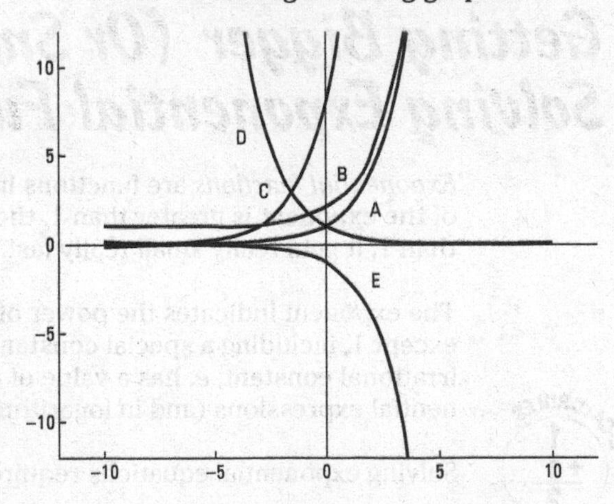

1. Solve for x in $27^{x+3} = 81^{3x-9}$.

Solve It

2. Solve for x in $e^{2x-4} = e^{6x+8}$.

Solve It

3. Solve for x in $(5^{2x} - 1)(25^{x} - 125) = 0$.

Solve It

4. Solve for x in $3 \cdot 9^{x} - 8 = -7$.

Solve It

5. Sketch the graph of $y = -3^x - 4$.

Solve It

6. Sketch the graph of $y = -3e^{x-2}$.

Solve It

No Getting Stumped Here: Edgy Logarithm Solutions

Just as multiplying by the reciprocal is another way to write division, *logarithms* are simply another way to write exponents. Exponential and logarithmic functions are inverses of each other. In other words, logarithmic functions are really just another way of writing exponential functions. So you may ask, "Why do you need both?" Well, logarithms are extremely helpful for an immense number of practical applications. In fact, before the invention of computers, logarithms were the only way to perform many complex computations in physics, chemistry, astronomy, and engineering.

For solving and graphing logarithmic functions (logs), remember this inverse relationship and you'll be sawing . . . er, solving logs in no time! Here's this relationship in equation form (the double arrow means "if and only if"):

$$y = \log_b x \leftrightarrow b^y = x$$

Where b is greater than 0 and not equal to 1. Observe that $x = b^y > 0$.

Just as with exponential functions, the base can be any positive number except 1, including e. In fact, a base of e is so common in science and calculus that $\log_e$ has its own special name: ln. Thus, $\log_e x = \ln x$.

Similarly, $\log_{10}$ is so commonly used that it's often just written as log (without the written base).

Remember our review of domain from Chapter 3? The domain for the basic logarithm $y = \log_b x$ is $x > 0$. Therefore, when you're solving logarithmic functions, it's important to check for extraneous roots (review Chapter 1).

Here are more properties that are true for *any* logarithm:

$\log_b 1 = 0$

$\log_b b = 1$

The product rule: $\log_b(a \cdot c) = \log_b a + \log_b c$

The quotient rule: $\log_b\left(\frac{a}{c}\right) = \log_b a - \log_b c$

The power rules: $\log_b a^c = c \cdot \log_b a$; $\log_b b^x = x$

$b^{\log_b x} = x$, for $x > 0$

$\log_b a = \frac{\log_c a}{\log_c b}$

If $\log_b a = \log_b c$, then $a = c$

Using these properties, simplifying logarithmic expressions and solving logarithmic equations is a snap (we did say logs, not twigs, right?).

Q. Rewrite the following logarithmic expression to a single log: $3\log_5 x + \log_5(2x-1) - 2\log_5(3x+2)$.

A. $\log_5\left[\frac{x^3(2x-1)}{(3x+2)^2}\right]$

Using the properties of logs, begin by rewriting the coefficients as exponents: $3\log_5 x = \log_5 x^3$ and $2\log_5(3x+2) = \log_5(3x+2)^2$. Next, rewrite the addition of the first two logs as the log of the product of two functions: $\log_5 x^3 + \log_5(2x-1) = \log_5 x^3(2x-1)$. Last, rewrite the difference of these two logs as the log of the quotient: $\log_5 x^3(2x-1) - \log_5 x^3(3x+2)^2 =$

$$\log_5\left[\frac{x^3(2x-1)}{(3x+2)^2}\right].$$

Q. Sketch the graphs of (A) $y = \log_2 x$, (B) $y = 1 + \log_2 x$, (C) $y = \log_2(x+3)$, and (D) $y = -\log_2 x$, all on the same set of axes.

A. First, in the following figure, you can see that Graphs B–D are transformations of Graph A. Graph B is a shift of 1 up, Graph C is a shift of 3 to the left, and Graph D is a reflection of Graph A over the x-axis. Second (nifty trick here), these are all inverses of Graphs A–D in the preceding section on solving exponential functions. Another way to graph logarithms is to change the log to an exponential function. Using the properties of logarithms, find the inverse function by switching x and y, graph the inverse, and reflect every point over the line $y = x$. For a review of inverses, see Chapter 3. Here, we stick with transforming the parent graph.

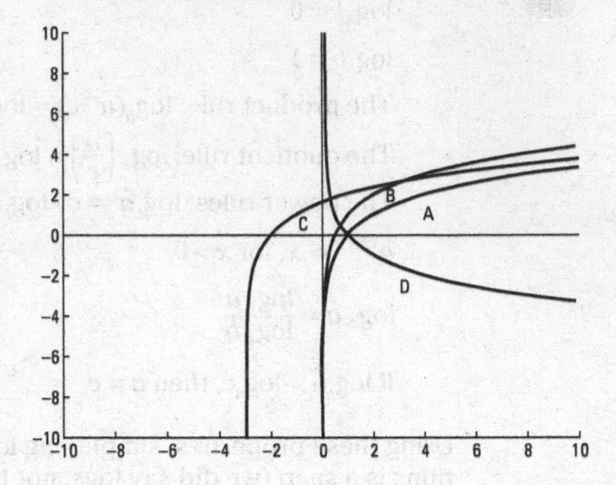

7. Rewrite the given expression as a single logarithm:
$\ln 4x + 3\ln(x-2) - 2(\ln 2x + \ln(3x-4))$.

Solve It

8. Solve for x in $\log_7 4 + \log_7(x+4) - 2\log_7 2 = \log_7(x-2) + \frac{1}{2}\log_7 9$.

Solve It

9. Solve for x in $\ln x + \ln(2x-1) = \ln 1$.

Solve It

10. Find $\log_b(48b)$ if $\log_b 2 = 0.36$ and $\log_b 3 = 0.56$.

Solve It

11. Sketch the graph of $y = -3 + \log(x + 2)$.

Solve It

12. Sketch the graph of $y = \ln(x - 2) + 4$.

Solve It

Putting Exponents and Logs Together

Now, we show you how to put these two lovely functions together. By keeping in mind the inverse relationship ($y = \log_b x \leftrightarrow b^y = x$), you can solve even more complex problems. Aren't you excited?!

A helpful key to remember when solving equations using exponents and logs is that if the variable is in the exponent, you convert the equation into logarithmic form. This is especially helpful if you use *natural log* (ln) or the *common log* ($\log_{10} x$), often referred to as just $\log x$, because you can plug the expression into your calculator to get a decimal approximation of the solution.

One pitfall to avoid when manipulating logs relates to the products and quotients of logs. Remember: $\log 18 - \log 3 = \log\left(\frac{18}{3}\right)$, not $\frac{\log 18}{\log 3}$. These are entirely different expressions. In fact, if you plug them into your calculator, you can see that $\log\left(\frac{18}{3}\right) = \log 6 \approx 0.778$, whereas $\left(\frac{\log 18}{\log 3}\right) \approx 2.631$. The same can be said for products and logs: $\log 6 + \log 7 = \log(6 \cdot 7)$, not $\log(6 + 7)$ nor $(\log 6)(\log 7)$.

Q. Solve for x in $\log(50x + 250) - \log x = 2$.

A. $x = 5$. Start by combining the logs as a quotient: $\log\left(\frac{50x+250}{x}\right) = 2$. Next, rewrite in exponential form (remember that log means $\log_{10}$): $\left(\frac{50x+250}{x}\right) = 10^2$. Because $10^2 = 100$, you can rewrite the equation as $\frac{50x+250}{x} = 100$. After cross-multiplying, you can then solve algebraically: $50x + 250 = 100x$; $250 = 50x$; $x = 5$.

Q. Solve for x in $3^x = 2^{x+2}$.

A. $x = \frac{\ln 4}{\ln\left(\frac{3}{2}\right)} \approx 3.419$. First, recognize that the variable is in the exponent of each term, so you can easily remedy that by taking either log or ln of both sides. We use ln, but it really doesn't make a difference. So $3^x = 2^{x+2}$ becomes $\ln 3^x = \ln 2^{(x+2)}$. Then, you can use properties of logarithms to solve. Start by changing the exponents to coefficients: $x \cdot \ln 3 = (x + 2)\ln 2$. Using algebra, you can distribute the ln2 across $(x + 2)$: $x \cdot \ln 2 + 2 \cdot \ln 2$. Still using algebra, get the terms with the variable on the same side by subtracting $x \cdot \ln 2$ to the opposite side: $x \cdot \ln 3 - x \cdot \ln 2 = 2 \cdot \ln 2$. Then, using the distributive property again, remove the x as a greatest common factor: $x(\ln 3 - \ln 2) = 2 \cdot \ln 2$. Finally, isolate x by dividing $\ln 3 - \ln 2$ from both sides: $x = \frac{\ln 2^2}{\ln 3 - \ln 2}$. Last, use the quotient rule to simplify and get $x = \frac{\ln 4}{\ln\left(\frac{3}{2}\right)} \approx 3.419$.

13. Solve for x in $\log(x + 6) - \log(x - 3) = 1$.

Solve It

14. Solve for x in $3^x = 5$.

Solve It

15. Solve for x in $4^x - 4 \cdot 2^x = -3$.

Solve It

16. Solve for x in $3^x = 5^{2x-3}$.

Solve It

Using Exponents and Logs in Practical Applications

When will I ever use this? Well, in addition to being used in mathematics courses, exponential functions actually have many practical applications. Common uses of exponential functions include figuring compound interest, computing population growth, and doing radiocarbon dating (no, that's not some new online matchmaking system). In fact, these uses are so common, many teachers make you memorize their formulas. If you need nonstandard formulas to do a problem, they're provided in the question itself.

Here are formulas for interest rate and half-life:

- **Compound interest formula:** $A = P\left(1+\frac{r}{n}\right)^{nt}$ where A is the amount after t time in years compounded n times per year if P dollars are invested at annual interest rate r.
- **Continuous compound interest formula:** $A = Pe^{r \cdot t}$, where A is the amount after t time in years if P dollars are invested at interest rate r with interest compounded continuously throughout the year.
- **Formula for the remaining mass of a radioactive element:** $M(x) = c \cdot 2^{-x/h}$ where $M(x)$ is the mass at the time x, c is the original mass of the element, and h is the half-life of the element.

Q. If you deposit $600 at 5.5% interest compounded continuously, what will your balance be in 10 years?

A. $1,039.95. Because this is continuous compound interest, you use the formula $A = Pe^{rt}$ when you're solving for A: $A = \$600e^{(0.055)(10)}$. Plugging this into a calculator, you get approximately $1,039.95.

Q. How old is a piece of bone that has lost 60 percent of its carbon-14? (The half-life of carbon-14 is 5,730 years.)

A. Approximately 7,575 years old. You can figure out this problem using the formula for half-life. First, because 60 percent of the carbon-14 is gone, the mass of carbon remaining is 40 percent, so you can write the present mass as $0.40c$. Therefore, the equation will be: $0.40c = c \cdot 2^{-x/5,730}$. You can start solving this by cancelling c from both sides: $0.40 = 2^{-x/5,730}$. Taking the natural log of both sides allows you to move the variable from the exponent position: $\ln 0.40 = \ln 2^{-x/5,730}$; $\ln 0.40 = \left(\frac{-x}{5,730}\right)\ln 2$. From here, you can solve algebraically:

$$\frac{\ln 0.40}{\ln 2} = \frac{x}{-5,730};\ -5,730\frac{\ln 0.40}{\ln 2} \approx 7,575 \text{ years.}$$

17. If you deposit $3,000 at 8-percent interest per year compounded quarterly, in approximately how many years will the investment be worth $10,500?

Solve It

18. The half-life of Krypton-85 is 10.4 years. How long will it take for 600 grams to decay to 15 grams?

Solve It

19. The deer population in a certain area in year t is approximately $P(t) = \frac{3{,}000}{1 + 299 \cdot e^{-0.56t}}$. When will the deer population reach 2,000?

Solve It

20. If you deposit $20,000 at 6.5-percent interest compounded continuously, how long will it take for you to have $1 million?

Solve It

Answers to Problems on Exponential and Logarithmic Functions

Following are the answers to problems dealing with exponential and logarithmic functions. We also provide guidance on getting the answers if you need to review where you went wrong.

1 Solve for x in $27^{x+3} = 81^{3x-9}$. The answer is $x = 5$.

First, rewrite 27 as 3^3 and 81 as 3^4. Simplify the power to get $3^{3(x+3)} = 3^{4(3x-9)}$. Now that the bases are the same, set the two exponents equal to each other: $3(x + 3) = 4(3x - 9)$, and then solve for x: $3x + 9 = 12x - 36$; $-9x = -45$; $x = 5$.

2 Solve for x in $e^{2x-4} = e^{6x+8}$. The answer is $x = -3$.

Start by setting the exponents equal to each other: $2x - 4 = 6x + 8$; then solve algebraically: $4x = -12$; $x = -3$.

3 Solve for x: $(5^{2x} - 1)(25^x - 125) = 0$. The answer is $x = 0, \frac{3}{2}$.

Using the fact that $25 = 5^2$, replace 25^x with 5^{2x} to get $(5^{2x} - 1)(5^{2x} - 125) = 0$. Next, set each factor equal to 0 using the zero product property (see Chapter 4 for a review) and solve: First, $5^{2x} - 1 = 0$, which implies $5^{2x} = 1$, and because anything to the power of 0 equals 1, $5^{2x} = 5^0$. Therefore, $2x = 0$ and thus $x = 0$. Second, $5^{2x} - 125 = 0$, which implies $5^{2x} = 125$, and because 125 is equal to 5^3, rewrite the second equation as $5^{2x} = 5^3$. Set the exponents equal to each other, $2x = 3$, and solve for $x = \frac{3}{2}$. Both solutions work.

4 Solve for x in $3 \cdot 9^x - 8 = -7$. The answer is $x = -\frac{1}{2}$.

Start by isolating the exponential expression: $3 \cdot 9^x = 1$; $9^x = \frac{1}{3}$. Next, replace 9^x with 3^{2x} and $\frac{1}{3}$ with 3^{-1}, so $3^{2x} = 3^{-1}$. Set the exponents equal to each other: $2x = -1$, and solve for $x = -\frac{1}{2}$.

5 Sketch the graph of $y = -3^x - 4$. See the graph for the answer.

The y-intercept is $(0, -5)$. The graph of this function is the basic exponential graph of $y = 3^x$ reflected across the x-axis first and then shifted 4 units down.

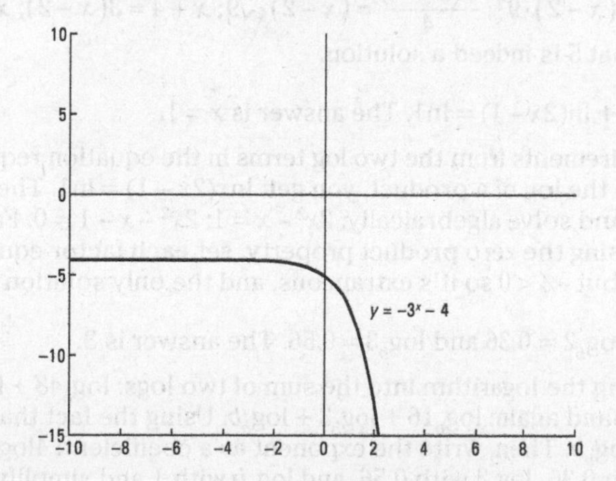

6 Sketch the graph of $y = -3e^{x-2}$. See the graph for the answer.

The y-intercept is approximately $(0, -0.406)$. The graph of this function is the basic exponential graph of $y = e^x$ shifted 2 units to the right, reflected across the x-axis, and followed with a vertical stretch by a factor of 3.

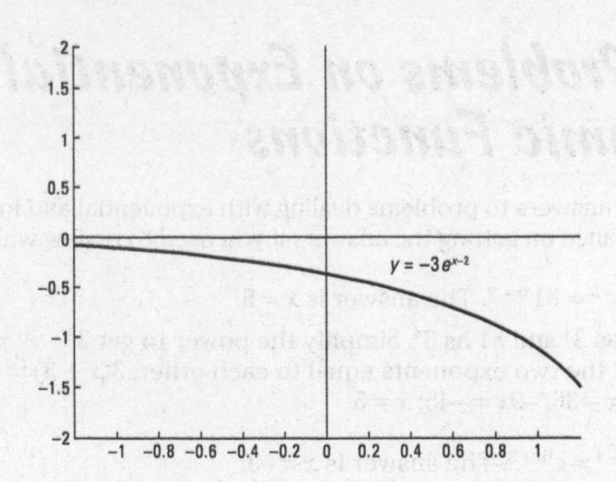

7 Rewrite the given expression as a single logarithm: $\ln 4x + 3\ln(x-2) - 2(\ln 2x + \ln(3x-4))$. The answer is $\ln\left[\frac{(x-2)^3}{x(3x-4)^2}\right]$.

Begin by rewriting coefficients as exponents: $\ln 4x + \ln(x-2)^3 - \left(\ln(2x)^2 + \ln(3x-4)^2\right)$. Next, rewrite the first two logarithms as a single product: $\ln 4x(x-2)^3 - \left(\ln 4x^2 + \ln(3x-4)^2\right)$. Then use the product rule in the parentheses to get $\ln 4x(x-2)^3 - \ln 4x^2(3x-4)^2$. Finally, write the difference of logarithms as the log of a quotient and reduce the 4 in the numerator and denominator: $\ln\left[\frac{(x-2)^3}{x(3x-4)^2}\right]$.

8 Solve for x in: $\log_7 4 + \log_7(x+4) - 2\log_7 2 = \log_7(x-2) + \frac{1}{2}\log_7 9$. The answer is $x = 5$.

The first step is to write the coefficients as exponents: $\log_7 4 + \log_7(x+4) - \log_7 2^2 = \log_7(x-2) + \log_7 9^{1/2}$. Next, rewrite the sums and differences of logs as the logs of products and quotients: $\log_7\frac{4(x+4)}{2^2} = \log_7(x-2)\cdot 9^{\frac{1}{2}}$. Using the rules of logarithms, set $\frac{4(x+4)}{2^2} = (x-2)\cdot 9^{\frac{1}{2}}$. Solve algebraically: $\frac{4(x+4)}{2^2} = (x-2)\cdot 9^{\frac{1}{2}}$; $\frac{4(x+4)}{4} = (x-2)\cdot\sqrt{9}$; $x+4 = 3(x-2)$; $x+4 = 3x-6$; $2x = 10$; $x = 5$. It's easy to see that 5 is indeed a solution.

9 Solve for x in $\ln x + \ln(2x-1) = \ln 1$. The answer is $x = 1$.

The domain requirements from the two log terms in the equation require $x > ½$. Rewriting the sum of natural logs as the log of a product, you get: $\ln x(2x-1) = \ln 1$. Then, using rules of logarithms, set $x(2x-1) = 1$ and solve algebraically: $2x^2 - x = 1$; $2x^2 - x - 1 = 0$. Factor the quadratic: $(2x+1)(x-1) = 0$ and, using the zero product property, set each factor equal to 0. The solutions are $x = -½$ and $x = 1$, but $-½ < 0$ so it's extraneous, and the only solution is $x = 1$.

10 Find $\log_b(48b)$ if $\log_b 2 = 0.36$ and $\log_b 3 = 0.56$. The answer is 3.

Start by expanding the logarithm into the sum of two logs: $\log_b 48 + \log_b b$. Next, factoring the 48 into $16 \cdot 3$, expand again: $\log_b 16 + \log_b 3 + \log_b b$. Using the fact that $16 = 2^4$, write the first log: $\log_b 2^4 + \log_b 3 + \log_b b$. Then, write the exponent as a coefficient: $4\log_b 2 + \log_b 3 + \log_b b$. Last, replace $\log_b 2$ with 0.36, $\log_b 3$ with 0.56, and $\log_b b$ with 1 and simplify: $4(0.36) + 0.56 + 1 = 3$.

11 Sketch the graph of $y = -3 + \log(x+2)$. See the graph for the answer.

The y-intercept is approximately $(0, -2.699)$. The graph of this function is the basic logarithmic graph of $y = \log x$ shifted 2 units to the left and 3 units down.

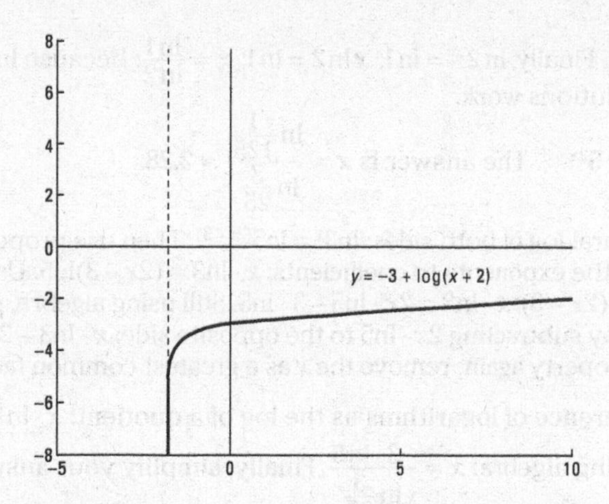

12 Sketch the graph of $y = \ln(x-2)+4$. See the graph for the answer.

There's a vertical asymptote at $x = 2$. The graph of this function is the basic logarithmic graph of $y = \ln x$ shifted 2 units to the right and 4 units up.

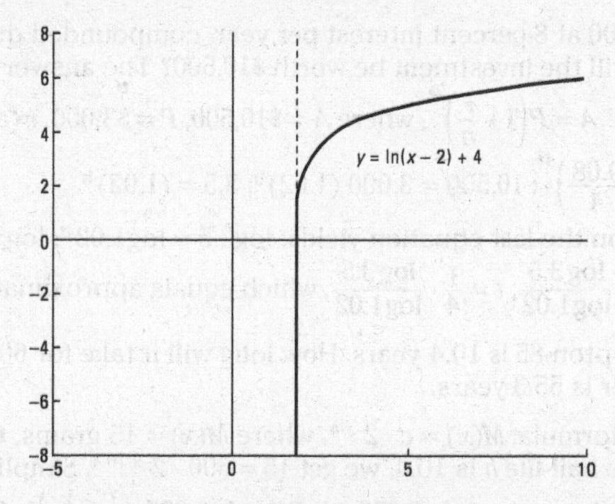

13 Solve for x in $\log(x+6)-\log(x-3)=1$. The answer is $x = 4$.

Begin by writing the difference of logs as the log of a quotient: $\log\frac{x+6}{x-3}$. Next, rewrite the logarithm as an exponent: $\frac{x+6}{x-3}=10^1$. Then, solve algebraically: $x+6=10(x-3)$; $x+6=10x-30$; $36=9x$; $x=4$. You see 4 is indeed a solution.

14 Solve for x in $3^x = 5$. The answer is $x=\frac{\ln 5}{\ln 3}$ or $x \approx 1.465$.

First, take the natural log of both sides: $\ln 3^x = \ln 5$. Then use the power rule to simplify: $x \cdot \ln 3 = \ln 5$. Last, divide both sides by ln3: $x=\frac{\ln 5}{\ln 3}$ or $x \approx 1.464$.

15 Solve for x in $4^x - 4\cdot 2^x = -3$. The answer is $x=\frac{\ln 3}{\ln 2}\approx 1.585$ and $x = 0$.

Start by using the fact that $4 = 2^2$ and rewrite 4^x as 2^{2x}: $2^{2x}-4\cdot 2^x=-3$. Add 3 to both sides: $2^{2x}-4\cdot 2^x+3=0$. Notice that this is the same thing as $(2^x)^2-4\cdot 2^x+3=0$. So you can substitute y for 2^x to arrive at $y^2-4y+3=0$. Now, you can factor and solve using the zero product property: $(y-3)(y-1)=0$; $y=3$ and $y=1$. Then, resubstitute 2^x for y: $2^x=3$ and $2^x=1$. Taking the natural log of each side, you can solve for x by using the rules of logarithms: $\ln 2^x=\ln 3$; $x\ln 2=\ln 3$; $x=\frac{\ln 3}{\ln 2}$, your first solution. Finally, $\ln 2^x = \ln 1$; $x \ln 2 = \ln 1$; $x = \frac{\ln 1}{\ln 2}$; because $\ln 1 = 0$, $\frac{\ln 1}{\ln 2} = 0$, your second solution. Both solutions work.

16 Solve for x in $3^x = 5^{2x-3}$. The answer is $x = \frac{\ln \frac{1}{125}}{\ln \frac{3}{25}} \approx 2.28$.

First, take the natural log of both sides: $\ln 3^x = \ln 5^{(2x-3)}$. Then use properties of logarithms to solve. Start by changing the exponents to coefficients: $x \cdot \ln 3 = (2x-3)\ln 5$. Using algebra, you can distribute the ln5 across $(2x-3)$: $x \cdot \ln 3 = 2x \cdot \ln 5 - 3 \cdot \ln 5$. Still using algebra, get the terms with the variable on the same side by subtracting $2x \cdot \ln 5$ to the opposite side: $x \cdot \ln 3 - 2x \cdot \ln 5 = -3 \cdot \ln 5$. Then, using the distributive property again, remove the x as a greatest common factor: $x(\ln 3 - 2\ln 5) = -3 \cdot \ln 5$.

Combine the difference of logarithms as the log of a quotient: $x \cdot \ln\left(\frac{3}{5^2}\right) = -3 \cdot \ln 5$.

Next, isolate x using algebra: $x = \frac{-3 \cdot \ln 5}{\ln \frac{3}{25}}$. Finally, simplify your answer using the rules of exponents:

$\frac{\ln 5^{-3}}{\ln \frac{3}{25}} = \frac{\ln \frac{1}{125}}{\ln \frac{3}{25}}$, which equals approximately 2.28.

17 If you deposit \$3,000 at 8-percent interest per year, compounded quarterly, in approximately how many years will the investment be worth \$10,500? The answer is approximately 15.82 years.

Using the equation: $A = P\left(1+\frac{r}{n}\right)^{nt}$, where A = \$10,500, P = \$3,000, r (as a decimal) = 0.08, and $n = 4$:

$10{,}500 = 3{,}000\left(1+\frac{0.08}{4}\right)^{4t}$; $10{,}500 = 3{,}000\,(1.02)^{4t}$; $3.5 = (1.02)^{4t}$.

Using logarithms on the last equation yields: $\log 3.5 = \log 1.02^{4t}$; $\log 3.5 = 4t \log 1.02$; finally, solve algebraically: $4t = \frac{\log 3.5}{\log 1.02}$; $t = \frac{1}{4} \cdot \frac{\log 3.5}{\log 1.02}$, which equals approximately 15.82 years.

18 The half-life of Krypton-85 is 10.4 years. How long will it take for 600 grams to decay to 15 grams? The answer is 55.3 years.

Using the half-life formula: $M(x) = c \cdot 2^{-x/h}$, where $M(x)$ = 15 grams, the original mass c is 600 grams, and the half-life h is 10.4, we get $15 = 600 \cdot 2^{-x/10.4}$. Simplifying to $0.025 = 2^{-x/10.4}$, we can solve using logarithms: $\ln 0.025 = \ln 2^{\frac{-x}{10.4}}$, $\ln 0.025 = \left(\frac{-x}{10.4}\right)\ln 2$, $\frac{\ln 0.025}{\ln 2} = \frac{-x}{10.4}$ and which approximates to 55.3 years.

19 The deer population in a certain area in year t is approximately $P(t) = \frac{3{,}000}{1+299 \cdot e^{-0.56t}}$. When will the deer population reach 2,000? The answer is approximately 11.4 years.

Here, you simply plug in 2,000 for $P(t)$ and solve: $2{,}000 = \frac{3{,}000}{1+299 \cdot e^{-0.56t}}$; $2{,}000\left(1+299 \cdot e^{-0.56t}\right) = 3{,}000$; $1+299 \cdot e^{-0.56t} = 1.5$; $299 \cdot e^{-0.56t} = 0.5$; $e^{-0.56t} = 0.00167$; $-0.56t = \ln 0.00167$; $t = \frac{\ln 0.00167}{-0.56}$, which equals approximately 11.4 years.

20 If you deposit \$20,000 at 6.5-percent interest compounded continuously, how long will it take for you to have \$1 million? The answer is approximately 60.2 years.

Using the equation for continuous compound interest, $A = Pe^{rt}$, where the amount A is \$1,000,000, the initial investment P is \$20,000, and the interest rate r in decimal form is 0.065, you get: $1{,}000{,}000 = 20{,}000e^{0.065t}$. Simplify and use logarithms to solve: $50 = e^{0.065t}$; $\ln 50 = 0.065t$; $\frac{\ln 50}{0.065} = t$, which equals approximately 60.2 years.

Part II

Trig Is the Key: Basic Review, the Unit Circle, and Graphs

"Periodic functions?! I can tell you all about periodic functions. My whole life is about periodic functions!"

In this part . . .

You should be familiar with the basics of trigonometry from your geometry class — right triangles, trig ratios, and angles, for example. But your Algebra II course may or may not have expanded on those ideas to prepare you for the direction that pre-calc is going to take you. For this reason, we assume that you've never seen this stuff before. We don't want to leave you behind on our journey.

This part begins with trig ratios and word problems and then moves on to the unit circle: how to build it and how to use it. We solve some trig equations and make and measure arcs. Graphing trig functions is a major component of pre-calc, so we walk you through how to graph each of the six functions.

Chapter 6

Basic Trigonometry and the Unit Circle

In This Chapter

- Working with the six trigonometric ratios
- Making use of right triangles to solve word problems
- Using the unit circle to find points, angles, and right triangle ratios
- Isolating trig terms to solve trig equations
- Calculating arc lengths

Ah . . . trigonometry, the math of triangles! Invented by the ancient Greeks, trigonometry is used to solve problems in navigation, astronomy, and surveying. Think of a sailor lost at sea. All he has to do is triangulate his position against two other objects, such as two stars, and calculate his position using — you guessed it — trigonometry!

In this chapter, we review the basics of right triangle trigonometry. Then we show you how to apply that knowledge to the unit circle, a very useful tool for graphically representing trigonometric ratios and relationships. From there, you can solve trig equations. Finally, we combine these concepts so that you can apply them to arcs. The ancient Greeks didn't know what they started with trigonometry, but the modern applications are endless!

It's All Right-Triangle Trig — Finding the Six Trigonometric Ratios

Dude! Did you see that? He just did a 2π on his board! Huh? Oh . . . we mean a 360. In geometry, angles are measured in degrees (°), with 360° describing a full circle on a coordinate plane (or skateboard). However, in pre-calculus, you also use another measure for angles: *radians*. Radians, from the word radius, are usually designated without a symbol for units. Because both radians and degrees are used often in pre-calc, you see both in use here.

To convert radians to degrees and vice versa, you use the fact that $360° = 2\pi$ radians, or $180° = \pi$. Therefore, to convert degrees to radians, you simply multiply by the ratio $\pi/180$. Similarly, to convert radians to degrees, you can multiply by the ratio $180/\pi$.

When solving right triangles or finding all the sides and angles (θ), it's important to remember the six basic trigonometric ratios: sine ($\sin\theta$), cosine ($\cos\theta$), tangent ($\tan\theta$), cosecant ($\csc\theta$), secant ($\sec\theta$), and cotangent ($\cot\theta$). The first three are the most important to remember, because the second three are reciprocals of the first three. In other words, $\sin\theta = 1/\csc\theta$, $\cos\theta = 1/\sec\theta$, and $\tan\theta = 1/\cot\theta$.

Acronyms! We love 'em . . . LOL — otherwise known as Laugh Out Loud! One of the most famous acronyms in math is SOHCAHTOA. It helps you remember the first three trigonometric ratios:

$$\text{Sine} = \frac{\text{Opposite}}{\text{Hypotenuse}}$$

$$\text{Cosine} = \frac{\text{Adjacent}}{\text{Hypotenuse}}$$

$$\text{Tangent} = \frac{\text{Opposite}}{\text{Adjacent}}$$

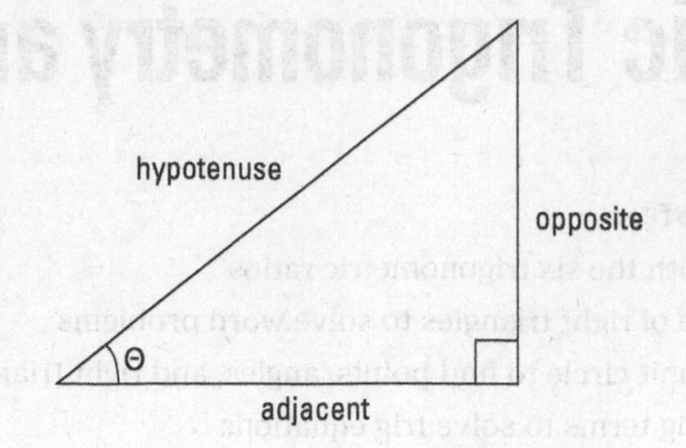

Q. Given $\triangle KLM$ in the following figure, find $\sin\angle K$.

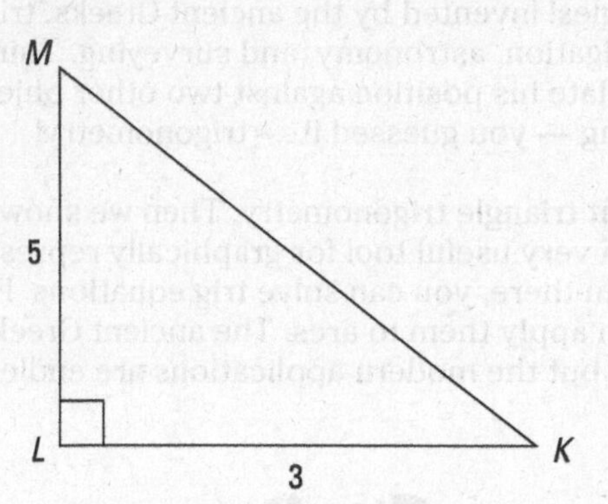

A. $\sin\angle K = \frac{5\sqrt{34}}{34}$. Because $\sin\angle K = {}^{\text{opp}}\!/_{\text{hyp}}$, you first need to find the hypotenuse. To do this, you need to use the Pythagorean Theorem, which says that $(\text{leg})^2 + (\text{leg})^2 = (\text{hypotenuse})^2$. Using this, you find that $3^2 + 5^2 = (\text{hyp})^2$, and therefore $34 = (\text{hyp})^2$, so the hypotenuse is $\sqrt{34}$. Plugging this into your sine ratio, you get $\sin\angle K = \frac{5}{\sqrt{34}}$. You can rationalize the denominator and get $\sin\angle K = \frac{5\sqrt{34}}{34}$.

Q. Solve $\triangle RST$, referring to the following figure.

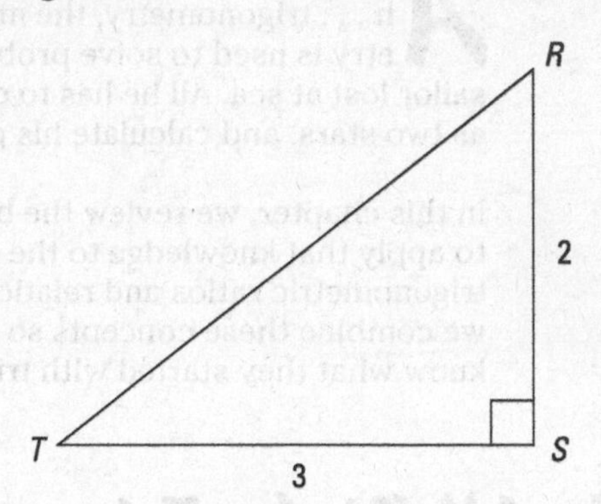

A. $RT = \sqrt{13}$, $\angle T \approx 33.7°$, $\angle R \approx 56.3°$.

Remember, solving a triangle means finding all the angles and sides. So you start by using the Pythagorean Theorem to find the hypotenuse: $2^2 + 3^2 = (\text{hyp})^2$, $\text{hyp} = \sqrt{13}$. Next, use any trigonometric ratio to find an angle. You can use $\sin\angle T = \frac{2}{\sqrt{13}}$. To get the angle by itself, you use the fact that the inverse operation of sin is $\sin^{-1}$, or arcsine. Thus, you get $\angle T = \sin^{-1}\left(\frac{2}{\sqrt{13}}\right)$, which you can find using your calculator to be 33.7° (not in radians). Or, if you want to use radians, $\angle T$ is 0.59. We prefer degrees for now. Lastly, using the fact that the angles of a triangle add up to 180°, you can find $\angle R$: $180 - (90 + 33.7°) = 56.3°$.

1. Find $\cos A$ in $\triangle ABC$.

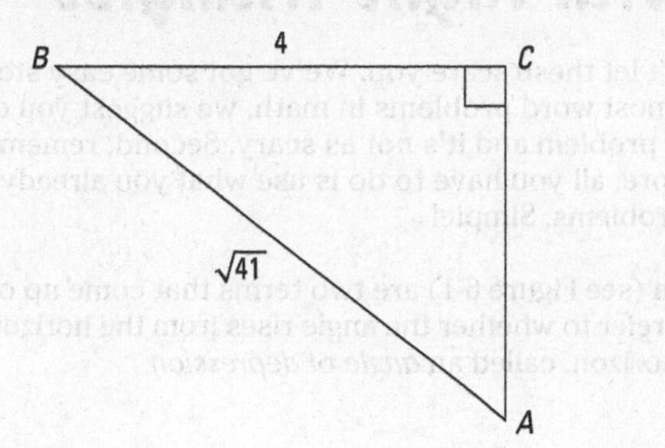

Solve It

2. Solve $\triangle DEF$.

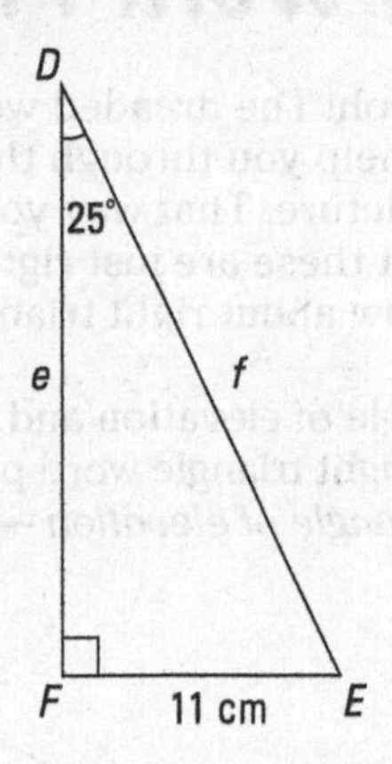

Solve It

3. Find $\angle Q$ in $\triangle QRS$ (round to the nearest tenth).

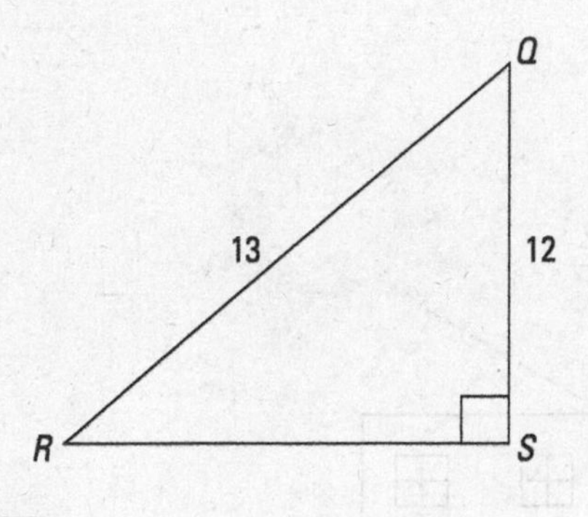

Solve It

4. Find the six trigonometric ratios of $\angle R$ in $\triangle QRS$ from Question 3.

Solve It

Solving Word Problems with Right Triangles

Uh-oh! The dreaded word problems! Don't let these scare you. We've got some easy steps to help you through them. First, as with most word problems in math, we suggest you draw a picture. That way you can visualize the problem and it's not as scary. Second, remember that these are just right triangles. Therefore, all you have to do is use what you already know about right triangles to solve the problems. Simple!

Angle of elevation and angle of depression (see Figure 6-1) are two terms that come up often in right triangle word problems. They just refer to whether the angle rises from the horizon — an *angle of elevation* — or falls from the horizon, called an *angle of depression.*

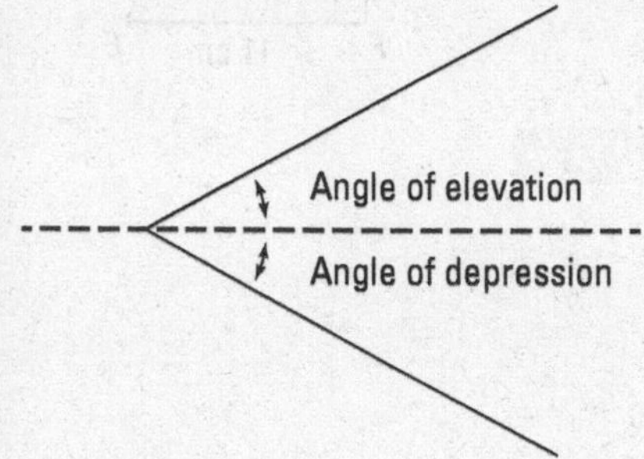

Figure 6-1: The ups and downs: angle of elevation and angle of depression.

Q. When the sun is at an angle of elevation of 32°, a building casts a shadow that's 152 feet from its base. What is the height of the building?

A. The building is approximately 95 feet tall. Okay, remember your steps. Step one, draw a picture:

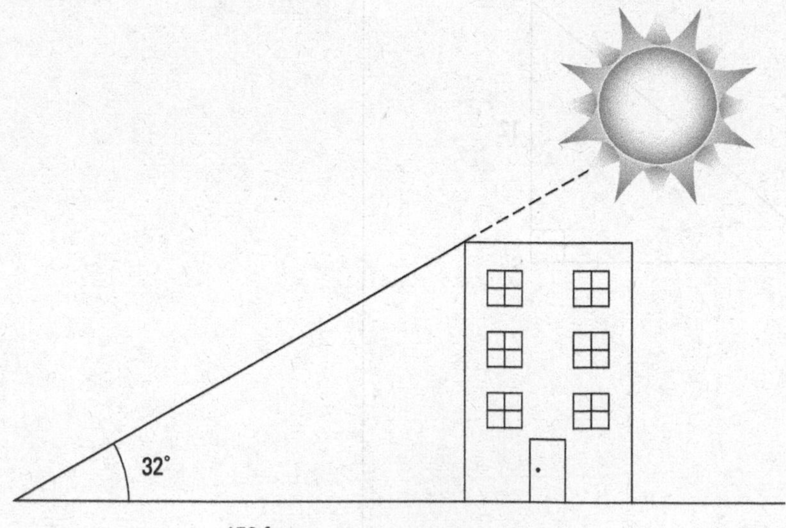

Step two, recall what you know about right triangles. Because you want to find the building's height, x, which is opposite the angle, and you have the shadow length, which is adjacent to the angle, you can use the tangent ratio. Setting up your ratio, you get $\tan 32° = x/152$, or $x = 152 \cdot \tan 32°$. Using a calculator, you find that the building height is approximately 95 feet.

Q. Two boat captains whose boats are in a straight line from a lighthouse look up to the top of the lighthouse at the same time. The captain of Boat A sees the top of the 40-foot lighthouse from an angle of elevation of 45°, while the captain of Boat B sees the top of the lighthouse from an angle of elevation of 30°. How far are the boats from each other, to the nearest foot?

A. The boats are 29 feet apart. Ooh . . . this is a tricky one! But don't let it scare you — it's completely doable! First, remember to draw a picture. In this case, you may want to draw three: one for the lighthouse and both boats, and two separate pictures with one boat each:

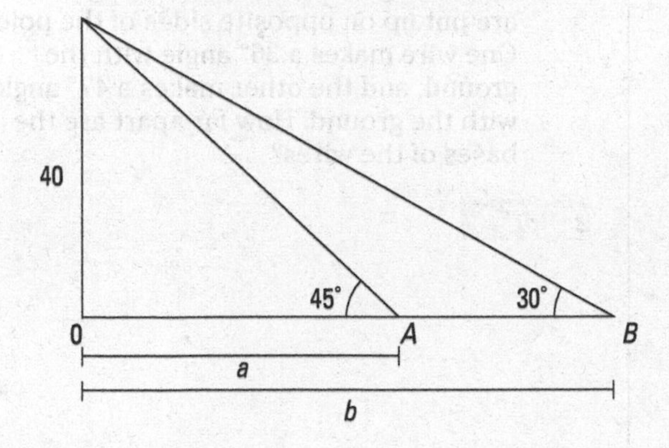

From the picture, you can see that to find the distance between the boats, you need to find the distance that each boat is from the base of the lighthouse and subtract Boat A's distance from the distance of Boat B. Because the angle of elevation is 45° for Boat A, you can set up the trigonometric ratio: $\tan 45° = 40/a$. Solving for a, you find that the distance from Boat A to the base of the lighthouse is 40 feet. Similarly, you can set up a trigonometric ratio for Boat B's distance: $\tan 30° = 40/b$. Solving, you get that $b = 69$ feet. Subtracting these two distances, you find that the distance between the boats is 29 feet. Whew!

5. Romero wants to deliver a rose to his girlfriend, Jules, who is sitting on her balcony 24 feet above the street. If Romero has a 28-foot ladder, at what angle must he place the bottom of the ladder to reach his love, Jules?

Solve It

6. Sam needs to cross a river. He spies a bridge directly ahead of him. Looking across the river, he sees that he's 27° below the bridge from the other side. How far must he walk on his side of the river to reach the bridge if the bridge length is 40 feet?

Solve It

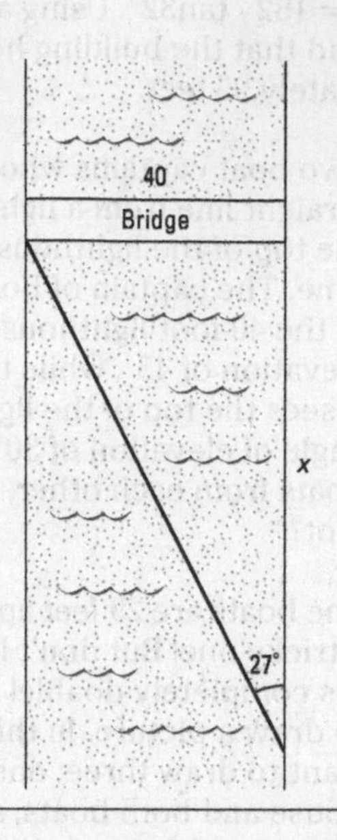

7. Paul, a 6-foot-tall man, is holding the end of a kite string 5 feet above the ground. The string of the kite is 75 feet long at 35° of elevation. Paulette, Paul's 5-foot-tall daughter, is directly underneath the kite. How far above Paulette's head is the kite?

Solve It

8. To hold up a 100-foot pole, two guide wires are put up on opposite sides of the pole. One wire makes a 36° angle with the ground, and the other makes a 47° angle with the ground. How far apart are the bases of the wires?

Solve It

Unit Circle and the Coordinate Plane: Finding Points and Angles

The unit circle is a very useful tool in pre-calculus. The information it provides can help you solve problems very quickly. Essentially, the *unit circle* is a circle with a radius (r) of one unit, centered on the origin of a coordinate plane. Thinking of the trigonometric ratios you've been dealing with in terms of x and y values, where x is adjacent to the angle, y is opposite the angle, and r is the hypotenuse, allows you to make a right triangle by using a point on the unit circle and the x-axis. This is often called *point-in-plane,* and it results in an alternate definition of the six trigonometric ratios:

$\sin\theta = y/r$ $\csc\theta = r/y$

$\cos\theta = x/r$ $\sec\theta = r/x$

$\tan\theta = y/x$ $\cot\theta = x/y$

When graphing on a coordinate plane, how you measure your angles is important. In pre-calculus, the angle always begins on the positive side of the x-axis, called the *initial side.* Any angle in this position is in *standard position.* The angle can extend to anywhere on the plane, ending on what's called the *terminal side.* Any angles that have different measures but have the same terminal side are called *co-terminal angles.* These can be found by adding or subtracting 360° or 2π to any angle.

From the initial side, an angle that moves in the counterclockwise direction has a *positive measure,* and an angle that moves in the clockwise direction has a *negative measure.*

Q. Find three co-terminal angles of 520°.

A. Sample answer: 160°, –200°, and 880°, but other answers are possible. To get these angles, you simply add or subtract multiples of 360° from 520°. $520° - 360° = 160°$; $520° - 2 \cdot 360° = -200°$; and $520° + 360° = 880°$.

Q. Evaluate the six trigonometric ratios of the point (2, –3).

A. $\sin\theta = \frac{-3\sqrt{13}}{13}$, $\cos\theta = \frac{2\sqrt{13}}{13}$, $\tan\theta = -\frac{3}{2}$, $\csc\theta = \frac{\sqrt{13}}{-3}$, $\sec\theta = \frac{\sqrt{13}}{2}$, $\cot\theta = -\frac{2}{3}$.

Start by finding the radius using the Pythagorean Theorem: $2^2 + (-3)^2 = r^2$, $4 + 9 = r^2$, $13 = r^2$, $\sqrt{13} = r$. Then, simply plug the known values into the trigonometric ratios given: $x = 2$, $y = -3$, and $r = \sqrt{13}$. Don't forget to rationalize any radicals in the denominator!

$\sin\theta = \frac{-3\sqrt{13}}{13}$, $\cos\theta = \frac{2\sqrt{13}}{13}$, $\tan\theta = -\frac{3}{2}$,

$\csc\theta = \frac{\sqrt{13}}{-3}$, $\sec\theta = \frac{\sqrt{13}}{2}$, $\cot\theta = -\frac{2}{3}$.

9. Find three co-terminal angles of $\pi/5$.

Solve It

10. Find two positive co-terminal angles of $-775°$.

Solve It

11. Evaluate the six trigonometric ratios of the point (3, 4).

Solve It

12. Evaluate the six trigonometric ratios of the point (–5, –7).

Solve It

13. Evaluate the six trigonometric ratios of the point $\left(-2, 2\sqrt{3}\right)$.

Solve It

14. Evaluate the six trigonometric ratios of the point $\left(6, -3\sqrt{5}\right)$.

Solve It

Isn't That Special? Finding Right Triangle Ratios on the Unit Circle

Well, isn't that special? Yes, it *is* special — special right triangles that is! Remember your geometry teacher drilling in 30-60-90 and 45-45-90 triangles? Well, they're back! And with good reason, because they're very common in pre-calculus, and they're the foundation of the unit circle. Using these two special triangles, you can find the specific trig values that you see on the completed unit circle in Figure 6-2.

One important point to remember about the unit circle is that the radius is 1. Therefore, the hypotenuse of any right triangle drawn from a point to the x-axis is 1. Thus, for any point, (x, y), you know that $x^2 + y^2 = 1$.

Recalling 30-60-90 triangles, the sides are in the ratio of $1 : \sqrt{3} : 2$. Therefore, if you want the hypotenuse to be 1, as it is in the unit circle, divide each side by 2. Similarly, the sides of 45-45-90 triangles are in the ratio of $1 : 1 : \sqrt{2}$. Converting to a unit circle, the values are $\frac{\sqrt{2}}{2} : \frac{\sqrt{2}}{2} : 1$.

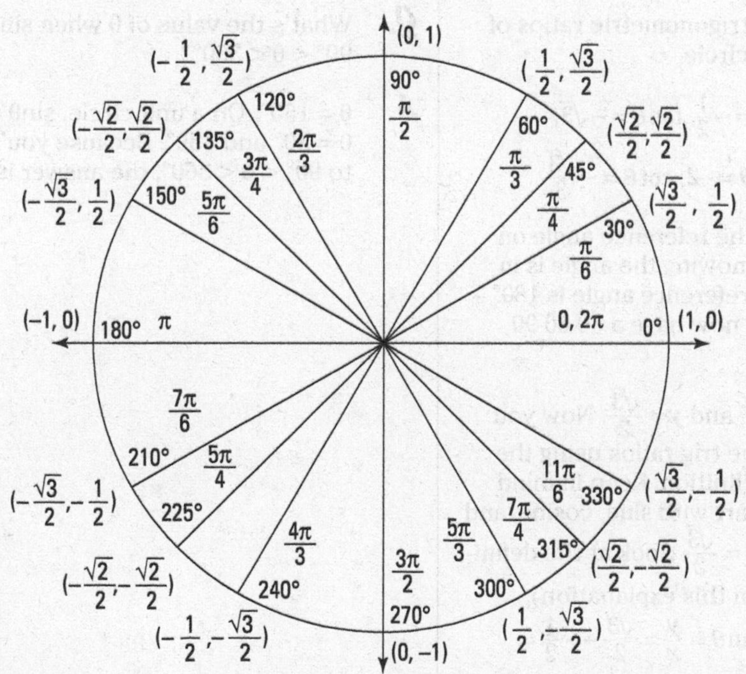

Figure 6-2: The whole unit circle.

Now, using the point-in-plane definition, the six trigonometric ratios are easy to find. In fact, because the hypotenuse is now 1, $\sin\theta = y/r$ becomes $\sin\theta = y$. Similarly, $\cos\theta = x/r$ becomes $\cos\theta = x$. Thus, any point on the unit circle has the coordinates $(\cos\theta, \sin\theta)$. Imagine the possibilities!

If you don't have a unit circle handy, you can always use *reference angles* to find your solutions. A reference angle is the angle between the x-axis and the terminal side of an angle.

It's different for each quadrant (see Figure 6-3). If the original angle is θ, then the reference angle in Quadrant I is θ. In Quadrant II, the reference angle is 180° – θ. For Quadrant III, the reference angle is θ – 180°. Lastly, Quadrant IV's reference angle is 360° – θ.

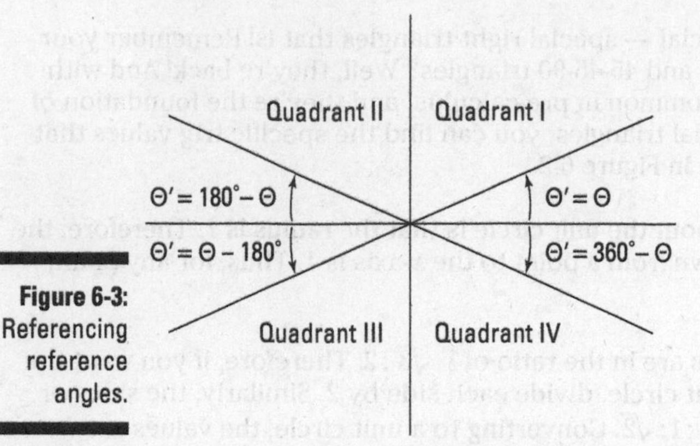

Figure 6-3: Referencing reference angles.

Q. Evaluate the six trigonometric ratios of 120° on the unit circle.

A. $\sin\theta = \frac{\sqrt{3}}{2}$, $\cos\theta = -\frac{1}{2}$, $\tan\theta = -\sqrt{3}$, $\csc\theta = \frac{2\sqrt{3}}{3}$, $\sec\theta = -2$, $\cot\theta = -\frac{\sqrt{3}}{3}$

Start by finding the reference angle on the unit circle. Knowing the angle is in Quadrant II, the reference angle is 180° – 120°, or 60°. You now have a 30-60-90 triangle!

Therefore, $x = -½$ and $y = \frac{\sqrt{3}}{2}$. Now you can easily find the trig ratios using the point-in-plane definition. Keep in mind that $r = 1$. You start with sine, cosine, and tangent: $\sin\theta = y = \frac{\sqrt{3}}{2}$ (look at the definition of y earlier in this explanation); $\cos\theta = x = -\frac{1}{2}$; $\tan\theta = \frac{y}{x} = \frac{\sqrt{3}}{2} \div -\frac{1}{2} = \frac{\sqrt{3}}{2} \cdot -\frac{2}{1} = -\sqrt{3}$

Next, do the reciprocal ratios: $\csc\theta = \frac{1}{y} = \frac{2}{\sqrt{3}} = \frac{2\sqrt{3}}{3}$; $\sec\theta = \frac{1}{x} = 1 \div -\frac{1}{2} = 1 \cdot -\frac{2}{1} = -2$; and $\cot\theta = \frac{x}{y} = -\frac{1}{2} \div \frac{\sqrt{3}}{2} = -\frac{1}{2} \cdot \frac{2}{\sqrt{3}} = -\frac{1}{\sqrt{3}} = -\frac{\sqrt{3}}{3}$

Q. What's the value of θ when sinθ = ½ and 90° < θ < 360°?

A. θ = 150°. On a unit circle, sinθ = ½ when θ = 30° and 150°. Because you're limited to 90° < θ < 360°, the answer is just 150°.

15. Evaluate the six trigonometric ratios of 225°.

Solve It

16. Find all values of θ satisfying $\cos\theta = \frac{\sqrt{3}}{2}$ and $0° < \theta < 360°$.

Solve It

17. Evaluate the six trigonometric ratios of 330°.

Solve It

18. Find all values of θ satisfying $\tan\theta = \frac{\sqrt{3}}{3}$ and $0° < \theta < 360°$.

Solve It

Solving Trig Equations

Solving trigonometric equations is just like solving regular algebraic equations, with a twist! The twist is the trigonometric term. Instead of isolating the variable, you need to isolate the trigonometric term. From there, you can use the handy unit circle to find your solution. For a complete unit circle, refer to Figure 6-2.

Given what you already know about co-terminal angles, you know that any given equation may have an infinite number of solutions. Therefore, for these examples, stick with angles that are within one positive rotation of the unit circle $0 \le x \le 2\pi$. But make sure that you check for multiple solutions within that unit circle!

Q. Solve $2\sin x = 1$ in terms of radians.

A. $x = \frac{\pi}{6}$, or $\frac{5\pi}{6}$. Because you already know how to solve $2y = 1$, you also know how to solve $2\sin x = 1$: It's $\sin x = \frac{1}{2}$. The question is what to do with it from there. Well, now you need to find the angle or angles that make the equation true. Here's where that unit circle comes in handy! Remembering that $\sin\theta = y$, you can look at the unit circle to find which angles have $y = \frac{1}{2}$. The two angles are $\frac{\pi}{6}$ and $\frac{5\pi}{6}$.

Q. Solve $2\cos^2 x - \cos x = 1$, giving answers in terms of degrees.

A. $x = 0°$, $120°$, $240°$, and $360°$. Don't let this one trip you up! Just keep in mind your amazing basic algebra skills. Observe that $\cos^2 x = (\cos x)^2$; you can think of this as $2y^2 - y = 1$. You see that it's a simple quadratic that you need to try and factor and then solve using the zero product property: $2\cos^2 x - \cos x - 1 = 0$ factors into $(2\cos x + 1)(\cos x - 1) = 0$. Using the zero product property, $2\cos x + 1 = 0$, so $2\cos x = -1$; hence, $\cos x = -\frac{1}{2}$; or $\cos x - 1 = 0$, so $\cos x = 1$. Now it's time to use those reference angles! Ask yourself this: When is $\cos x = -\frac{1}{2}$? Well, considering that $\cos x = -\frac{1}{2}$ on the unit circle, it's clear that your reference angle (θ) is $60°$, and your answer falls in Quadrants II and III. Therefore, the resulting angles are in Quad II ($180° - 60° = 120°$) and Quad III ($180° + 60° = 240°$). For your second equation, $\cos x = 1$ and $x = 0°$ or $360°$. Therefore, your four solutions are $0°$, $120°$, $240°$, and $360°$.

19. Solve for θ in $3\tan\theta - 1 = 2$.

Solve It

20. Solve for θ in $\sin^2\theta = \sin\theta$.

Solve It

21. Solve for θ in $2\cos^2\theta - 1 = 0$.

Solve It

22. Solve for θ in $4\sin^2\theta + 3 = 4$.

Solve It

23. Solve for θ in $4\sin^4\theta - 7\sin^2\theta + 3 = 0$ in degrees.

Solve It

24. Solve for θ in $\tan^2\theta - \tan\theta = 0$.

Solve It

Making and Measuring Arcs

If someone asks you how far an ant on the edge of a 6-inch CD travels if the CD spins at 120°, you probably wonder why it matters. You may even be thinking that the ant is probably messing up your CD player! But wacky math teachers love coming up with questions like that, so we're here to help you solve them.

To calculate the measure of an *arc,* a portion of the circumference of a circle like the path that pesky ant is taking, you need to remember that arcs can be measured in two ways: as an angle and as a length. As an angle, there's nothing to calculate — the measure of the arc is simply the same as the measure of the central angle. As a length, the measure of the arc is directly proportional to the circumference of the circle. If θ is measured in degrees, r is the radius, and s is the arc length facing the angle with θ; then the ratio of s over the circumference of the circle is the same as the angle value in degrees over 360. This gives you the nifty formula $s = \frac{\theta}{360} \cdot 2\pi r$. When θ is expressed in radians, the formula becomes $s = \theta \cdot r$. Use Figure 6-4 as your guide.

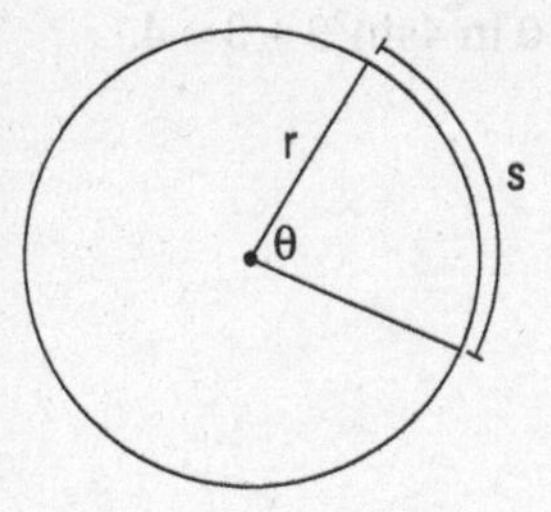

Figure 6-4: Calculating arc length and the variables involved.

Q. Back to that ant! A pesky ant is on the edge of a 6-inch CD. How far does the ant travel if the CD spins 120°?

A. The ant travels approximately 6.3 inches. We can use both formulas. However, to use the second formula, the angle needs to be in radians, so multiply the 120° by $\pi/180°$ to get $2\pi/3$. The diameter is 6 inches, so the radius is 3 inches. Using the formula $s = \theta \cdot r = 2\pi/3 \cdot 3 = 2\pi$ inches, which is approximately 6.3 inches.

25. Find the length of an arc in a circle with a radius of 4 feet if the central angle is $\pi/6$.

Solve It

26. Find the length of an arc in a circle with a diameter of 16 centimeters if the central angle is $7\pi/4$.

Solve It

27. Find the length of an arc in a circle with a radius of 18 feet if the angle is 210°.

Solve It

28. Find the radius of a circle in which an angle of 2 radians cuts an arc with a length of 42 inches.

Solve It

Answers to Problems on Basic Trig and the Unit Circle

This section contains the answers for the practice problems presented in this chapter. We suggest you read the following explanations if your answers don't match up with ours (or if you just want a refresher on solving a particular type of problem).

1 Find cos$A$ in $\triangle ABC$. The answer is $\cos A = \frac{5\sqrt{41}}{41}$.

Because $\cos\theta = {}^{\text{adj}}/_{\text{hyp}}$, you need to find the adjacent side using the Pythagorean Theorem: $x^2 + 4^2 = \left(\sqrt{41}\right)^2$, $x^2 + 16 = 41$, $x^2 = 25$, $x = 5$. Therefore, $\cos A = \frac{5}{\sqrt{41}}$. Rationalizing the denominator, $\cos A = \frac{5\sqrt{41}}{41}$.

2 Solve $\triangle DEF$. The answer is $\angle E = 65°$, side $e = 23.6$ cm, and side $f = 26$ cm.

First, because you know $\angle D$ and $\angle F$, you can find $\angle E$ by subtracting the sum of $\angle D$ and $\angle F$ from 180°: $180° - (25° + 90°) = 65°$. To find side e, you can use $\tan 65° = e/11$. Multiplying both sides by 11, you get $e = 11 \cdot \tan 65°$, which is approximately 23.6. To find side f, you can use $\sin 25° = 11/f$. Multiply both sides by f: $f \cdot \sin 25° = 11$. Divide by sin 25° to get $f = 11/\sin 25°$, which is approximately 26 cm.

3 Find $\angle Q$ in $\triangle QRS$ (round to the nearest tenth). The answer is approximately 22.6°.

Because you have the adjacent side to $\angle Q$ and the hypotenuse, you use cosine: $\cos Q = (12/13)$. To solve, take the inverse cosine of each side: $\angle Q = \cos^{-1}(12/13)$, which is approximately 22.6°.

4 Find the six trigonometric ratios of $\angle R$ in $\triangle QRS$ from Question 3. The answer is $\sin R = 12/13$, $\cos R = 5/13$, $\tan R = 12/5$, $\csc R = 13/12$, $\sec R = 13/5$, and $\cot R = 5/12$.

Start by using the Pythagorean Theorem to find the third side: $12^2 + q^2 = 13^2$, $144 + q^2 = 169$, $q^2 = 25$, $q = 5$. Then, plug the sides into the trigonometric ratios: $\sin R = 12/13$, $\cos R = 5/13$, $\tan R = 12/5$, $\csc R = 13/12$, $\sec R = 13/5$, $\cot R = 5/12$.

5 Romero wants to deliver a rose to his girlfriend, Jules, who is sitting on her balcony 24 feet above the street. If Romero has a 28-foot ladder, at what angle must he place the bottom of the ladder to reach his love, Jules? The answer is 59°.

To solve, draw a picture:

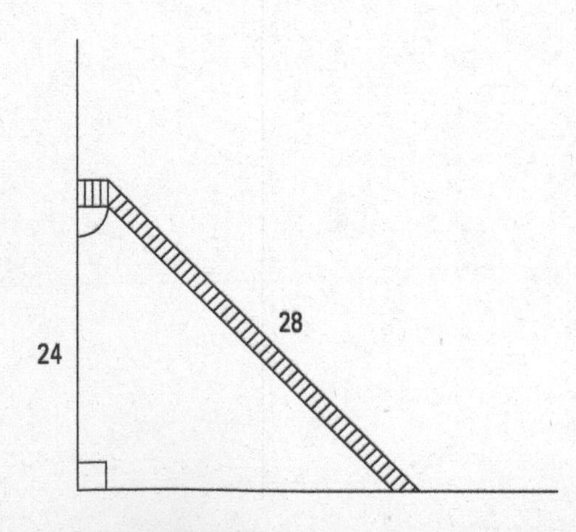

With the picture, you can see that you have the opposite side, 24 feet, opposite the angle you want and the hypotenuse, 28 feet. Therefore, you can use the sine ratio to solve: $\sin\theta = (^{24}/_{28})$. To isolate the angle, use inverse sine: $\sin\theta = \sin^{-1}(^{24}/_{28})$, which is approximately 59°.

6 Sam needs to cross a river. He spies a bridge directly ahead of him. Looking across the river, he sees that he's 27° below the bridge from the other side. How far must he walk on his side of the river to reach the bridge if the bridge length is 40 feet? The answer is 79 more feet.

First, consider the picture:

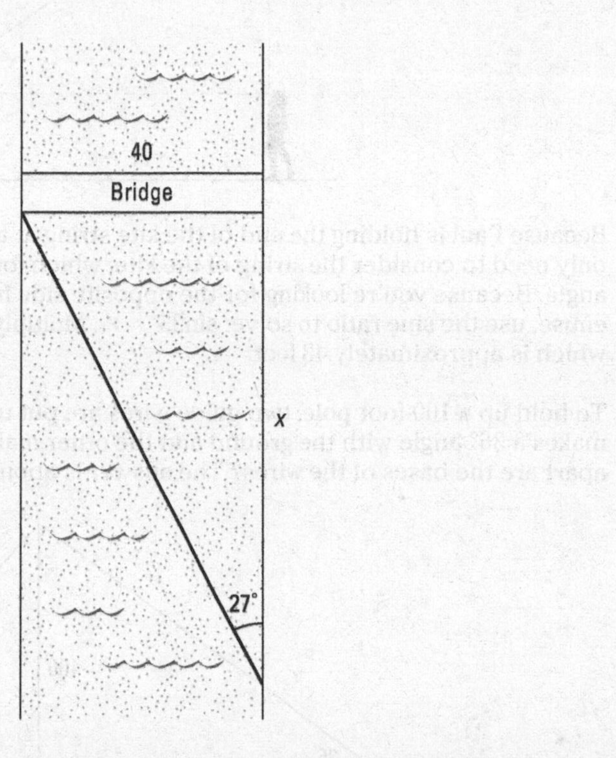

Considering that you have the opposite side from the angle, 40 feet, and you're looking for the adjacent side, you can use the tangent ratio: $\tan 27° = ^{40}/_{x}$. Multiplying both sides by x, you get $x \cdot \tan 27° = 40$. Dividing 40 by tan27°, you get that x equals approximately 79 feet.

7 Paul, a 6-foot-tall man, is holding the end of a kite string 5 feet above the ground. The string of the kite is 75-feet long at 35° of elevation. Paulette, Paul's 5-foot-tall daughter, is directly underneath the kite. How far above Paulette's head is the kite? The answer is about 43 feet.

Begin by (you guessed it!) drawing a picture:

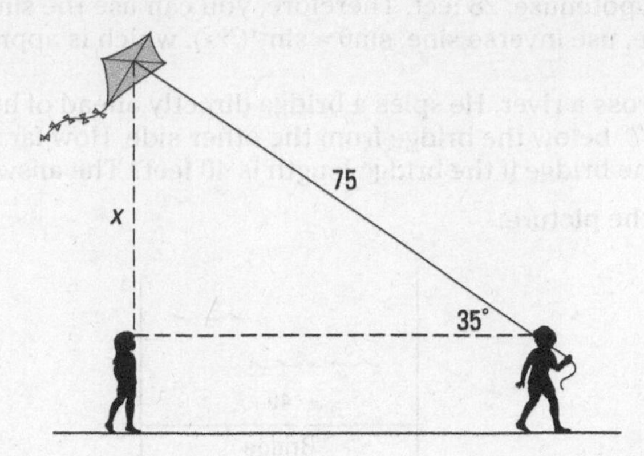

Because Paul is holding the end of the kite string at the same height as Paulette's head, you only need to consider the string of the kite, which forms the hypotenuse of the triangle and the angle. Because you're looking for the opposite side from the angle and you have the hypotenuse, use the sine ratio to solve: $\sin 35° = \frac{x}{75}$. Multiplying both sides by 75, you get $x = 75 \sin 35°$, which is approximately 43 feet.

8 To hold up a 100-foot pole, two guide wires are put up on opposite sides of the pole. One wire makes a 36° angle with the ground and the other makes a 47° angle with the ground. How far apart are the bases of the wires? The answer is about 231 feet apart.

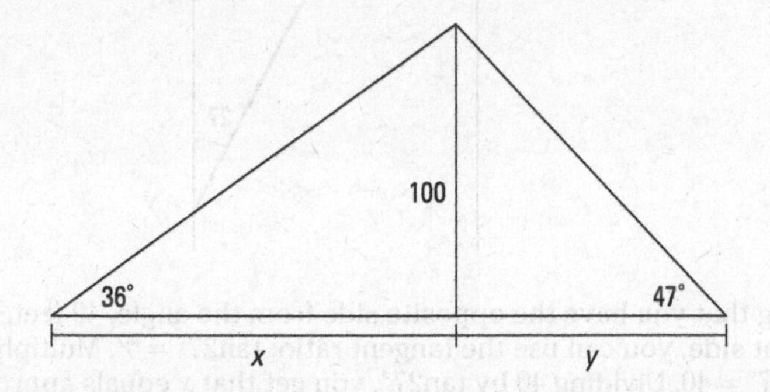

Using your picture, you can set up two tangent ratios: $\tan 36° = \frac{100}{x}$ and $\tan 47° = \frac{100}{y}$. Multiply both sides by x and y respectively: $x \cdot \tan 36° = 100$ and $y \cdot \tan 47° = 100$. Isolate the variables by dividing: $x = 100 \div \tan 36°$ and $y = 100 \div \tan 47°$. Therefore, x is approximately 137.6 feet and y is approximately 93.3. Add these together to get that the total distance apart is about 231 feet.

9 Find three co-terminal angles of $\frac{\pi}{5}$. Although there are multiple answers, three possible answers are $\frac{11\pi}{5}$, $\frac{-9\pi}{5}$, and $\frac{21\pi}{5}$.

Simply add or subtract multiples of 2π: $\frac{\pi}{5} + 2\pi = \frac{11\pi}{5}$; $\frac{\pi}{5} - 2\pi = \frac{-9\pi}{5}$; $\frac{\pi}{5} + 2 \cdot 2\pi = \frac{21\pi}{5}$.

10 Find two positive co-terminal angles of –775°. Two possible answers are 305° and 665°.

Here, just add multiples of 360° to –775° until you get two positive co-terminal angles: $-775° + 3 \cdot 360° = 305°$; $-775° + 4 \cdot 360° = 665°$.

11 Evaluate the six trigonometric ratios of the point (3, 4). The answers are $\sin\theta = \frac{4}{5}$, $\cos\theta = \frac{3}{5}$, $\tan\theta = \frac{4}{3}$, $\csc\theta = \frac{5}{4}$, $\sec\theta = \frac{5}{3}$, and $\cot\theta = \frac{3}{4}$.

First, find the radius using the Pythagorean Theorem: $3^2 + 4^2 = r^2$, $9 + 16 = r^2$, $25 = r^2$, $5 = r$. Using this and $x = 3$, $y = 4$, plug into the trigonometric ratios: $\sin\theta = \frac{4}{5}$, $\cos\theta = \frac{3}{5}$, $\tan\theta = \frac{4}{3}$, $\csc\theta = \frac{5}{4}$, $\sec\theta = \frac{5}{3}$, and $\cot\theta = \frac{3}{4}$.

12 Evaluate the six trigonometric ratios of the point (–5, –7). The answers are $\sin\theta = -\frac{7\sqrt{74}}{74}$, $\cos\theta = -\frac{5\sqrt{74}}{74}$, $\tan\theta = \frac{7}{5}$, $\csc\theta = -\frac{\sqrt{74}}{7}$, $\sec\theta = -\frac{\sqrt{74}}{5}$, and $\cot\theta = \frac{5}{7}$.

Find your radius: $(-5)^2 + (-7)^2 = r^2$, $25 + 49 = r^2$, $74 = r^2$, $\sqrt{74} = r$. Using this and the point (x, y), plug into the trig ratios and rationalize if necessary:

$\sin\theta = \frac{-7}{\sqrt{74}} = -\frac{7\sqrt{74}}{74}$; $\cos\theta = \frac{-5}{\sqrt{74}} = -\frac{5\sqrt{74}}{74}$; $\tan\theta = \frac{-7}{-5} = \frac{7}{5}$; $\csc\theta = -\frac{\sqrt{74}}{7}$; $\sec\theta = -\frac{\sqrt{74}}{5}$;

and $\cot\theta = \frac{-5}{-7} = \frac{5}{7}$.

13 Evaluate the six trigonometric ratios of the point $\left(-2, 2\sqrt{3}\right)$. The answers are $\sin\theta = \frac{\sqrt{3}}{2}$, $\cos\theta = \frac{-1}{2}$, $\tan\theta = -\sqrt{3}$, $\csc\theta = \frac{2\sqrt{3}}{3}$, $\sec\theta = -2$, and $\cot\theta = -\frac{\sqrt{3}}{3}$.

Begin by finding your radius: $(-2)^2 + \left(2\sqrt{3}\right)^2 = r^2$, $4 + 12 = r^2$, $16 = r^2$, $4 = r$. Now plug into your trig ratios and rationalize if necessary: $\sin\theta = \frac{2\sqrt{3}}{4} = \frac{\sqrt{3}}{2}$; $\cos\theta = \frac{-2}{4} = \frac{-1}{2}$; $\tan\theta = \frac{2\sqrt{3}}{-2} = -\sqrt{3}$;

$\csc\theta = \frac{4}{2\sqrt{3}} = \frac{2}{\sqrt{3}} = \frac{2\sqrt{3}}{3}$; $\sec\theta = \frac{4}{-2} = -2$ and $\cot\theta = \frac{-2}{2\sqrt{3}} = -\frac{1}{\sqrt{3}} = -\frac{\sqrt{3}}{3}$.

14 Evaluate the six trigonometric ratios of the point $\left(6, -3\sqrt{5}\right)$. The answers are $\sin\theta = -\frac{\sqrt{5}}{3}$, $\cos\theta = \frac{2}{3}$, $\tan\theta = -\frac{\sqrt{5}}{2}$, $\csc\theta = -\frac{3\sqrt{5}}{5}$, $\sec\theta = \frac{3}{2}$, and $\cot\theta = -\frac{2\sqrt{5}}{5}$.

Start by finding the radius: $6^2 + \left(-3\sqrt{5}\right)^2 = r^2$, $36 + 45 = r^2$, $81 = r^2$, $9 = r$. Plug into your trig ratios and rationalize if necessary: $\sin\theta = \frac{-3\sqrt{5}}{9} = -\frac{\sqrt{5}}{3}$; $\cos\theta = \frac{6}{9} = \frac{2}{3}$; $\tan\theta = \frac{-3\sqrt{5}}{6} = -\frac{\sqrt{5}}{2}$;

$\csc\theta = \frac{9}{-3\sqrt{5}} = -\frac{3}{\sqrt{5}} = \frac{-3\sqrt{5}}{5}$; $\sec\theta = \frac{9}{6} = \frac{3}{2}$; and $\cot\theta = \frac{6}{-3\sqrt{5}} = -\frac{2}{\sqrt{5}} = \frac{-2\sqrt{5}}{5}$.

15 Evaluate the six trigonometric ratios of 225°. The answers are $\sin\theta = -\frac{\sqrt{2}}{2}$, $\cos\theta = -\frac{\sqrt{2}}{2}$, $\tan\theta = 1$, $\csc\theta = -\sqrt{2}$, $\sec\theta = -\sqrt{2}$, and $\cot\theta = 1$.

Using reference angles, you can see that you're dealing with a 45-45-90 triangle (225° – 180°). Therefore, $x = -\frac{\sqrt{2}}{2}$ and $y = -\frac{\sqrt{2}}{2}$. Now, by using the point-in-plane definition, you can find the six trigonometric ratios: $\sin\theta = y = -\frac{\sqrt{2}}{2}$, $\cos\theta = x = -\frac{\sqrt{2}}{2}$, $\tan\theta = \frac{y}{x} = -\frac{\sqrt{2}}{2} \div -\frac{\sqrt{2}}{2} = 1$,

$\csc\theta = \frac{1}{y} = 1 \div -\frac{\sqrt{2}}{2} = -\sqrt{2}$, $\sec\theta = \frac{1}{x} = 1 \div -\frac{\sqrt{2}}{2} = -\sqrt{2}$, and $\cot\theta = \frac{x}{y} = -\frac{\sqrt{2}}{2} \div -\frac{\sqrt{2}}{2} = 1$.

16 Find θ when $\cos\theta = \frac{\sqrt{3}}{2}$ and 0° < θ < 360°. The answer is 30° and 330°.

Looking at the special right triangles, you can see that $\cos\theta = \frac{\sqrt{3}}{2}$ when θ is 30°. Because cosine is equal to the x value on the unit circle and x is positive in Quadrants I and IV, the answer is 30° and 330°.

17 Evaluate the six trigonometric ratios of 330°. The answers are $\sin\theta = -\frac{1}{2}$, $\cos\theta = \frac{\sqrt{3}}{2}$, $\tan\theta = -\frac{\sqrt{3}}{3}$, $\csc\theta = -2$, $\sec\theta = \frac{2\sqrt{3}}{3}$, and $\cot\theta = -\sqrt{3}$.

Considering that 330° is in Quadrant IV, using reference angles (360° – 330°), you find that you're dealing with a 30-60-90 triangle. Using the point-in-plane definition, you get $\sin\theta = y = -\frac{1}{2}$,

$\cos\theta = x = \frac{\sqrt{3}}{2}$, $\tan\theta = \frac{y}{x} = -\frac{1}{2} \div \frac{\sqrt{3}}{2} = -\frac{1}{2} \cdot \frac{2}{\sqrt{3}} = -\frac{1}{\sqrt{3}} = -\frac{\sqrt{3}}{3}$, $\csc\theta = \frac{1}{y} = 1 \div -\frac{1}{2} = -2$,

$\sec\theta = \frac{1}{x} = 1 \div \frac{\sqrt{3}}{2} = \frac{2}{\sqrt{3}} = \frac{2\sqrt{3}}{3}$, and $\cot\theta = \frac{x}{y} = \frac{\sqrt{3}}{2} \div -\frac{1}{2} = -\frac{\sqrt{3}}{2} \cdot \frac{2}{1} = -\sqrt{3}$.

18 Find θ when $\tan\theta = \frac{\sqrt{3}}{3}$ and 0° < θ < 360°. The answer is θ = 30° and 210°.

To solve this, use special right triangles. You can see that if θ is 30°, then $y = \frac{1}{2}$ and $x = \frac{\sqrt{3}}{2}$.

This means that because $\tan\theta = \frac{y}{x}$, $\tan\theta = \frac{1}{2} \div \frac{\sqrt{3}}{2} = \frac{1}{2} \cdot \frac{2}{\sqrt{3}} = \frac{1}{\sqrt{3}} = \frac{\sqrt{3}}{3}$, which is what you want.

Because the tangent value is positive, both sine and cosine must be the same sign, which occurs in Quadrants I and III. Therefore, θ = 30° and 210°.

19 Solve for θ in 3tanθ – 1 = 2. The answer is θ = 45° and 225°.

Begin by using algebra to isolate tanθ: 3tanθ – 1 = 2, 3tanθ = 3, and tanθ = 1. Because $\tan\theta = \frac{y}{x}$, sine and cosine must be the same value for tanθ to equal 1. This occurs when θ = 45°. Because the answer is positive, both sine and cosine must be the same sign, which occurs in Quadrants I and III. Therefore, using reference angles, for Quadrant I, θ = 45°, and for Quadrant III, 180° + 45° = 225°.

20 Solve for θ in $\sin^2\theta = \sin\theta$. The answer is θ = 0, $\frac{\pi}{2}$, π, and 2π.

To solve, think of $\sin^2\theta = \sin\theta$ as $x^2 = x$, which can be solved by bringing both terms to the same side and factoring: $x^2 - x = 0$, $x(x - 1) = 0$. Similarly, $\sin^2\theta = \sin\theta$, $\sin^2\theta - \sin\theta = 0$, $\sin\theta(\sin\theta - 1) = 0$. Therefore, sinθ = 0, or sinθ – 1 = 0, which means sinθ = 1. Knowing that sinθ = y on the unit circle, sinθ = 0 at 0, and π, 2π and sinθ = 1 at $\frac{\pi}{2}$, you have your answers!

21 Solve for θ in $2\cos^2\theta - 1 = 0$. The answer is θ = $\frac{\pi}{4}$, $\frac{3\pi}{4}$, $\frac{5\pi}{4}$, and $\frac{7\pi}{4}$.

First, isolate the cosine term using algebra: $2\cos^2\theta - 1 = 0$, $2\cos^2\theta = 1$, $\cos^2\theta = \frac{1}{2}$. Now, take the square root of each side: $\cos\theta = \pm\sqrt{\frac{1}{2}}$. Thus, $\cos\theta = \pm\frac{1}{\sqrt{2}} = \pm\frac{\sqrt{2}}{2}$. This occurs at four angles on the unit circle: $\frac{\pi}{4}$, $\frac{3\pi}{4}$, $\frac{5\pi}{4}$, and $\frac{7\pi}{4}$.

22 Solve for θ in $4\sin^2\theta + 3 = 4$. The answer is $\theta = \frac{\pi}{6}, \frac{5\pi}{6}, \frac{7\pi}{6}$, and $\frac{11\pi}{6}$.

Begin by using algebra to isolate the sine term: $4\sin^2\theta + 3 = 4$, $4\sin^2\theta = 1$, $\sin^2\theta = \frac{1}{4}$. Taking the square root of each side, you get: $\sin\theta = \pm\sqrt{\frac{1}{4}}$, $\sin\theta = \pm \frac{1}{2}$. This means that $\theta = \frac{\pi}{6}, \frac{5\pi}{6}, \frac{7\pi}{6}$, and $\frac{11\pi}{6}$.

23 Solve for θ in $4\sin^4\theta - 7\sin^2\theta + 3 = 0$ in degrees. The answer is $\theta = 60°, 90°, 120°, 240°, 270°$, and $300°$.

Start by thinking of $4\sin^4\theta - 7\sin^2\theta + 3 = 0$ as $4x^4 - 7x^2 + 3 = 0$, which factors into $(4x^2 - 3)(x^2 - 1) = 0$. Similarly, $4\sin^4\theta - 7\sin^2\theta + 3 = 0$ factors into $(4\sin^2\theta - 3)(\sin^2\theta - 1) = 0$. Set each factor equal to zero and take the square root of each side to find $\sin\theta$: $4\sin^2\theta - 3 = 0$, $4\sin^2\theta = 3$, $\sin^2\theta = \frac{3}{4}$, $\sin\theta = \pm\sqrt{\frac{3}{4}}$, $\sin\theta = \pm\frac{\sqrt{3}}{2}$. Therefore, $\theta = 60°, 120°, 240°$, and $300°$. Or, $\sin^2\theta - 1 = 0$, $\sin^2\theta = 1$, $\sin\theta = \pm\sqrt{1}$, $\sin\theta = \pm 1$. Therefore, $\theta = 90°$ and $270°$.

24 Solve for θ in $\tan^2\theta - \tan\theta = 0$. The answer is $\theta = 0, \pi, 2\pi, \frac{\pi}{4}$, and $\frac{5\pi}{4}$.

Notice that this problem is similar to Question 20. You can factor the same way: $\tan^2\theta - \tan\theta = 0$, $\tan\theta(\tan\theta - 1) = 0$. Set each factor to zero: $\tan\theta = 0$ or $\tan\theta - 1 = 0$, so $\tan\theta = 1$. These occur at 0, π, 2π, $\frac{\pi}{4}$, and $\frac{5\pi}{4}$.

25 Find the length of an arc in a circle with a radius of 4 feet if the central angle is $\frac{\pi}{6}$. The answer is $s = \frac{2\pi}{3}$ feet, or approximately 2.1 feet.

Use the formula $s = \theta \cdot r = \frac{\pi}{6} \cdot 4 = \frac{2\pi}{3}$ feet, which is approximately 2.1 feet.

26 Find the length of an arc in a circle with a diameter of 16 centimeters if the central angle is $\frac{7\pi}{4}$. The answer is $s = 14\pi$ cm ≈ 44 cm.

Start by finding the radius by dividing the diameter by two: 8 cm. Next, plug into the arc length formula: $s = \theta \cdot r = \frac{7\pi}{4} \cdot 8 = 14\pi$ cm ≈ 44 cm.

27 Find the length of an arc in a circle with a radius of 18 feet if the angle is 210°. The answer is $s = 21\pi$ feet, which is approximately 66 feet.

Begin by changing degrees to radians by multiplying 210° by $\frac{\pi}{180}$: $210° \cdot \frac{\pi}{180} = \frac{7\pi}{6}$. Now, plug the radius and angle into the arc length formula: $s = \theta \cdot r = \frac{7\pi}{6} \cdot 18 = 21\pi$ feet, which is approximately 66 feet.

28 Find the radius of a circle in which an angle of 2 radians cuts an arc of length 42 inches. The answer is 21 inches.

Just plug 'em in! $s = \theta \cdot r$, $42 = 2 \cdot r$. Dividing both sides by 2, you find $r = 21$ inches.

Chapter 7

Graphing and Transforming Trig Functions

In This Chapter

- Exploring period graphs
- Graphing sine and cosine
- Picturing tangent and cotangent
- Charting secant and cosecant

Graphing a trig function is similar to graphing any other function. You simply insert values into the input to find the output. In this case, the input is typically θ and the output is typically *y*. And, just like graphing any other function, knowing the *parent trig graph* — the most basic, unshifted graph — makes graphing more complex graphs easier. In this chapter, we show you the parent graph of each trig function and its transformations.

Getting a Grip on Periodic Graphs

Periodic graphs are like that everlasting bunny we all know . . . they keep going, and going, and going. If you remember that trig functions are periodic graphs, the steps to graphing them are easy! Because they repeat their values over and over again, you just need to figure out one period (or cycle), and then you can repeat it as many times as you like.

The key is to graph one period. To do this, you need to start by graphing the parent graph, and then make transformations (just like you do in Chapter 3 for other types of graphs). As with other graphs, transformations can be applied to trig graphs:

- For trig functions, vertical stretches and shrinks are achieved by simply multiplying the parent function by a constant. For example, $f(\theta) = 2\sin\theta$ is the same as the parent graph, only its wave goes up to a value of 2 and down to –2. Multiplying a negative constant to the parent graph simply flips the graph upside down, or reflects it over the *x*-axis.
- Horizontal stretches and compressions occur by changing the period of the graph. For sine and cosine parent graphs, the period is 2π. The same is true for cosecant and secant graphs. For tangent and cotangent graphs, the period is π. Multiplying the angle in the function by a constant transforms the period. For example, ($f(\theta) = \cos 2\theta$) results in a graph that repeats itself twice in the amount of space the parent graph would occupy.

Vertical and horizontal translations shift the parent graph up, down, left, or right:

- Just as we show you in Chapter 3, vertical and horizontal shifts just change the location of the parent graph: up, down, left, or right.
- The general equation for these shifts for sine, for example, is $f(\theta) = \sin(\theta - h) + v$, where h represents the horizontal shift left or right and v represents the vertical shift up or down.
- To find the horizontal shift of a function, simply set the inside parentheses equal to 0. For example, the horizontal shift for $\sin(\theta + 3)$ is –3 because $\theta + 3 = 0$, so $\theta = -3$.

We suggest that when you put the transformations together, you follow this simple order:

1. **Change the amplitude, if applicable.**
2. **Change the period, if applicable.**
3. **Shift the graph horizontally, if applicable.**
4. **Shift the graph vertically, if applicable.**

Parent Graphs and Transformations: Sine and Cosine

Sine and cosine graphs look like waves. These waves, or *sinusoids* in math speak, keep going and going like our bunny friend. To graph these sinusoids, you need to start with checking out the parent graphs (see Figure 7-1).

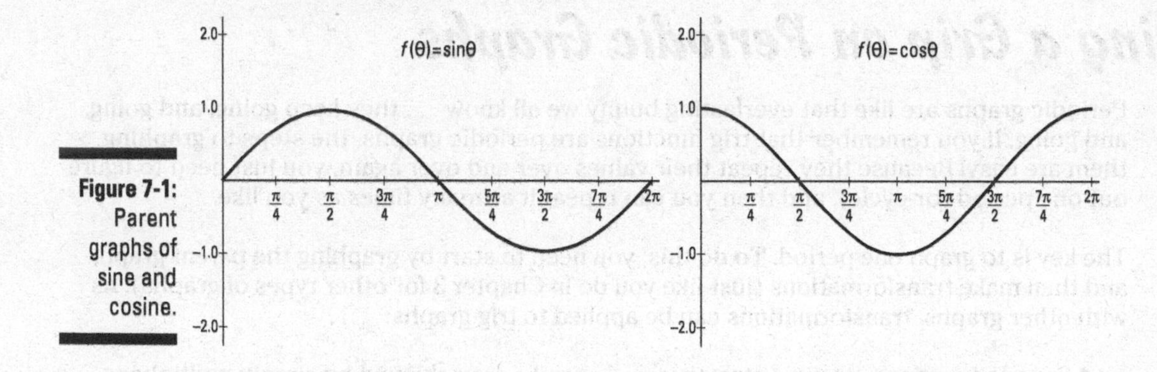

Figure 7-1: Parent graphs of sine and cosine.

Notice that the periods of both the sine and cosine graphs are the same: 2π. Similarly, they both have an amplitude (or height) of 1. You use this information for your transformations.

Putting together all the transformation information into one equation, you get

$$f(\theta) = a \cdot \sin[p(\theta - h)] + v$$

$$f(\theta) = a \cdot \cos[p(\theta - h)] + v$$

where a is the amplitude, h is the horizontal shift, v is the vertical shift, and you divide 2π by p to get the period.

Q. Graph $f(\theta) = 2\sin\theta + 3$

A. Starting with amplitude, you can see that $a = 2$, so your amplitude is 2. The period is 2π because the period doesn't change from the parent equation. The vertical shift is positive 3 because $v = 3$. The graph is shown in the following figure.

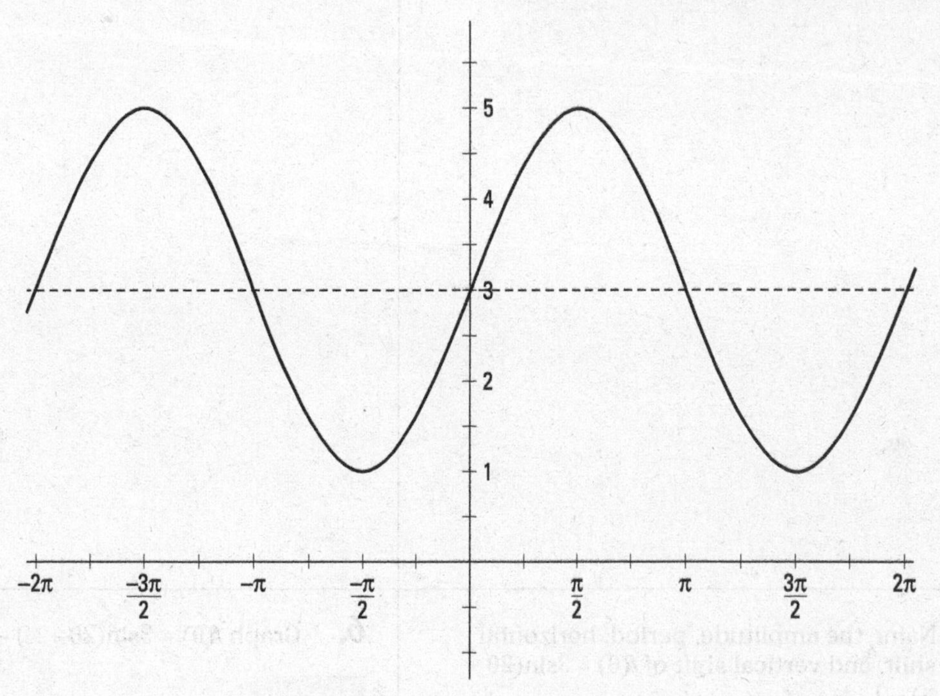

1. Graph $f(\theta) = -\frac{1}{2}\cos\theta$.

Solve It

2. Graph $f(\theta) = \cos\frac{1}{2}\theta$.

Solve It

3. Graph $f(\theta) = \sin(\theta + \pi/4)$.

Solve It

4. Graph $f(\theta) = \cos\frac{1}{3}\theta + 2$.

Solve It

5. Name the amplitude, period, horizontal shift, and vertical shift of $f(\theta) = 3\sin(2\theta + \pi/2) - 1$.

Solve It

6. Graph $f(\theta) = 3\sin(2\theta + \pi/2) - 1$.

Solve It

Moms, Pops, and Children: Tangent and Cotangent

Tangent and cotangent are both periodic, but they're not wavelike like sine and cosine. Instead, they have vertical asymptotes that break up their graphs. As we discuss in Chapter 3, a *vertical asymptote* is where the function is undefined. Because tangent $\tan\theta = \frac{\sin\theta}{\cos\theta}$ and cotangent $\cot\theta = \frac{\cos\theta}{\sin\theta}$ are ratios, they both have values that are undefined where their denominators are equal to 0. For tangent, this occurs on the unit circle at $\frac{\pi}{2}$ and $-\frac{\pi}{2}$. For cotangent, this occurs at 0, π, and $-\pi$ on the unit circle. Therefore, these are the locations of their asymptotes (see Figure 7-2).

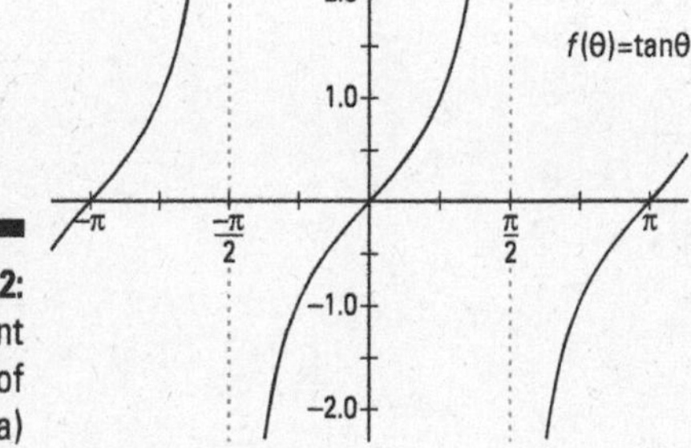

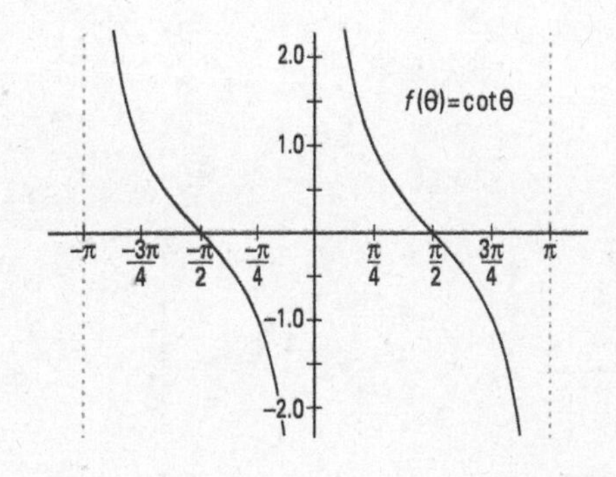

Figure 7-2: Parent graphs of tangent (a) and cotangent (b).

Notice that the periods of both the tangent and cotangent graphs are the same: π. The x-intercepts for tangent are 0, π, and $-\pi$. For cotangent, the x-intercepts are $\frac{\pi}{2}$ and $-\frac{\pi}{2}$. Using this information, you can make your transformations.

Putting together all the transformation information from earlier in this chapter into one equation, you get

$$f(\theta) = a \cdot \tan[p(\theta - h)] + v$$

$$f(\theta) = a \cdot \cot[p(\theta - h)] + v$$

where a is the vertical transformation (no amplitude with tangent and cotangent), h is the horizontal shift, v is the vertical shift, and you divide π by p to get the period.

Q. Graph $f(\theta) = \frac{1}{2} \cdot \tan 2\theta$.

A. Starting with the vertical transformation (shrink), you can see that it's ½. Next, find the period by dividing π by 2, which is π/2. Because there are no horizontal or vertical shifts, you're ready to graph:

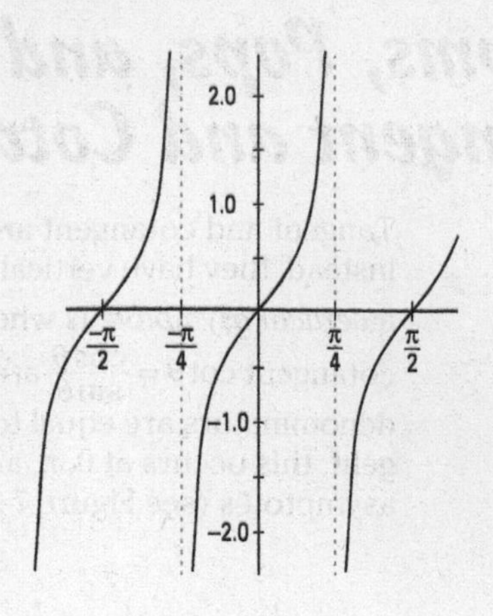

7. Graph $f(\theta) = \cot\frac{1}{2}\theta$.

Solve It

8. Graph $f(\theta) = \tan\theta + 2$.

Solve It

9. Graph $f(\theta) = \frac{1}{3}\cot\theta$.

Solve It

10. Graph $f(\theta) = \tan(\theta - \frac{\pi}{2})$.

Solve It

11. Name the vertical stretch or shrink, period, horizontal shift, and vertical shift of $f(\theta) = 2 \cdot \tan(\theta + \frac{\pi}{4}) - 1$.

Solve It

12. Graph $f(\theta) = 2\tan(\theta + \frac{\pi}{4}) - 1$.

Solve It

Generations: Secant and Cosecant

To graph cosecant and secant, it's important to remember that they're the reciprocals of sine and cosine, respectively: $\csc\theta = \frac{1}{\sin\theta}$ and $\sec\theta = \frac{1}{\cos\theta}$. Using this fact, the easiest way to graph cosecant or secant is to start by graphing sine or cosine — the graphs of the reciprocals are easily found from there.

Given that cosecant and secant are reciprocal functions of sine and cosine, respectively, as sine and cosine approach zero, their reciprocals get larger and larger without limit. So, yep, you guessed it — cosecant and secant graphs have asymptotes. These occur wherever their reciprocal functions (sine or cosine) have a value of 0. To graph, follow these easy steps:

1. **Graph the sine graph with transformations to graph a cosecant graph, or graph the cosine graph with transformations to get the secant graph.**
2. **Draw vertical asymptotes where the sine or cosine functions are equal to 0.**
3. **Sketch the reciprocal graph of sine or cosine between each pair of asymptotes.**

   For example, if the sine graph gets bigger, the cosecant graph gets smaller.

The parent graphs of cosecant and secant are shown in Figure 7-3. The parent sine and cosine graphs are also shown so you can see where they came from.

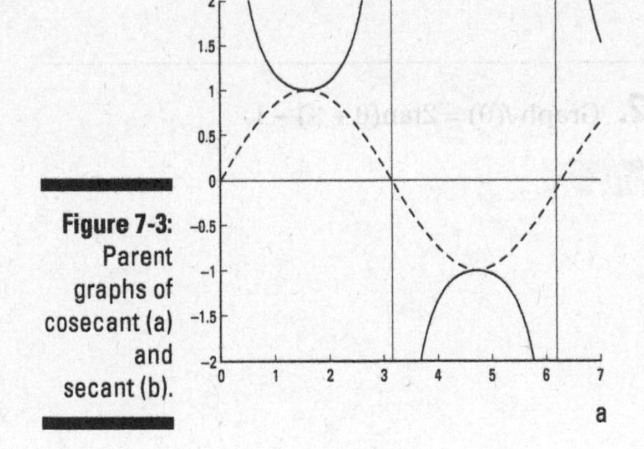

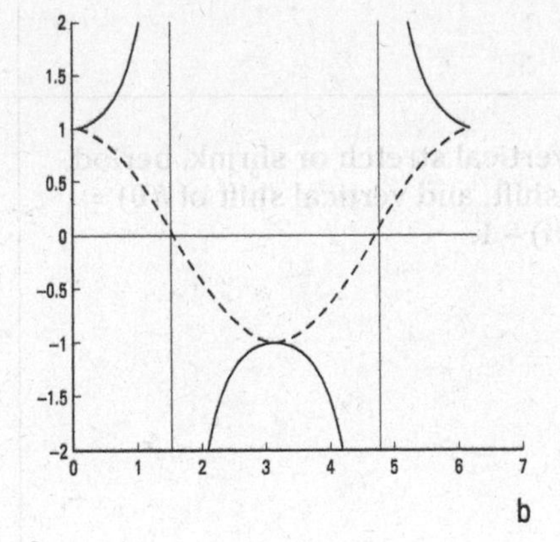

Figure 7-3: Parent graphs of cosecant (a) and secant (b).

Q. Graph $f(\theta) = \csc\theta + 1$.

A. Here, there's only a vertical shift of 1, which means you shift the parent graph up one (see the following figure).

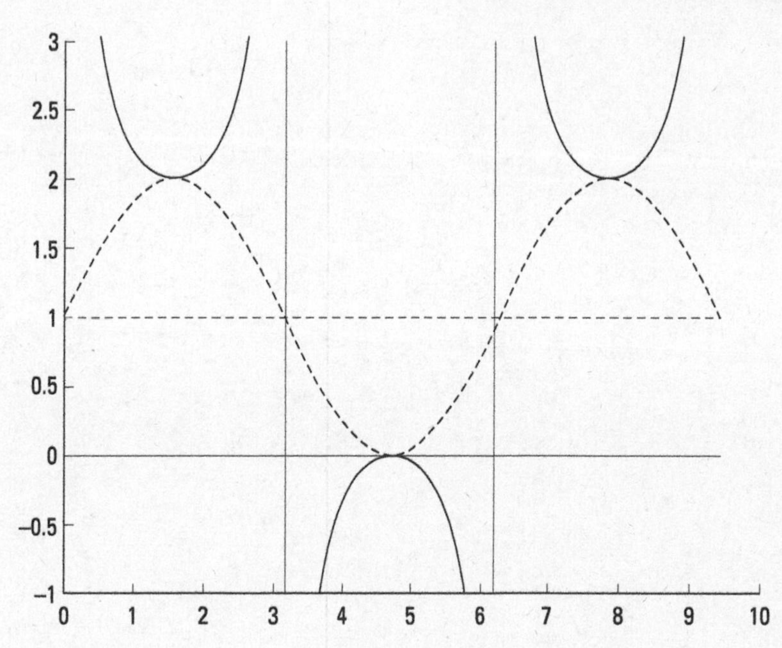

13. Graph $f(\theta) = -\csc\theta - 1$.

Solve It

14. Graph $f(\theta) = \sec 2\theta + 1$.

Solve It

15. Name the asymptotes for one period and the horizontal and vertical shifts of $f(\theta) = \frac{1}{4} \cdot \csc\left(\theta - \frac{\pi}{2}\right) - 1$

Solve It

16. Graph $f(\theta) = \frac{1}{4} \cdot \csc\left(\theta - \frac{\pi}{2}\right) - 1$

Solve It

17. Name the period, horizontal shift, and vertical shift of $f(\theta) = 2 \cdot \sec\left(\frac{1}{2}\theta + \frac{\pi}{4}\right) + 1$

Solve It

18. Graph $f(\theta) = 2 \cdot \sec\left(\frac{1}{2}\theta + \frac{\pi}{4}\right) + 1$

Solve It

Answers to Problems on Graphing and Transforming Trig Functions

Following are the answers to problems dealing with trig functions. We also provide guidance on getting the answers if you need to review where you went wrong.

1 Graph $f(\theta) = -\frac{1}{2}\cos\theta$. See the graph for the answer.

Because the cosine function is multiplied by $-\frac{1}{2}$, the graph is inverted with an amplitude of $\frac{1}{2}$. The period doesn't change, and there are no shifts.

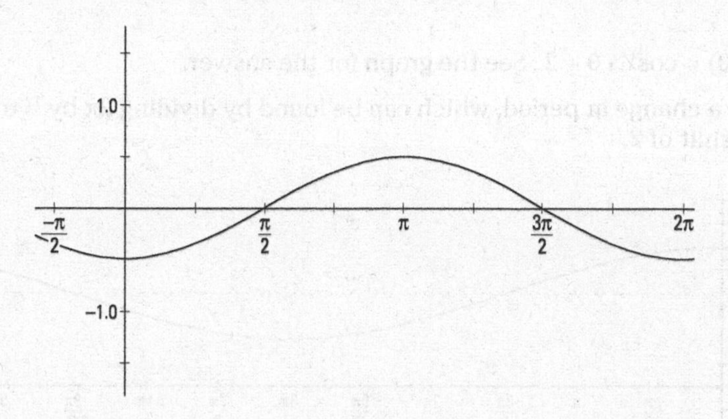

2 Graph $f(\theta) = \cos\frac{1}{2}\theta$. See the graph for the answer.

Here, the amplitude doesn't change, but the period does. The new period is found by dividing 2π by $\frac{1}{2}$, which is 4π. There are no vertical or horizontal shifts.

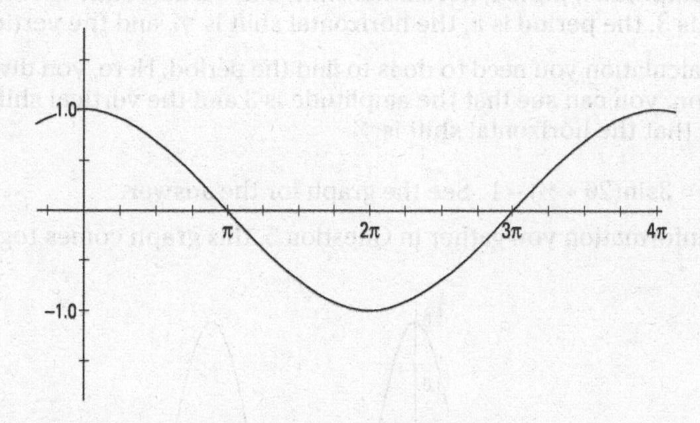

3 Graph $f(\theta) = \sin(\theta + \frac{\pi}{4})$. See the graph for the answer.

This graph has a horizontal shift. To find it, set what's inside the parentheses to the starting value of the parent graph: $\theta + \frac{\pi}{4} = 0$, so $\theta = \frac{-\pi}{4}$. There are no other changes from the parent graph.

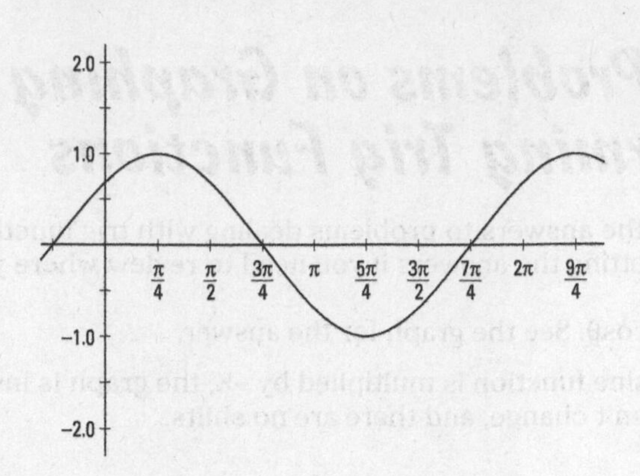

4 Graph $f(\theta) = \cos\frac{1}{3}\cdot\theta + 2$. See the graph for the answer.

This has a change in period, which can be found by dividing 2π by $\frac{1}{3}$ to get 6π. There's also a vertical shift of 2.

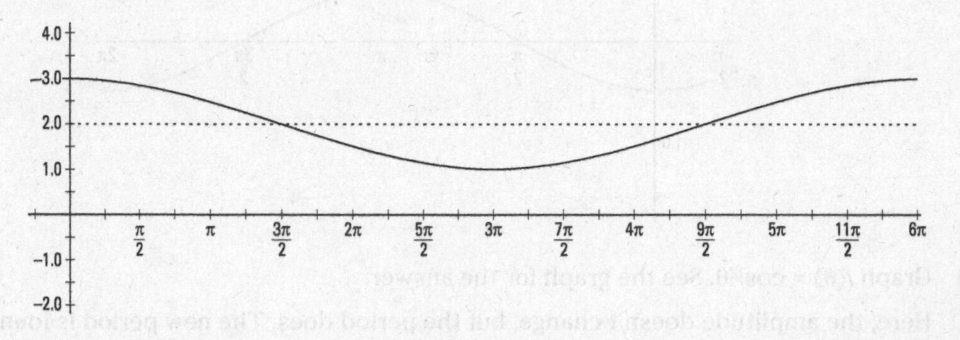

5 Name the amplitude, period, horizontal shift, and vertical shift of of $f(\theta) = 3\sin(2\theta + \frac{\pi}{2}) - 1$. The amplitude is 3, the period is π, the horizontal shift is $-\frac{\pi}{4}$, and the vertical shift is –1.

The only calculation you need to do is to find the period. Here, you divide 2π by 2 to get π. From the equation, you can see that the amplitude is 3 and the vertical shift is –1. By setting $2\theta + \frac{\pi}{2} = 0$, you find that the horizontal shift is $-\frac{\pi}{4}$.

6 Graph $f(\theta) = 3\sin(2\theta + \frac{\pi}{2}) - 1$. See the graph for the answer.

Using the information you gather in Question 5, this graph comes together quickly.

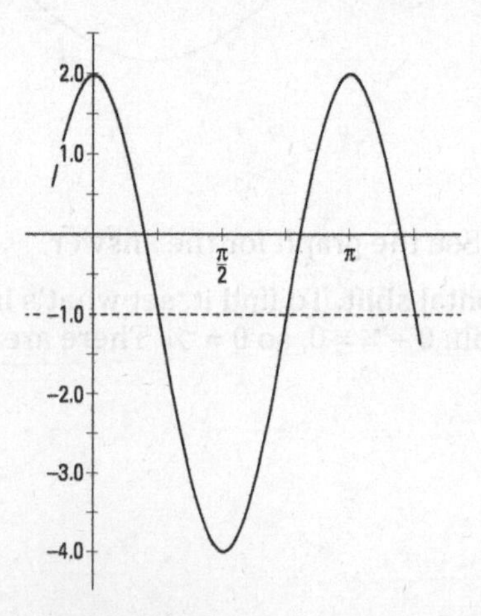

7 Graph $f(\theta) = \cot\frac{1}{2}\theta$. See the graph for the answer.

For this cotangent graph, the period has a change, which can be found by dividing π by ½ (you get 2π). There are no other changes to the parent cotangent graph.

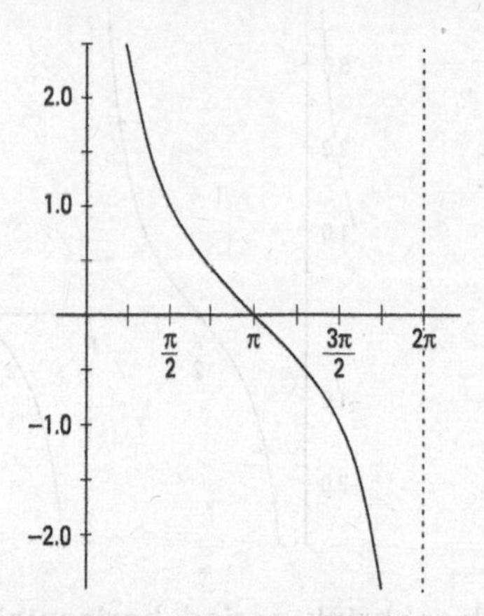

8 Graph $f(\theta) = \tan\theta + 2$. See the graph for the answer.

Here, there's only a vertical shift of 2. The period is the same as the parent graph.

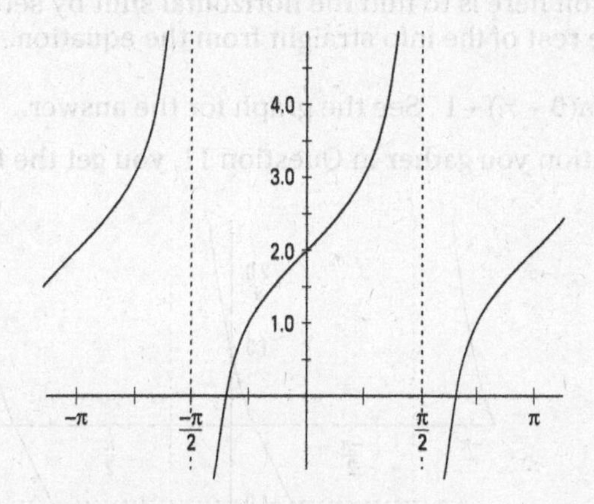

9 Graph $f(\theta) = \frac{1}{3}\cot\theta$. See the graph for the answer.

This graph shows a vertical transformation of ⅓.

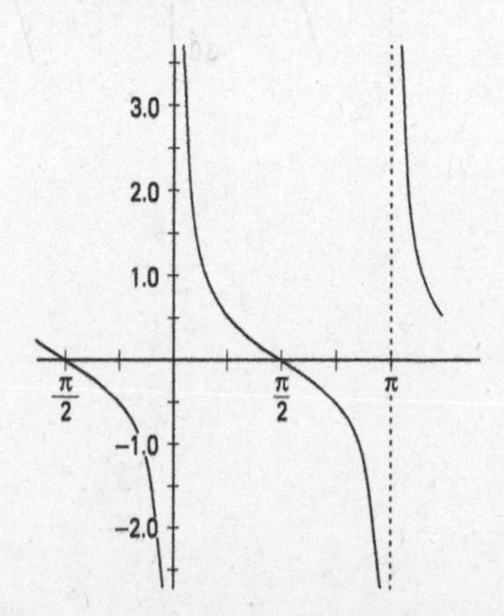

10 Graph $f(\theta) = \tan(\theta - \pi/2)$. See the graph for the answer.

This tangent graph has a horizontal shift of $\pi/2$. There are no other changes to the parent graph.

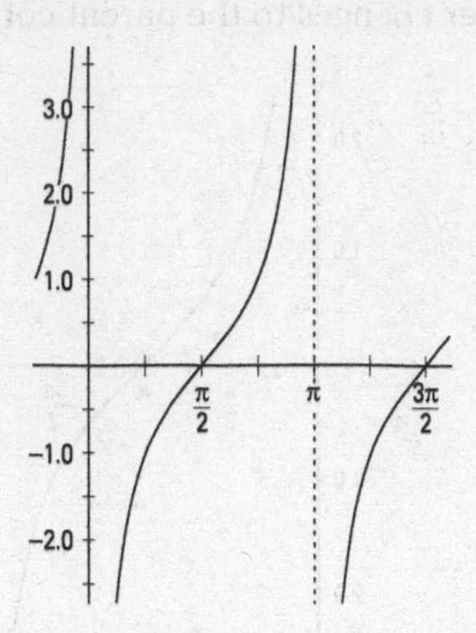

11 Name the vertical stretch or shrink, period, horizontal shift, and vertical shift of $f(\theta) = 2 \cdot \tan(\theta + \pi/4) - 1$. The vertical stretch is 2, the period is π, the horizontal shift is $-\pi/4$, and the vertical shift is -1.

The only calculation here is to find the horizontal shift by setting $\theta + \pi/4 = 0$. This gives you $\theta = -\pi/4$. You get the rest of the info straight from the equation.

12 Graph $f(\theta) = 2 \cdot \tan(\theta + \pi/4) - 1$. See the graph for the answer.

Using the information you gather in Question 11, you get the following graph:

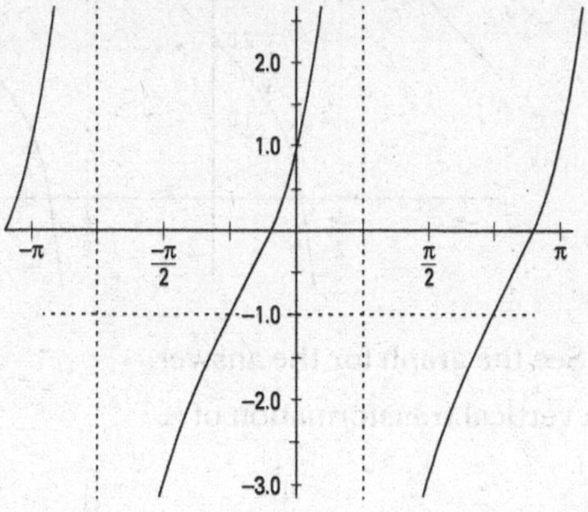

13 Graph $f(\theta) = -\csc\theta - 1$. See the graph for the answer.

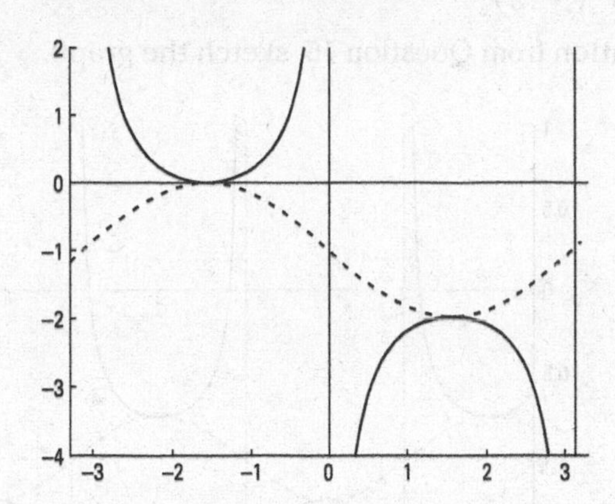

REMEMBER For these problems, you start by graphing the sine graph to find the asymptotes, and then sketch the reciprocal function of cosecant.

14 Graph $f(\theta) = \sec 2\theta + 1$. See the graph for the answer.

Again, we include the reciprocal cosine graph. This has a change in period, which you find by dividing 2π by 2 to get π. It also has a vertical shift of 1. Then you draw the asymptotes and sketch the secant graph.

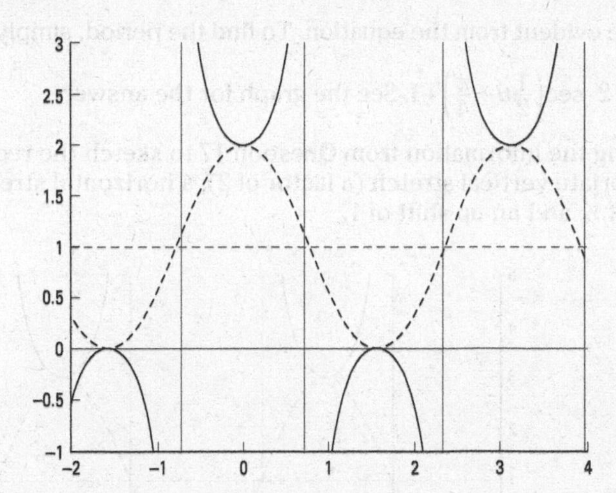

15 Name the asymptotes from 0 to 2π, and the horizontal and vertical shifts of $f(\theta) = \frac{1}{4} \cdot \csc\left(\theta - \frac{\pi}{2}\right) - 1$.

The asymptotes are at $\frac{\pi}{2}$ and $\frac{3\pi}{2}$. The horizontal shift is $\frac{\pi}{2}$, and the vertical shift is -1.

To find the asymptotes, you need to first look for any shifts or changes in period that would affect the parent graph. The shifts are evident from the equation, where the horizontal shift is $\frac{\pi}{2}$ and the vertical shift is -1. This makes the zeros of the reciprocal sine graph at $\frac{\pi}{2}$ and $\frac{3\pi}{2}$. This, then, is where the asymptotes will be.

16 Graph $f(\theta)=\frac{1}{4}\cdot\csc\left(\theta-\frac{\pi}{2}\right)-1$. **See the graph for the answer.**

Using the information from Question 15, sketch the graph.

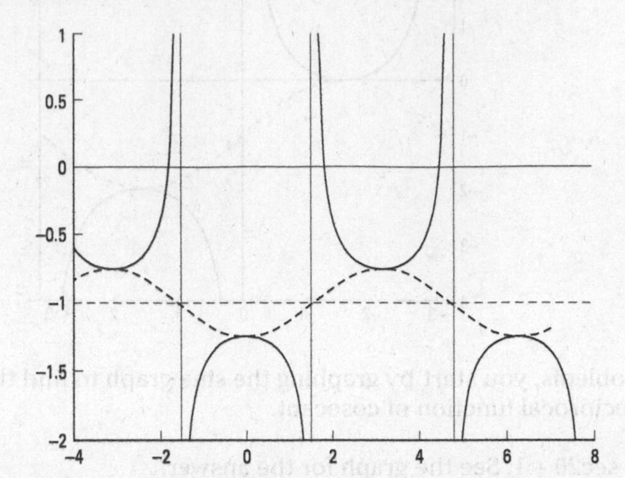

17 Name the period, horizontal shift, and vertical shift of $f(\theta)=2\cdot\sec\left(\frac{1}{2}\theta+\frac{\pi}{4}\right)+1$. **The period is 4π, the horizontal shift is $-\pi/2$, and the vertical shift is 1.**

The shifts are evident from the equation. To find the period, simply divide 2π by $1/2$ and you get 4π.

18 Graph $f(\theta)=2\cdot\sec\left(\frac{1}{2}\theta+\frac{\pi}{4}\right)+1$. **See the graph for the answer.**

Begin by using the information from Question 17 to sketch the reciprocal of the cosine graph. Make appropriate vertical stretch (a factor of 2), a horizontal stretch (also a factor of 2), a left translation of $\pi/2$, and an up-shift of 1.

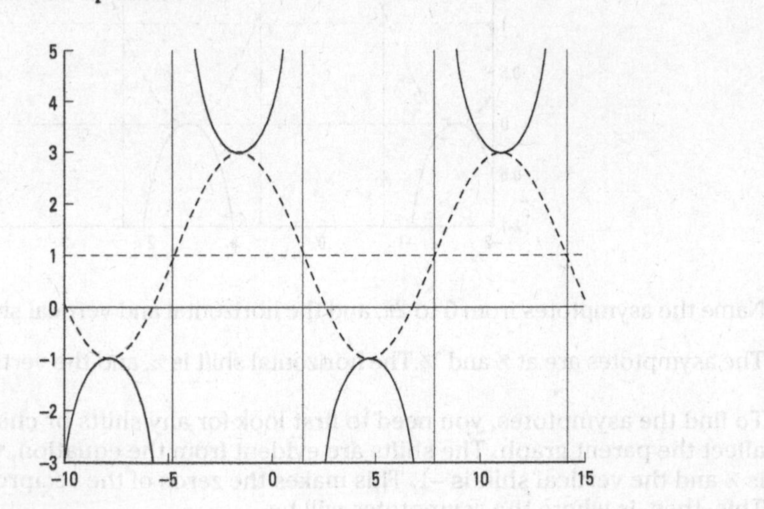

Part III

Digging into Advanced Trig: Identities, Theorems, and Applications

"He was there a minute ago. I just stepped out as he was running some trig equations in order to calculate the area of the Bermuda Triangle."

In this part . . .

The concepts of trig keep building, and we're here to help you follow along. These chapters move into identities — they're like formulas, but they're true all the time, no matter what you put in for the variable(s). These identities are used to simplify expressions and solve equations, and they're even used in trig proofs (and you thought you were done with proofs in geometry!). This part also includes some very specific and powerful identities that can be used to solve oblique triangles.

Chapter 8

Basic Trig Identities

In This Chapter

- Simplifying with reciprocal, Pythagorean, and co-function identities
- Finding patterns with even-odd identities
- Discovering periodicity identities
- Tackling trigonometric proofs

Ever want to pretend you were someone else — change your identity? Well, trig expressions have the opportunity to do that all the time. In this chapter, we cover basic *identities,* or statements that are always true. We use these identities to simplify problems and then to complete trigonometric proofs. Each section builds upon the previous one, so we recommend you spend some time reviewing the identities in each section before jumping to the end to practice proofs.

Using Reciprocal Identities to Simplify Trig Expressions

We actually introduce some trigonometry expressions back in Chapter 6, but now we're going to use reciprocal identities to simplify more complicated trig expressions. Because these identities are all review, we also include the ratios of tangent and cotangent — the *ratio identities* that we introduce in Chapter 6. The reciprocal (and ratio) identities are:

$$\sin x = \frac{1}{\csc x} \qquad \csc x = \frac{1}{\sin x}$$

$$\cos x = \frac{1}{\sec x} \qquad \sec x = \frac{1}{\cos x}$$

$$\tan x = \frac{1}{\cot x} = \frac{\sin x}{\cos x} \qquad \cot x = \frac{1}{\tan x} = \frac{\cos x}{\sin x}$$

Because each pair of expressions is mathematically equivalent, you can substitute one for another in a given expression and watch things simplify. Typically, changing a given expression to all sines and cosines causes a whole lot of canceling! Try it and see . . . we dare you!

Q. Use reciprocal identities to rewrite $\frac{\sin x \cdot \sec x}{\tan x}$

A. **The answer is 1. Start by using reciprocal and ratio identities to rewrite sec *x* and tan *x* in terms of sine and cosine (we dared you). Next, use your knowledge of fractions to rewrite the expression as a division problem. Then, multiply by the reciprocal and cancel where you can. Here's what it should look like:**

$$\frac{\sin x \cdot \sec x}{\tan x} = \frac{\sin x \cdot \frac{1}{\cos x}}{\frac{\sin x}{\cos x}} = \sin x \cdot \frac{1}{\cos x} \div \frac{\sin x}{\cos x} = \sin x \cdot \frac{1}{\cos x} \cdot \frac{\cos x}{\sin x} = \cancel{\sin x} \cdot \frac{1}{\cancel{\cos x}} \cdot \frac{\cancel{\cos x}}{\cancel{\sin x}} = 1$$

1. Simplify $\cot x \cdot \sec x$.

Solve It

2. Simplify $\sin x \cdot \sec x$.

Solve It

3. Simplify $\sin^3 x \cdot \csc^2 x + \tan x \cdot \cos x$.

Solve It

4. Simplify $\cot x \cdot \sin x \cdot \tan x$.

Solve It

Simplifying with Pythagorean Identities

Pythagorean identities are extremely helpful for simplifying complex trig expressions. These identities are derived from those right triangles on a unit circle (turn to Chapter 6 for a review if you need to). Remember that $\cos\theta$ = the x leg of a triangle, $\sin\theta$ = the y leg of a triangle, and the hypotenuse of the triangle on that unit circle is 1. Given the fact that $\text{leg}^2 + \text{leg}^2 = \text{hypotenuse}^2$, we get the first Pythagorean identity. The other two are derived from that (check out *Pre-Calculus For Dummies* by Krystle Rose Forseth, Christopher Burger, and Michelle Rose Gilman [Wiley] if you want to see how this works!). These identities are especially helpful when simplifying expressions that have a term that has been squared ($\sin^2 x$, $\cos^2 x$, and so on). Here are the Pythagorean identities (and some variations):

$\sin^2 x + \cos^2 x = 1$ *or* $\cos^2 x = 1 - \sin^2 x$ *or* $\sin^2 x = 1 - \cos^2 x$

$\tan^2 x + 1 = \sec^2 x$ *or* $\tan^2 x = \sec^2 x - 1$ *or* $1 = \sec^2 x - \tan^2 x$

$1 + \cot^2 x = \csc^2 x$ *or* $\cot^2 x = \csc^2 x - 1$ *or* $1 = \csc^2 x - \cot^2 x$

Q. Simplify $(\sec x + \tan x)(1 - \sin x)(\cos x)$.

A. $\cos^2 x$. Start by changing everything to sine and cosine using the reciprocal and ratio identities from the preceding section. Then add the resulting fractions (the common denominator is cosine) and cancel the cosine in the numerator and denominator. This leaves you with two terms that you can FOIL. Recognize this last term as a Pythagorean identity? We hoped you might! Substitute it in and you have your answer. The steps look like this:

$$(\sec x + \tan x)(1 - \sin x)(\cos x)$$

$$= \left(\frac{1}{\cos x} + \frac{\sin x}{\cos x}\right)(1 - \sin x)(\cos x)$$

$$= \left(\frac{1 + \sin x}{\cancel{\cos x}}\right)(1 - \sin x)(\cancel{\cos x})$$

$$= (1 + \sin x)(1 - \sin x)$$

$$= 1 - \sin^2 x$$

$$= \cos^2 x$$

5. Simplify $\frac{\cos x}{\sin x}(\tan x + \cot x)$

Solve It

6. Simplify $1 - \frac{\sin x \cdot \tan x}{\sec x}$

Solve It

7. Simplify $\sin x \cdot \cot^2 x + \sin x$.

Solve It

8. Simplify $(\sin^2 x - 1)(\tan^2 x + 1)$.

Solve It

Discovering Even-Odd Identities

All functions, including trig functions, can be described as being even, odd, or neither (see Chapter 3 for review). Knowing whether a trig function is even or odd can actually help you simplify an expression. These even-odd identities are helpful when you have an expression where the variable inside the trig function is negative (such as $-x$). The even-odd identities are:

$\sin(-x) = -\sin x$ $\quad$ $\csc(-x) = -\csc x$

$\cos(-x) = \cos x$ $\quad$ $\sec(-x) = \sec x$

$\tan(-x) = -\tan x$ $\quad$ $\cot(-x) = -\cot x$

Q. Simplify $-\tan^2(-x) + \sec^2(-x)$.

A. The answer is 1. Using the even-odd identities, start by substituting for the negative angles: $-(-\tan x)^2 + (\sec x)^2 = -\tan^2 x + \sec^2 x$. Using the commutative property of addition (from Chapter 1), rewrite the expression: $\sec^2 x - \tan^2 x$. Recognize this from our Pythagorean identities from the preceding section? This expression equals 1.

9. Simplify $\sec(-x) \cdot \cot(-x)$.

Solve It

10. Simplify $\sin x \cdot [\csc x + \sin(-x)]$.

Solve It

11. Simplify $\dfrac{\csc x}{\cot(-x)}$

Solve It

12. Simplify $\dfrac{-\cot^2(-x)-1}{-\cot^2(x)}$

Solve It

Solving with Co-Function Identities

Ever notice that the graphs of sine and cosine look exactly alike, only shifted? (See Chapter 7 for a visual.) There are also similarities between tangent and cotangent (there is a refection in addition to a shift), as well as between secant and cosecant. Because these functions have the same values, only shifted, we can define them as being *co-functions.* We can write them as co-function identities and use them to simplify expressions. The co-function identities are:

$\sin x = \cos(\pi/2 - x)$ $\qquad \cos x = \sin(\pi/2 - x)$

$\tan x = \cot(\pi/2 - x)$ $\qquad \cot x = \tan(\pi/2 - x)$

$\csc x = \sec(\pi/2 - x)$ $\qquad \sec x = \csc(\pi/2 - x)$

Q. Simplify $\dfrac{\cot\left(\frac{\pi}{2} - x\right)}{\sec x}$

A. $\sin x$. Start by using the co-function identity to replace $\cot(\pi/2 - x)$ with $\tan x$. Next, rewrite the fraction as a division problem. Then, rewrite in terms of sine and cosine using reciprocal and ratio identities. Finally, simplify by using the multiplicative inverse, canceling any common terms. The steps look like this:

$$\frac{\cot\left(\frac{\pi}{2} - x\right)}{\sec x} = \frac{\tan x}{\sec x} = \tan x \div \sec x = \frac{\sin x}{\cos x} \div \frac{1}{\cos x} = \frac{\sin x}{\cancel{\cos x}} \cdot \frac{\cancel{\cos x}}{1} = \sin x$$

13. Simplify $\sin\left(\frac{\pi}{2} - x\right) \cdot \cot\left(\frac{\pi}{2} - x\right)$

Solve It

14. Simplify $\sin\left(\frac{\pi}{2} - x\right) + \cot\left(\frac{\pi}{2} - x\right) \cdot \cos\left(\frac{\pi}{2} - x\right)$

Solve It

15. Simplify $\dfrac{\cos\left(\frac{\pi}{2} - x\right)}{\sin\left(\frac{\pi}{2} - x\right)} \cdot \cot\left(\frac{\pi}{2} - x\right)$

Solve It

16. Simplify

$$\left[\sec\left(\frac{\pi}{2} - x\right) + \tan\left(\frac{\pi}{2} - x\right)\right]\left[1 - \sin\left(\frac{\pi}{2} - x\right)\right]$$

Solve It

Moving with Periodicity Identities

Recall that horizontal transformations change the period of a graph and horizontal shifts move it left or right (see Chapter 7). If you shift the graph by one whole period to the left or right, you end up with the same function. This is the idea behind *periodicity identities*. Because the periods of sine, cosine, cosecant, and secant repeat every 2π, and tangent and cotangent repeat every π, the periodicity identities are as follows:

$$\sin(x + 2\pi) = \sin x$$

$$\cos(x + 2\pi) = \cos x$$

$$\tan(x + \pi) = \tan x$$

$$\cot(x + \pi) = \cot x$$

$$\csc(x + 2\pi) = \csc x$$

$$\sec(x + 2\pi) = \sec x$$

Q. Simplify $1 - \sin(2\pi + x) \cdot \cot(\pi + x) \cdot \cos(2\pi + x)$.

A. $\sin^2 x$. Begin by rewriting the trig terms using periodicity identities: $1 - (\sin x \cdot \cot x \cdot \cos x)$. Next, rewrite cotangent in terms of sine and cosine: $1-\left(\sin x \cdot \frac{\cos x}{\sin x} \cdot \cos x\right)$. Then, cancel the sine from the numerator and denominator, leaving you with $1 - \cos^2 x$. Using Pythagorean identities, this is the same as $\sin^2 x$.

17. Simplify $\cos(2\pi + x) + \sin(2\pi + x) \cdot \cot(\pi + x)$

Solve It

18. Simplify $\frac{\cos(x+4\pi)}{\cot(x+2\pi)}$

Solve It

19. Simplify $\frac{\sec(x+2\pi)}{\csc(x+2\pi)}$

Solve It

20. Simplify $[\sec(x - 2\pi) - \tan(x - \pi)] \cdot [\sec(x + 2\pi) + \tan(x + \pi)]$

Solve It

Tackling Trig Proofs

Proofs?!? You thought you left those behind in geometry. Nope, sorry. Don't worry, though — we walk you through pre-calc's version of proofs, which are trig proofs. One thing to remember is that you're just using what you've already practiced in this chapter. These proofs are composed of two sides of an equation. Your job is to make one side look like the other. Here are some hints on how to solve these:

- **Deal with fractions using basic fraction rules.** The same rules apply to simplifying trig expressions as any other expression. Two key rules to remember:
  - Dividing an expression by another expression is the same as multiplying by the reciprocal of that expression.
  - Use the lowest common denominator (or LCD) when adding or subtracting fractions.
- **Factor when you can.** Keep an eye out for factorable terms, including factoring out the greatest common factor (GCF) and factoring trinomials (see Chapter 4).
- **Square square roots.** When you have a square root in a proof, you probably have to square both sides of the equation.
- **Work on the more complicated side first.** Because the goal is to make one side look like the other, starting on the more complicated side first is generally easier. If you get stuck, try working on the other side for a while. You can then work backward to simplify the first side.

Q. Prove $\frac{1}{\cot^2 x} - \frac{1}{\csc^2 x} = \frac{\sin^4 x}{\cos^2 x}$

A. The left side is more complicated, so we work on that side. Start by finding the common denominator and add the fractions.

$$\frac{1}{\cot^2 x} - \frac{1}{\csc^2 x} = \frac{\csc^2 x}{\cot^2 x \csc^2 x} - \frac{\cot^2 x}{\cot^2 x \csc^2 x} = \frac{\csc^2 x - \cot^2 x}{\cot^2 x \csc^2 x}$$

From there, notice that you have a Pythagorean identity. Rewrite the resulting fraction in terms of sine and cosine using ratio identities, and multiply terms to complete the proof.

$$\frac{1}{\cot^2 x} \cdot \frac{1}{\csc^2 x} = \tan^2 x \cdot \sin^2 x = \frac{\sin^2 x}{\cos^2 x} \cdot \sin^2 x = \frac{\sin^4 x}{\cos^2 x}$$

21. Prove $\frac{\cot x - 1}{1 + \tan(-x)} = \cot x$ for x in $(\pi/4, \pi + \pi/4)$

Solve It

22. Prove $\frac{\csc x + \tan x}{\cot\left(\frac{\pi}{2} - x\right)\csc\left(\frac{\pi}{2} - x\right)} = \cot^2 x + \cos x$

Solve It

23. Prove $\cot x = \frac{\csc^2 x - 1}{\cot x}$

Solve It

24. Prove $\frac{1 - \sin x}{\csc x} - \frac{\tan x \cos x}{1 + \sin x} = -\frac{\sin^2 x}{\csc x + 1}$

Solve It

25. Prove $\sqrt{\left[\cos x \cdot \sin\left(\frac{\pi}{2} - x\right) \cdot \csc(2\pi + x) \cdot \sec\left(\frac{\pi}{2} - x\right)\right] + 1} = |\csc x|$

Solve It

26. Prove $\sec x - \cos x = \sin x \cdot \tan x$.

Solve It

Answers to Problems on Basic Trig Identities

This section contains the answers for the practice problems presented in this chapter. We suggest you read the following explanations if your answers don't match up with ours (or if you just want a refresher on solving a particular type of problem).

1 Simplify $\cot x \cdot \sec x$. The answer is $\csc x$.

Start by using ratio identities to show the expression in terms of sine and cosine. Cancel the cosine in the numerator and denominator and write as a single fraction. Finally, rewrite using reciprocal identities.

$$\cot x \cdot \sec x = \frac{\cos x}{\sin x} \cdot \frac{1}{\cos x} = \frac{\cancel{\cos x}}{\sin x} \cdot \frac{1}{\cancel{\cos x}} = \frac{1}{\sin x} = \csc x$$

2 Simplify $\sin x \cdot \sec x$. The answer is $\tan x$.

Begin by rewriting $\sec x$ using reciprocal identities. Then multiply the terms to get a single fraction. Finally, rewrite as a single expression using ratio identities.

$$\sin x \cdot \sec x = \sin x \cdot \frac{1}{\cos x} = \frac{\sin x}{\cos x} = \tan x$$

3 Simplify $\sin^3 x \cdot \csc^2 x + \tan x \cdot \cos x$. The answer is $2\sin x$.

Start by using reciprocal and ratio identities to rewrite the expression in terms of sine and cosine. Next, cancel any terms you can. Finally, combine the like terms.

$$\sin^3 x \cdot \csc^2 x + \tan x \cdot \cos x = \sin^3 x \cdot \frac{1}{\sin^2 x} + \frac{\sin x}{\cos x} \cdot \cos x =$$

$$\sin^{\cancel{3}\,1} x \cdot \frac{1}{\cancel{\sin^2 x}} + \frac{\sin x}{\cancel{\cos x}} \cdot \cancel{\cos x} = \sin x + \sin x = 2\sin x$$

Be careful to write $\sin x + \sin x = 2\sin x$, not $\sin 2x$; the latter expression is the sin of a double angle (which we introduce in the next chapter).

4 Simplify $\cot x \cdot \sin x \cdot \tan x$. The answer is $\sin x$.

Here, you may want to start by rewriting the tangent term using ratio identities. Then, cancel the like terms in the numerator and denominator, giving you your answer.

$$\cot x \cdot \sin x \cdot \tan x = \cot x \cdot \sin x \cdot \frac{1}{\cot x} = \cancel{\cot x} \cdot \sin x \cdot \frac{1}{\cancel{\cot x}} = \sin x$$

5 Simplify $\frac{\cos x}{\sin x}(\tan x + \cot x)$. The answer is $\csc^2 x$.

Begin by using ratio identities to rewrite the ratio of cosine and sine as cotangent. Then distribute the cotangent term and rewrite the tangent term using ratio identities, canceling terms in the numerator and denominator. Notice that the resulting expression can be simplified using a Pythagorean identity.

$$\frac{\cos x}{\sin x}(\tan x + \cot x) = \cot x(\tan x + \cot x) = \cot x \cdot \tan x + \cot x \cdot \cot x =$$

$$\cot x \cdot \frac{1}{\cot x} + \cot^2 x = \cancel{\cot x} \cdot \frac{1}{\cancel{\cot x}} + \cot^2 x = 1 + \cot^2 x = \csc^2 x$$

6 Simplify $1-\frac{\sin x\cdot\tan x}{\sec x}$. The answer is $\cos^2 x$.

Again, begin by using reciprocal identities to rewrite tangent and secant in terms of sine and cosine. Next, rewrite the fraction as a division problem; then multiply by the reciprocal. Cancel terms in the numerator and denominator. Finally, use Pythagorean identities to rewrite the expression as a single term.

$$1-\frac{\sin x\cdot\tan x}{\sec x}=1-\frac{\sin x\cdot\frac{\sin x}{\cos x}}{\frac{1}{\cos x}}=1-\sin x\cdot\frac{\sin x}{\cos x}\div\frac{1}{\cos x}=$$

$$1-\sin x\cdot\frac{\sin x}{\cos x}\cdot\cos x=1-\sin x\cdot\frac{\sin x}{\cancel{\cos x}}\cdot\cancel{\cos x}=1-\sin^2 x=\cos^2 x$$

7 Simplify $\sin x\cdot\cot^2 x+\sin x$. The answer is $\csc x$.

Factor $\sin x$ from both terms. Then, using Pythagorean identities, simplify the expression. Next, rewrite the cosecant term using reciprocal identities. Cancel the like terms in the numerator and denominator. Finally, use reciprocal identities to simplify the final term.

$$\sin x\cdot\cot^2 x+\sin x=\sin x(\cot^2 x+1)=\sin x\cdot\csc^2 x=$$

$$\sin x\cdot\frac{1}{\sin^2 x}=\cancel{\sin x}\cdot\frac{1}{\sin^{\cancel{2}1} x}=\frac{1}{\sin x}=\csc x$$

8 Simplify $(\sin^2 x-1)(\tan^2 x+1)$. The answer is -1.

Start by replacing the sine term using Pythagorean identities. Next, distribute the cosine term across the tangent expression. Then, using ratio identities, rewrite the tangent term. You can then cancel cosine from the numerator and denominator. The resulting expression is another Pythagorean identity.

$$(\sin^2 x-1)(\tan^2 x+1)=-\cos^2 x(\tan^2 x+1)=-\cos^2 x\cdot\tan^2 x-\cos^2 x=$$

$$-\cos^2 x\cdot\frac{\sin^2 x}{\cos^2 x}-\cos^2 x=-\cancel{\cos^2 x}\cdot\frac{\sin^2 x}{\cancel{\cos^2 x}}-\cos^2 x=-\sin^2 x-\cos^2 x=-1$$

9 Simplify $\sec(-x)\cdot\cot(-x)$. The answer is $-\csc x$.

Begin by using even-odd identities to get rid of all the $-x$ values inside the trig expressions. Next, use reciprocal and ratio identities to rewrite the expression in terms of sine and cosine. Then, cancel any like terms in the numerator and denominator. Last, rewrite the resulting fraction using reciprocal identities.

$$\sec(-x)\cdot\cot(-x)=\sec x\cdot(-\cot x)=\frac{1}{\cos x}\cdot\left(-\frac{\cos x}{\sin x}\right)=$$

$$\frac{1}{\cancel{\cos x}}\left(-\frac{\cancel{\cos x}}{\sin x}\right)=-\frac{1}{\sin x}=-\csc x$$

10 Simplify $\sin x\cdot[\csc x+\sin(-x)]$. The answer is $\cos^2 x$.

Use even-odd identities to replace the $-x$ value; then you can distribute the $\sin x$ term. Using reciprocal identities, rewrite the cosecant term and cancel where you can. The resulting expression can be simplified using Pythagorean identities.

$$\sin x\cdot[\csc x+\sin(-x)]=\sin x\cdot[\csc x-\sin x]=\sin x\cdot\csc x-\sin x\cdot\sin x=$$

$$\sin x\cdot\frac{1}{\sin x}-\sin^2 x=\cancel{\sin x}\cdot\frac{1}{\cancel{\sin x}}-\sin^2 x=1-\sin^2 x=\cos^2 x$$

11 Simplify $\frac{\csc x}{\cot(-x)}$. The answer is $-\sec x$.

Start by replacing the $-x$ value using even-odd identities. Then, using reciprocal and ratio identities, rewrite everything in terms of sine and cosine. You can then rewrite the big division bar with a division sign and invert the fraction so you can multiply. Cancel any terms you can. Finally, simplify the resulting expression using reciprocal identities.

$$\frac{\csc x}{\cot(-x)} = \frac{\csc x}{-\cot x} = \frac{\frac{1}{\sin x}}{-\frac{\cos x}{\sin x}} = \frac{1}{\sin x} \div \left(-\frac{\cos x}{\sin x}\right) =$$

$$\frac{1}{\sin x} \cdot \left(-\frac{\sin x}{\cos x}\right) = \frac{1}{\cancel{\sin x}} \cdot \left(-\frac{\cancel{\sin x}}{\cos x}\right) = -\frac{1}{\cos x} = -\sec x$$

12 Simplify $\frac{-\cot^2(-x)-1}{-\cot^2(x)}$. The answer is $\sec^2 x$.

Again, you can start by replacing the $-x$ using even-odd identities. Then you can factor out a negative from the numerator. This results in a Pythagorean identity, which can be simplified. At this point the negatives in the numerator and denominator cancel each other out, leaving you with a positive expression. Next, using reciprocal and ratio identities, change everything into terms using sine and cosine. Rewrite the big fraction as a division problem, which can then be changed to multiplication by inverting the fraction. Cancel what you can and use reciprocal identities to simplify. *Voilà!*

$$\frac{-\cot^2(-x)-1}{-\cot^2 x} = \frac{\cot^2 x+1}{\cot^2 x} = \frac{\csc^2 x}{\cot^2 x} = \frac{\frac{1}{\sin^2 x}}{\frac{\cos^2 x}{\sin^2 x}} =$$

$$\frac{1}{\sin^2 x} \div \frac{\cos^2 x}{\sin^2 x} = \frac{1}{\sin^2 x} \cdot \frac{\sin^2 x}{\cos^2 x} = \frac{1}{\cancel{\sin^2 x}} \cdot \frac{\cancel{\sin^2 x}}{\cos^2 x} = \frac{1}{\cos^2 x} = \sec^2 x$$

13 Simplify $\sin\left(\frac{\pi}{2}-x\right) \cdot \cot\left(\frac{\pi}{2}-x\right)$. The answer is $\sin x$.

Here you can start by using co-function identities to rewrite the sine and cotangent terms. Next, use ratio identities to rewrite the tangent term as a rational function of sine and cosine. Cancel the cosine terms, leaving you with your answer.

$$\sin\left(\frac{\pi}{2}-x\right) \cdot \cot\left(\frac{\pi}{2}-x\right) = \cos x \cdot \tan x = \cos x \cdot \frac{\sin x}{\cos x} = \cancel{\cos x} \cdot \frac{\sin x}{\cancel{\cos x}} = \sin x$$

14 Simplify $\sin\left(\frac{\pi}{2}-x\right) + \cot\left(\frac{\pi}{2}-x\right) \cdot \cos\left(\frac{\pi}{2}-x\right)$. The answer is $\sec x$.

Begin by using co-function identities to replace any terms with $\pi/2$ in them. Then rewrite tangent using ratio identities. Next, you need to find a common denominator in order to add the resulting terms. Rewriting the resulting fraction, you can see that the numerator can be simplified using a Pythagorean identity. Finally, use reciprocal identities to simplify the resulting expression.

$$\sin\left(\frac{\pi}{2}-x\right) + \cot\left(\frac{\pi}{2}-x\right) \cdot \cos\left(\frac{\pi}{2}-x\right) = \cos x + \tan x \cdot \sin x = \cos x + \frac{\sin x}{\cos x} \cdot \sin x =$$

$$\cos x + \frac{\sin^2 x}{\cos x} = \frac{\cos^2 x}{\cos x} + \frac{\sin^2 x}{\cos x} = \frac{\cos^2 x + \sin^2 x}{\cos x} = \frac{1}{\cos x} = \sec x$$

15 Simplify $\dfrac{\cos\left(\frac{\pi}{2}-x\right)}{\sin\left(\frac{\pi}{2}-x\right)}\cdot\cot\left(\frac{\pi}{2}-x\right)$. The answer is $\tan^2 x$.

Begin by using co-function identities to replace each term. Then use ratio identities to replace the sine/cosine fraction with tangent. Finally, multiply the tangent terms.

$$\frac{\cos\left(\frac{\pi}{2}-x\right)}{\sin\left(\frac{\pi}{2}-x\right)}\cdot\cot\left(\frac{\pi}{2}-x\right)=\frac{\sin x}{\cos x}\cdot\tan x=\tan x\cdot\tan x=\tan^2 x$$

16 Simplify $\left[\sec\left(\frac{\pi}{2}-x\right)+\tan\left(\frac{\pi}{2}-x\right)\right]\left[1-\sin\left(\frac{\pi}{2}-x\right)\right]$. The answer is $\sin x$.

Start by replacing terms using co-function identities. Next, use reciprocal and ratio identities to rewrite the terms using sine and cosine. Then, because the first two terms have a common denominator of sine, you can write them as a single fraction. Next, you can multiply using FOIL. The resulting numerator can be simplified using a Pythagorean identity. Finally, cancel sine from the numerator and denominator.

$$\left[\sec\left(\frac{\pi}{2}-x\right)+\tan\left(\frac{\pi}{2}-x\right)\right]\left[1-\sin\left(\frac{\pi}{2}-x\right)\right]=(\csc x+\cot x)(1-\cos x)=$$

$$\left(\frac{1}{\sin x}+\frac{\cos x}{\sin x}\right)(1-\cos x)=\left(\frac{1+\cos x}{\sin x}\right)\left(\frac{1-\cos x}{1}\right)=\frac{1-\cos^2 x}{\sin x}=\frac{\sin^2 x}{\sin x}=\frac{\sin^{\not 2} x}{\not{\sin x}}=\sin x$$

17 Simplify $\cos(2\pi+x)+\sin(2\pi+x)\cdot\cot(\pi+x)$. The answer is $2\cos x$.

Using periodicity identities, replace each term. Then, using ratio identities, replace the cotangent with a ratio of cosine and sine. Cancel the sine from the numerator and denominator and add the two cosines.

$$\cos(2\pi+x)+\sin(2\pi+x)\cdot\cot(\pi+x)=\cos x+\sin x\cdot\cot x=$$

$$\cos x+\sin x\cdot\frac{\cos x}{\sin x}=\cos x+\not{\sin x}\cdot\frac{\cos x}{\not{\sin x}}=\cos x+\cos x=2\cos x$$

18 Simplify $\dfrac{\cos(x+4\pi)}{\cot(x+2\pi)}$. The answer is $\sin x$.

Ah! Don't let this one trick you. 4π is a multiple of 2π, so adding that to x also gives you a periodicity identity. Replacing the terms with the appropriate periodicity identity, you can easily simplify this problem. The next step is to use ratio identities to replace the cotangent term with a ratio of cosine and sine. Then, simplify the complex fraction by multiplying the numerator by the reciprocal of the denominator. Cancel where you can to get a simplified expression.

$$\frac{\cos(x+4\pi)}{\cot(x+2\pi)}=\frac{\cos x}{\cot x}=\frac{\cos x}{\frac{\cos x}{\sin x}}=\cos x+\frac{\cos x}{\sin x}=\cos x\cdot\frac{\sin x}{\cos x}=\not{\cos x}\cdot\frac{\sin x}{\not{\cos x}}=\sin x$$

19 Simplify $\frac{\sec(x+2\pi)}{\csc(x+2\pi)}$. The answer is $\tan x$.

Start by replacing both terms using periodicity identities. Then, you can rewrite the fraction as a division problem. Next, rewrite the division problem by multiplying by the reciprocal. Then use reciprocal identities to replace both terms. Multiply to write as a single term, which can be simplified using ratio identities.

$$\frac{\sec(x+2\pi)}{\csc(x+2\pi)}=\frac{\sec x}{\csc x}=\sec x\div\csc x=\sec x\cdot\frac{1}{\csc x}=\frac{1}{\cos x}\cdot\sin x=\frac{\sin x}{\cos x}=\tan x$$

20 Simplify $[\sec(x-2\pi)-\tan(x-\pi)][\sec(x+2\pi)+\tan(x+\pi)]$. The answer is 1.

Simplify by replacing every term using periodicity identities. Then you can FOIL. Finally, use a Pythagorean identity.

$$[\sec(x-2\pi)-\tan(x-\pi)][\sec(x+2\pi)+\tan(x+\pi)]=$$
$$(\sec x-\tan x)(\sec x+\tan x)=\sec^2 x-\tan^2 x=1$$

21 Prove $\frac{\cot x-1}{1+\tan(-x)}=\cot x$ for x in $(\frac{\pi}{4}, \pi+\frac{\pi}{4})$.

For this proof, we start with the left side because it's more complicated. Begin by using even-odd identities to replace the $-x$ and use ratio identities to rewrite tangent and cotangent in terms of sine and cosine.

$$\frac{\cot x-1}{1+\tan(-x)}=\frac{\cot x-1}{1-\tan x}=\frac{\frac{\cos x}{\sin x}-1}{1-\frac{\sin x}{\cos x}}$$

Now multiply the numerator and denominator by the LCD to simplify the complex fraction.

$$\frac{\left(\frac{\cos x}{\sin x}-1\right)(\sin x\cdot\cos x)}{\left(1-\frac{\sin x}{\cos x}\right)(\sin x\cdot\cos x)}=\frac{\frac{\cos x}{\sin x}\sin x\cdot\cos x-\sin x\cdot\cos x}{\sin x\cdot\cos x-\frac{\sin x}{\cos x}\sin x\cdot\cos x}=$$

$$\frac{\frac{\cos x}{\cancel{\sin x}}\cancel{\sin x}\cdot\cos x-\sin x\cdot\cos x}{\sin x\cdot\cos x-\frac{\sin x}{\cancel{\cos x}}\sin x\cdot\cancel{\cos x}}=\frac{\cos^2 x-\sin x\cdot\cos x}{\sin x\cdot\cos x-\sin^2 x}$$

Finally, you can pull out a common factor on both the numerator and denominator, cancel any like terms, and simplify the resulting fraction using ratio identities.

$$\frac{\cos^2 x-\sin x\cdot\cos x}{\sin x\cdot\cos x-\sin^2 x}=\frac{\cos x(\cos x-\sin x)}{\sin x(\cos x-\sin x)}=\frac{\cos x(\cancel{\cos x-\sin x})}{\sin x(\cancel{\cos x-\sin x})}=\frac{\cos x}{\sin x}=\cot x$$

22 Prove $\frac{\csc x+\tan x}{\cot\left(\frac{\pi}{2}-x\right)\csc\left(\frac{\pi}{2}-x\right)}=\cot^2 x+\cos x$.

Start by using co-function identities to replace the terms with π.

$$\frac{\csc x+\tan x}{\cot\left(\frac{\pi}{2}-x\right)\csc\left(\frac{\pi}{2}-x\right)}=\frac{\csc x+\tan x}{\tan x\sec x}$$

Next, you can separate the fraction into two different fractions and then cancel any terms on both the numerator and denominator.

$$\frac{\csc x}{\tan x\sec x}+\frac{\tan x}{\tan x\sec x}=\frac{\csc x}{\tan x\sec x}+\frac{\cancel{\tan x}}{\cancel{\tan x}\sec x}=\frac{\csc x}{\tan x\sec x}+\frac{1}{\sec x}$$

Then, use reciprocal and ratio identities to rewrite all terms using sine and cosine. You can simplify the denominator of the complex fraction using multiplication.

$$\frac{\frac{1}{\sin x}}{\frac{\sin x}{\cos x}\cdot\frac{1}{\cos x}}+\cos x=\frac{\frac{1}{\sin x}}{\frac{\sin x}{\cos^2 x}}+\cos x$$

Next, get rid of the complex fraction by changing the large fraction bar to division and then multiplying by the reciprocal. The resulting fraction can also be simplified using ratio identities, giving you the answer.

$$\frac{1}{\sin x}\div\frac{\sin x}{\cos^2 x}+\cos x=\frac{1}{\sin x}\cdot\frac{\cos^2 x}{\sin x}+\cos x=\frac{\cos^2 x}{\sin^2 x}+\cos x=\cot^2 x+\cos x$$

23 Prove $\cot x=\frac{\csc^2 x-1}{\cot x}$

In this proof, the right side is more complicated, so it's wise to start there. Notice the squared term? You can replace it using Pythagorean identities, which then cancels out the 1 in the numerator. Last, simply cancel a cotangent from both the numerator and denominator, and you're there!

$$\frac{\csc^2 x-1}{\cot x}=\frac{\left(\cot^2 x+1\right)-1}{\cot x}=\frac{\cot^2 x}{\cot x}=\frac{\cot^{\cancel{2}1}}{\cancel{\cot x}}=\cot x$$

24 Prove $\frac{1-\sin x}{\csc x}-\frac{\tan x\cos x}{1+\sin x}=-\frac{\sin^2 x}{\csc x+1}$

For this proof, you want to start by finding a common denominator for the two fractions and multiplying it through using FOIL for the first fraction. Next, you can simplify using a Pythagorean identity.

$$\frac{1-\sin x}{\csc x}-\frac{\tan x\cos x}{1+\sin x}=\frac{(1-\sin x)(1+\sin x)}{\csc x(1+\sin x)}-\frac{\tan x\cos x\csc x}{(1+\sin x)\csc x}=$$

$$\frac{1-\sin^2 x}{\csc x(1+\sin x)}-\frac{\tan x\cdot\cos x\cdot\csc x}{(1+\sin x)\csc x}=\frac{\cos^2 x}{\csc x(1+\sin x)}-\frac{\tan x\cdot\cos x\cdot\csc x}{(1+\sin x)\csc x}$$

After you have a single fraction, use reciprocal and ratio identities to change the numerator, allowing you to cancel many of the terms. Again, you have a Pythagorean identity that you can simplify.

$$\frac{\cos^2 x-\frac{\sin x}{\cos x}\cdot\cos x\cdot\frac{1}{\sin x}}{\csc x(1+\sin x)}=\frac{\cos^2 x-\frac{\cancel{\sin x}}{\cancel{\cos x}}\cdot\cancel{\cos x}\cdot\frac{1}{\cancel{\sin x}}}{\csc x(1+\sin x)}=$$

$$\frac{\cos^2 x-1}{\csc x(1+\sin x)}=\frac{-\sin^2 x}{\csc x(1+\sin x)}$$

Now that the numerator looks like your final answer, you can concentrate on the denominator. Distribute the cosecant and use reciprocal identities to simplify.

$$\frac{-\sin^2 x}{\csc x + \csc x \sin x} = \frac{-\sin^2 x}{\csc x + \frac{1}{\sin x} \cdot \sin x} = \frac{-\sin^2 x}{\csc x + \frac{1}{\cancel{\sin x}} \cdot \cancel{\sin x}} = \frac{-\sin^2 x}{\csc x + 1}$$

25 Prove $\sqrt{\left[\cos x \cdot \sin\left(\frac{\pi}{2} - x\right) \cdot \csc(2\pi + x) \cdot \sec\left(\frac{\pi}{2} - x\right)\right] + 1} = |\csc x|$

Start by dealing with the gigantic square root by squaring both sides. (***Note:*** This method of proof works here because both sides are positive.)

$$\sqrt{\left[\cos x \cdot \sin\left(\frac{\pi}{2} - x\right) \cdot \csc(2\pi + x) \cdot \sec\left(\frac{\pi}{2} - x\right)\right] + 1} = |\csc x|$$

$$\left[\cos x \cdot \sin\left(\frac{\pi}{2} - x\right) \cdot \csc(2\pi + x) \cdot \sec\left(\frac{\pi}{2} - x\right)\right] + 1 = \csc^2 x$$

Next, replace any terms with $\pi/2$ using co-function identities, and replace any terms with 2π using periodicity identities. Multiply the resulting trig terms.

$\cos x \cdot \cos x \cdot \csc x \cdot \csc x + 1 = \cos^2 x \cdot \csc^2 x + 1$

Replace the cosecant term using reciprocal identities and write as a single fraction that you can replace using ratio identities.

$$\cos^2 x \cdot \frac{1}{\sin^2 x} + 1 = \frac{\cos^2 x}{\sin^2 x} + 1 = \cot^2 x + 1$$

Finally, you have a Pythagorean identity that can be simplified to get the answer you want: $\csc^2 x$

26 Prove $\sec x - \cos x = \sin x \cdot \tan x$.

Start by replacing secant with its reciprocal identity. You end up with a fraction. Add the fractions using a common denominator of cosine and then simplify the resulting numerator with a Pythagorean identity. Finally, factor out a sine. You're left with a ratio of sine and cosine that you can replace using ratio identities to get the right side of the proof.

$$\sec x - \cos x = \frac{1}{\cos x} - \cos x = \frac{1}{\cos x} - \frac{\cos^2 x}{\cos x} = \frac{1 - \cos^2 x}{\cos x} =$$

$$\frac{\sin^2 x}{\cos x} = \sin x \cdot \frac{\sin x}{\cos x} = \sin x \tan x$$

Chapter 9

Advanced Identities

In This Chapter

- Using sum and difference identities
- Exploring double- and half-angle identities
- Tapping product-to-sum and sum-to-product identities
- Solving with power-reducing formulas

Okay . . . the training wheels are off! We're getting into the advanced stuff here — advanced identities, that is. This chapter builds on the basic identities you practice in Chapter 8.

In this chapter, we give you formulas that are essential for calculating precise values of trig functions at certain angles that you can't get any other way (even your calculator only gives approximate answers). These identities are essential for calculus and are well loved by pre-calc teachers, so it's time to get friendly with advanced identities.

Simplifying with Sum and Difference Identities

Here, those pesky mathematicians took simple concepts (addition and subtraction) and related them to a more complex one (trigonometric angles). These *sum and difference identities* allow you to write an angle that's not from the special triangles of 45-45-90 or 30-60-90 (see Chapter 6) as the sum or difference of those helpful angle measures. For example, you can rewrite the measure of 105° as the sum of 45° and 60°. The problems presented here (and in pre-calc books everywhere) can always be written using the angles you already have exact values for, even though these identities can be used for any value.

The sum and difference identities are:

$$\sin(a \pm b) = \sin a \cdot \cos b \pm \cos a \cdot \sin b$$

$$\cos(a \pm b) = \cos a \cdot \cos b \mp \sin a \cdot \sin b$$

$$\tan(a \pm b) = \frac{\tan a \pm \tan b}{1 \mp \tan a \cdot \tan b}$$

Q. Find $\tan \frac{7\pi}{12}$ using sum or difference identities.

A. $-2-\sqrt{3}$. Start by breaking up the fraction into a sum of two values that can be found on the unit circle: $\frac{7\pi}{12} = \frac{3\pi}{12} + \frac{4\pi}{12} = \frac{\pi}{4} + \frac{\pi}{3}$. Next, plug the angles into the sum identity for tangent:

$$\tan\left(\frac{\pi}{4}+\frac{\pi}{3}\right)=\frac{\tan\frac{\pi}{4}+\tan\frac{\pi}{3}}{1-\tan\frac{\pi}{4}\cdot\tan\frac{\pi}{3}}$$

Keep a close eye on the order of a plus or minus symbol in an equation. If it's inverted to be a minus or plus symbol, then you perform the opposite operation than the given problem. In this case, the problem involves addition, so we use addition in the numerator and subtraction in the denominator.

Finally, plug in the known values for the angles from the unit circle in Chapter 6 and simplify. Remember to rationalize the denominator using conjugates (see Chapter 2 for review). The steps are as follows:

$$\frac{\tan\frac{\pi}{4}+\tan\frac{\pi}{3}}{1-\tan\frac{\pi}{4}\cdot\tan\frac{\pi}{3}}=$$

$$\frac{1+\sqrt{3}}{1-(1)\left(\sqrt{3}\right)}=$$

$$\frac{1+\sqrt{3}}{1-\sqrt{3}}=$$

$$\left(\frac{1+\sqrt{3}}{1-\sqrt{3}}\right)\left(\frac{1+\sqrt{3}}{1+\sqrt{3}}\right)=$$

$$\frac{1+2\sqrt{3}+3}{1-3}=$$

$$\frac{4+2\sqrt{3}}{-2}=$$

$$-2-\sqrt{3}$$

1. Find $\cos 15°$ using sum or difference identities.

Solve It

2. Express $\tan(45° - x)$ in terms of $\tan x$.

Solve It

3. Prove $\frac{\sin(x+\pi)}{\cos(x+\pi)} = \tan x$.

Solve It

4. Prove $\frac{\sin(x+y)}{\sin x \cdot \cos y} = 1 + \cot x \cdot \tan y$.

Solve It

5. Find $\csc \frac{5\pi}{12}$ using sum or difference identities.

Solve It

6. Simplify $\sec(180° + x)$ using sum or difference identities.

Solve It

Using Double-Angle Identities

Double-angle identities help you find the trig value of twice an angle. These can be used to find an exact value if you know the original angle. They can also be used to prove theorems (see Chapter 8) or solve trig equations. Cosine has three double-angle identities created from the identities of Chapter 8. You have a choice as to which you want to use, depending on the problem.

The double-angle identities are:

$$\sin 2x = 2\sin x \cos x$$

$$\cos 2x = \cos^2 x - \sin^2 x = 2\cos^2 x - 1 = 1 - 2\sin^2 x$$

$$\tan 2x = \frac{2\tan x}{1-\tan^2 x}$$

Q. Solve $6\cos^2 x - 6\sin^2 x = 3$ for $0 < x < \pi$.

A. $x = \frac{\pi}{6}$ or $\frac{5\pi}{6}$. Start by factoring out the 6 from the left side of the equation $6(\cos^2 x - \sin^2 x) = 3$. Next, substitute using the appropriate double-angle identity: $6(\cos 2x) = 3$. Isolate the trigonometric term: $\cos 2x = \frac{1}{2}$. Then, take the inverse of cosine using the unit circle in Chapter 6: $2x = \frac{\pi}{3}$ or $\frac{5\pi}{3}$. Finally, solve for x by dividing both sides by 2: $x = \frac{\pi}{6}$ or $\frac{5\pi}{6}$.

7. Find the value of $\cos 2x$ if $\csc x = \frac{13}{5}$.

Solve It

8. Find the value $\tan 2x$ if $\cot x = \frac{1}{2}$.

Solve It

9. Prove $\frac{\cos 2x + 1}{\sin 2x} = \cot x$.

Solve It

10. Prove $\frac{\cos 2x}{\sin 2x} = \frac{1}{2}(\cot x - \tan x)$.

Solve It

11. Express $\tan 3x$ in terms of $\tan x$ by using double-angle identities.

Solve It

12. Solve $6 - 12\sin^2 x = 3\sqrt{3}$ for $0 < x < \pi$.

Solve It

Reducing with Half-Angle Identities

Similar to sum and difference identities, *half-angle identities* help you find exact values of unusual angles, namely ones that are half the value of ones you already know. For example, if you want to find a trig value of 22.5°, you use the half-angle identity on half of 45° because 22.5 is half of 45. Also, just like every other identity we've reviewed so far, half-angle identities can be used for proving theorems and solving trig equations.

The half-angle identities are:

$$\sin\left(\frac{a}{2}\right)=\pm\sqrt{\frac{1-\cos a}{2}}$$

$$\cos\left(\frac{a}{2}\right)=\pm\sqrt{\frac{1+\cos a}{2}}$$

$$\tan\left(\frac{a}{2}\right)=\frac{1-\cos a}{\sin a}=\frac{\sin a}{1+\cos a}$$

Q. Find $\cot\frac{5\pi}{12}$ using half-angle identities.

A. $2-\sqrt{3}$. First, because we didn't provide a half-angle formula for cotangent, you have to start by recognizing that $\cot\frac{5\pi}{12}$ is the reciprocal of tangent of the same angle. Therefore, you're going to find the value of $\tan\frac{5\pi}{12}$, and then take the reciprocal. The angle $\frac{5\pi}{12}$ can be rewritten as $\frac{\frac{5\pi}{6}}{2}$. Plugging that into the half-angle identity, you get

$$\tan\left(\frac{5\pi}{12}\right)=\frac{1-\cos\frac{5\pi}{6}}{\sin\frac{5\pi}{6}}$$

Replacing the trig expressions with the exact values from the unit circle, you get $\frac{1-\left(\frac{-\sqrt{3}}{2}\right)}{\frac{1}{2}}$. This simplifies to $2+\sqrt{3}$. But wait — you're not done! This is the value of $\tan\frac{5\pi}{12}$, and you need $\cot\frac{5\pi}{12}$. So you have to take the reciprocal, which, after rationalizing, you find to be $2-\sqrt{3}$.

13. Find $\tan\frac{3\pi}{8}$ using half-angle identities.

Solve It

14. Find $\sin\frac{7\pi}{12}$.

Solve It

15. Prove $\left[\tan\left(\frac{x}{2}\right)\right]\cdot\tan x=\frac{\sin^2 x}{\cos x+\cos^2 x}$.

Solve It

16. Find an approximate value of $\cos\frac{\pi}{24}$ using half-angle identities.

Solve It

Changing Products to Sums

You need to know three product-to-sum (or product-to-difference) identities: sin · cos, cos · cos, and sin · sin. They follow easily from the sum and difference identities. Here they are:

$\sin a\cdot\cos b=\frac{1}{2}[\sin(a+b)+\sin(a-b)]$

$\cos a\cdot\cos b=\frac{1}{2}[\cos(a+b)+\cos(a-b)]$

$\sin a\cdot\sin b=\frac{1}{2}[\cos(a-b)-\cos(a+b)]$

Q. Express $8\sin 3x \cdot \sin x$ as a sum or difference.

A. $4(\cos 2x - \cos 4x)$. Start by plugging in the appropriate product-to-sum identity:

$$8\left(\frac{1}{2}\left[\cos(3x-x)-\cos(3x+x)\right]\right).$$

Then, simplify using multiplication:

$$8\cdot\frac{1}{2}\left[\cos(2x)-\cos(4x)\right]=4(\cos 2x-\cos 4x)$$

17. Express $12\cos 6x \cdot \cos 2x$ as a sum or difference.

Solve It

18. Express $2\sin 5x \cdot \cos 2x$ as a sum or difference.

Solve It

19. Express $6\sin 6x \cdot \cos 3x$ as a sum or difference.

Solve It

20. Express $7\sin 8x \cdot \sin 3x$ as a sum or difference.

Solve It

Expressing Sums as Products

Although less frequently used than the other identities in this chapter, the *sum-to-product identities* are useful for finding exact answers for some trig expressions. In cases where the sum or difference of the two angles results in an angle from our special right triangles (Chapter 6), sum-to-product identities can be quite helpful. The sum-to-product (or difference-to-product) identities involve the addition or subtraction of either sine or cosine. They follow immediately from the product to sum/difference identities by letting $a = \frac{(x+y)}{2}$ and $b = \frac{(x-y)}{2}$.

The sum–to-product (or difference-to-product) identities are

$$\sin x + \sin y = 2\sin\left(\frac{x+y}{2}\right)\cos\left(\frac{x-y}{2}\right) \qquad \sin x - \sin y = 2\cos\left(\frac{x+y}{2}\right)\sin\left(\frac{x-y}{2}\right)$$

$$\cos x + \cos y = 2\cos\left(\frac{x+y}{2}\right)\cos\left(\frac{x-y}{2}\right) \qquad \cos x - \cos y = -2\sin\left(\frac{x+y}{2}\right)\sin\left(\frac{x-y}{2}\right)$$

Q. Find $\cos 165° + \cos 75°$.

A. $-\frac{\sqrt{2}}{2}$. Begin by using the sum-to-product identity to rewrite the expression: $2\cos\left(\frac{165+75}{2}\right)\cos\left(\frac{165-75}{2}\right)$. Simplify the results using unit circle values: $2\cos\left(\frac{240}{2}\right)\cos\left(\frac{90}{2}\right) = 2\cos 120° \cdot \cos 45° = 2\left(-\frac{1}{2}\right)\frac{\sqrt{2}}{2} = -\frac{\sqrt{2}}{2}$

21. Find $\sin 195° - \sin 75°$.

Solve It

22. Find $\cos 375° - \cos 75°$.

Solve It

23. Find $\sin \frac{7\pi}{12} + \sin \frac{\pi}{12}$.

Solve It

24. Find $\cos \frac{23\pi}{12} + \cos \frac{5\pi}{12}$.

Solve It

Powering Down: Power-Reducing Formulas

Power-reducing formulas can be used to simplify trig expressions with exponents and can be used more than once if you have a function that's raised to the fourth power or higher. These nifty formulas help you get rid of exponents and have many applications in calculus.

Here are the three power-reducing formulas:

$$\sin^2 u = \frac{1-\cos 2u}{2} \qquad \cos^2 u = \frac{1+\cos 2u}{2} \qquad \tan^2 u = \frac{1-\cos 2u}{1+\cos 2u}$$

Q. Express $\cos^4 x$ without exponents by using power-reducing formulas.

A. $\frac{1}{8}(3 + 4\cos 2x + \cos 4x)$. After rewriting $(\cos^2 x)^2$, you can see that you need to use the power-reducing formula twice. The first time gives you

$$\left(\frac{1+\cos 2x}{2}\right)^2 = \frac{(1+\cos 2x)(1+\cos 2x)}{4} = \frac{1+2\cos 2x+\cos^2 2x}{4} = \frac{1}{4}\left(1+2\cos 2x+\cos^2 2x\right).$$

Use the formula again on the remaining squared term and reduce:

$$\frac{1}{4}\left[1+2\cos 2x+\left(\frac{1+\cos 4x}{2}\right)\right] = \frac{1}{8}(2+4\cos 2x+1+\cos 4x) = \frac{1}{8}(3+4\cos 2x+\cos 4x)$$

25. Prove $\cos^2 3x - \sin^2 3x = \cos 6x$.

Solve It

26. Prove $(1 + \cos 2x) \cdot \tan^3 x = \tan x \cdot (1 - \cos 2x)$.

Solve It

27. Express $\sin^4 x - \cos^4 x$ without exponents.

Solve It

28. Express $\cot^2(\frac{x}{2})$ without exponents.

Solve It

Answers to Problems on Advanced Identities

Following are the answers to problems dealing with trig functions. We also provide guidance on getting the answers if you need to review where you went wrong.

1 Find $\cos 15°$ using sum or difference identities. The answer is $\frac{\sqrt{6}+\sqrt{2}}{4}$.

Start by rewriting the angle using special angles from the unit circle. We chose $15° = 45° - 30°$. Plugging these values into the difference formula, you get $\cos(45° - 30°) = \cos 45° \cdot \cos 30° + \sin 45° \cdot \sin 30°$. Use the unit circle to plug in the appropriate values and simplify: $\frac{\sqrt{2}}{2} \cdot \frac{\sqrt{3}}{2} + \frac{\sqrt{2}}{2} \cdot \frac{1}{2} = \frac{\sqrt{6}}{4} + \frac{\sqrt{2}}{4} = \frac{\sqrt{6}+\sqrt{2}}{4}$.

2 Express $\tan(45° - x)$ in terms of $\tan x$. The answer is $\frac{1-\tan x}{1+\tan x}$.

Rewrite the expression using the difference formula: $\frac{\tan 45° - \tan x}{1+\tan 45° \tan x}$. Then, replace $\tan 45°$ with its unit circle value of 1 and simplify the expression: $\frac{(1)-\tan x}{1+(1)\tan x} = \frac{1-\tan x}{1+\tan x}$

3 Prove $\frac{\sin(x+\pi)}{\cos(x+\pi)} = \tan x$.

Begin the proof by rewriting the left side using sum formulas. Next, substitute for unit circle values and simplify. Finally, replace the sine and cosine with tangent using ratio identities (from Chapter 8). The steps are as follows:

$$\frac{\sin(x+\pi)}{\cos(x+\pi)} = \frac{\sin x \cdot \cos\pi + \cos x \cdot \sin\pi}{\cos x \cdot \cos\pi - \sin x \cdot \sin\pi} = \frac{\sin x \cdot (-1) + \cos x \cdot (0)}{\cos x \cdot (-1) - \sin x \cdot (0)}$$

$$= \frac{-\sin x}{-\cos x} = \frac{\sin x}{\cos x} = \tan x$$

4 Prove $\frac{\sin(x+y)}{\sin x \cdot \cos y} = 1 + \cot x \cdot \tan y$.

Start by replacing the $\sin(x+y)$ term using the sum formula. Next, separate the fraction into the sum of two fractions and reduce terms. Last, rewrite using ratio identities from Chapter 8.

$$\frac{\sin(x+y)}{\sin x \cos y} = \frac{\sin x \cdot \cos y + \cos x \cdot \sin y}{\sin x \cos y} = \frac{\sin x \cdot \cos y}{\sin x \cos y} + \frac{\cos x \cdot \sin y}{\sin x \cos y} =$$

$$\frac{\cancel{\sin x} \cdot \cancel{\cos y}}{\cancel{\sin x}\,\cancel{\cos y}} + \frac{\cos x \cdot \sin y}{\sin x \cos y} = 1 + \frac{\cos x}{\sin x} \cdot \frac{\sin y}{\cos y} = 1 + \cot x \cdot \tan y$$

5 Find $\csc\frac{5\pi}{12}$ using sum or difference identities. The answer is $\sqrt{6} - \sqrt{2}$.

First, know that you can rewrite $\frac{5\pi}{12}$ as $\frac{2\pi}{12} + \frac{3\pi}{12}$, which is the same as $\frac{\pi}{6} + \frac{\pi}{4}$. Substitute this back into the problem: $\csc(\frac{\pi}{6} + \frac{\pi}{4})$. Because you don't have a sum identity for cosecant, you need to find $\frac{1}{\sin\left(\frac{\pi}{6}+\frac{\pi}{4}\right)}$. To ease this calculation, you start by finding $\sin(\frac{\pi}{6} + \frac{\pi}{4})$, and then invert it. Using sum identities, $\sin(\frac{\pi}{6} + \frac{\pi}{4}) = \sin\frac{\pi}{6} \cdot \cos\frac{\pi}{4} + \cos\frac{\pi}{6} \cdot \sin\frac{\pi}{4}$.

Replace the trig expressions with the appropriate values from the unit circle and simplify: $\frac{1}{2}\cdot\frac{\sqrt{2}}{2}+\frac{\sqrt{3}}{2}\cdot\frac{\sqrt{2}}{2}=\frac{\sqrt{2}}{4}+\frac{\sqrt{6}}{4}=\frac{\sqrt{2}+\sqrt{6}}{4}$.

The last step is to invert this solution and simplify by rationalizing the denominator:

$$\frac{4}{\sqrt{2}+\sqrt{6}}=\left(\frac{4}{\sqrt{2}+\sqrt{6}}\right)\cdot\frac{\left(\sqrt{2}-\sqrt{6}\right)}{\left(\sqrt{2}-\sqrt{6}\right)}=\frac{4\left(\sqrt{2}-\sqrt{6}\right)}{2-6}=\frac{4\left(\sqrt{2}-\sqrt{6}\right)}{-4}=-\left(\sqrt{2}-\sqrt{6}\right)=\sqrt{6}-\sqrt{2}$$

6 Simplify $\sec(180° + x)$ using sum or difference identities. The answer is $-\sec x$.

Because you don't have a sum identity for secant, you need to use the one for cosine and then find the reciprocal. Using this identity, you get $\cos(180° + x) = \cos 180° \cdot \cos x - \sin 180° \cdot \sin x$. Plugging in values from the unit circle and simplifying, you get $(-1)\cos x - (0)\sin x = -\cos x$. The reciprocal of this is $-\frac{1}{\cos x}$, which by reciprocal identities (from Chapter 8) is $-\sec x$.

7 Find the value of $\cos 2x$ if $\csc x = \frac{12}{5}$. The answer is $\frac{47}{72}$.

Because $\cos 2x = 1 - 2\sin^2 x$, you need to know $\sin x$ to plug it in. No problem! You have the reciprocal: $\csc x$. Therefore, $\sin x = \frac{5}{12}$. Plugging this in, you get $\cos 2x = 1 - 2(\frac{5}{12})^2 = 1 - 2(\frac{25}{144}) = 1 - \frac{25}{72} = \frac{47}{72}$.

8 Find the value of $\tan 2x$ if $\cot x = \frac{1}{2}$. The answer is $-\frac{4}{3}$.

Begin by finding $\tan x$ by taking the reciprocal of $\cot x$: $\tan x = \frac{2}{1} = 2$. Then plug it into the formula for $\tan 2x$: $\frac{2\tan x}{1-\tan^2 x}=\frac{2(2)}{1-(2)^2}=\frac{4}{1-4}=\frac{-4}{3}$.

9 Prove $\frac{\cos 2x+1}{\sin 2x}=\cot x$.

Working with the left side, change all the double angles using the double-angle identities. After combining like terms, cancel the 2 and cosine from the numerator and denominator. Finally, rewrite the result using ratio identities (see Chapter 8). The steps are as follows: $\frac{\cos 2x+1}{\sin 2x}=\frac{2\cos^2 x-1+1}{2\sin x\cos x}=\frac{2\cos^2 x}{2\sin x\cos x}=\frac{\cancel{2}\cos^{\cancel{2}} x}{\cancel{2}\sin x\cancel{\cos x}}=\frac{\cos x}{\sin x}=\cot x$. By reversing these steps, you complete the proof.

10 Prove $\frac{\cos 2x}{\sin 2x}=\frac{1}{2}(\cot x-\tan x)$.

Begin by using the double-angle identities for the left side: $\frac{\cos 2x}{\sin 2x}=\frac{\cos^2 x-\sin^2 x}{2\sin x\cos x}$. Next, separate the single fraction into the difference of two fractions: $\frac{\cos^2 x}{2\sin x\cos x}-\frac{\sin^2 x}{2\sin x\cos x}$. Then, cancel any terms in the numerator and denominator: $\frac{\cos^{\cancel{2}} x}{2\sin x\cancel{\cos x}}-\frac{\sin^{\cancel{2}} x}{2\cancel{\sin x}\cos x}=\frac{\cos x}{2\sin x}-\frac{\sin x}{2\cos x}$.

Finally, use ratio identities to replace the fractions with cotangent and tangent and factor out the GCF: $\frac{1}{2}\cot x - \frac{1}{2}\tan x = \frac{1}{2}(\cot x - \tan x)$.

11 Express $\tan 3x$ in terms of $\tan x$ by using double-angle identities. The answer is $\frac{3\tan x-\tan^3 x}{1-3\tan^2 x}$.

Start by separating out the angle into a single and double angle: $\tan(x + 2x)$. Next, use the sum identity for tangent: $\frac{\tan x+\tan 2x}{1-\tan x\tan 2x}$. Next, replace the double-angle terms using the double-angle formula: $\frac{\tan x+\frac{2\tan x}{1-\tan^2 x}}{1-\tan x\left(\frac{2\tan x}{1-\tan^2 x}\right)}$. Then, to simplify the complex fraction, multiply by the common denominator: $\frac{\tan x+\frac{2\tan x}{1-\tan^2 x}}{1-\tan x\left(\frac{2\tan x}{1-\tan^2 x}\right)}\cdot\frac{1-\tan^2 x}{1-\tan^2 x}$. Multiply through:

$$\frac{\tan x\left(1-\tan^2 x\right)+\frac{2\tan x}{\cancel{1-\tan^2 x}}\left(\cancel{1-\tan^2 x}\right)}{1\left(1-\tan^2 x\right)-\tan x\left(\frac{2\tan x}{\cancel{1-\tan^2 x}}\right)\left(\cancel{1-\tan^2 x}\right)}=\frac{\tan x-\tan^3 x+2\tan x}{1-\tan^2 x-2\tan^2 x}$$ and combine like terms: $\frac{3\tan x-\tan^3 x}{1-3\tan^2 x}$

12 Solve $6-12\sin^2 x=3\sqrt{3}$ for $0<x<\pi$. The answer is $\pi/12$ or $11\pi/12$.

Begin by factoring out 6 from both terms: $6\left(1-2\sin^2 x\right)=3\sqrt{3}$. Next, use double-angle identities to replace the trig term: $6(\cos 2x)=3\sqrt{3}$. Isolate the trig term by dividing both sides by 6 and simplifying: $\cos 2x=\frac{3\sqrt{3}}{6}=\frac{\cancel{3}\sqrt{3}}{\cancel{6}_2}=\frac{\sqrt{3}}{2}$. Then, to solve for x, take the inverse cosine of each side: $2x=\cos^{-1}\left(\frac{\sqrt{3}}{2}\right)=\frac{\pi}{6}$ or $\frac{11\pi}{6}$ and divide by 2: $\pi/12$ or $11\pi/12$.

13 Find $\tan 3\pi/8$ using half-angle identities. The answer is $\sqrt{2}+1$.

Start by rewriting the angle to find the half angle: $\tan\frac{\frac{3\pi}{4}}{2}$. Then, plug it into the half-angle identity: $\frac{1-\cos\frac{3\pi}{4}}{\sin\frac{3\pi}{4}}$. Find the exact values for the trigonometric terms on the unit circle: $\frac{1-\left(-\frac{\sqrt{2}}{2}\right)}{\frac{\sqrt{2}}{2}}$. Then simplify the complex fraction: $\frac{1+\frac{\sqrt{2}}{2}}{\frac{\sqrt{2}}{2}}=\frac{\frac{2+\sqrt{2}}{2}}{\frac{\sqrt{2}}{2}}=\frac{2+\sqrt{2}}{\cancel{2}}\cdot\frac{\cancel{2}}{\sqrt{2}}=\frac{2+\sqrt{2}}{\sqrt{2}}\left(\frac{\sqrt{2}}{\sqrt{2}}\right)=\frac{2\sqrt{2}+2}{2}=\sqrt{2}+1$.

14 Find $\sin 7\pi/12$. The answer is $\frac{1}{2}\sqrt{2+\sqrt{3}}$.

Begin by rewriting the angle: $\sin\frac{\frac{7\pi}{6}}{2}$. Plug the angle into the appropriate half-angle identity: $\sqrt{\frac{1-\cos\frac{7\pi}{6}}{2}}$. Because the angle is in Quadrant II, sine will be positive. The next step is to replace the trig terms with the exact values from the unit circle: $\sqrt{\frac{1-\left(-\frac{\sqrt{3}}{2}\right)}{2}}$. Simplify the complex fraction: $\sqrt{\frac{1+\frac{\sqrt{3}}{2}}{2}}=\sqrt{\frac{\frac{2+\sqrt{3}}{2}}{2}}=\sqrt{\frac{2+\sqrt{3}}{4}}=\frac{1}{2}\sqrt{2+\sqrt{3}}$.

15 Prove $\left[\tan\left(\frac{x}{2}\right)\right]\cdot\tan x=\frac{\sin^2 x}{\cos x+\cos^2 x}$.

Start with the left side by replacing the half-angle term using the appropriate identity: $\left(\frac{\sin x}{1+\cos x}\right)\tan x$. Next, use ratio identities to change everything to sine and cosine and multiply through to complete the proof: $\left(\frac{\sin x}{1+\cos x}\right)\frac{\sin x}{\cos x}=\frac{\sin^2 x}{\cos x+\cos^2 x}$

16 Find an approximate value of $\cos\frac{\pi}{24}$ using half-angle identities. The answer is ≈ .99.

Rewriting the angle using half-angle identities, remember that the angle is in Quadrant I, so the cosine is positive: $\cos\frac{\frac{\pi}{12}}{2}$. Replace it with the appropriate half-angle identity: $\sqrt{\frac{1+\cos\frac{\pi}{12}}{2}}$. Because you don't have a special right triangle value, you need to use the half-angle identities again: $\sqrt{\frac{1+\sqrt{\frac{1+\cos\frac{\pi}{6}}{2}}}{2}}$. Now you can replace the cosine term using the unit circle value (finally!): $\sqrt{\frac{1+\sqrt{\frac{1+\frac{\sqrt{3}}{2}}{2}}}{2}}$. Finish by plugging the complex fraction into your calculator: ≈ .99.

17 Express $12\cos 6x\cdot\cos 2x$ as a sum or difference. The answer is $6(\cos 8x+\cos 4x)$.

Start by plugging in the appropriate product-to-sum identity: $12\left(\frac{1}{2}\left[\cos(6x+2x)+\cos(6x-2x)\right]\right)$.

Then simplify using multiplication: $12\cdot\frac{1}{2}(\cos 8x+\cos 4x)=6(\cos 8x+\cos 4x)$.

18 Express $2\sin 5x\cdot\cos 2x$ as a sum or difference. The answer is $\sin 7x+\sin 3x$.

Begin by plugging in the appropriate product-to-sum identity: $2\left(\frac{1}{2}\left[\sin(5x+2x)+\sin(5x-2x)\right]\right)$.

Then simplify using multiplication: $2\cdot\frac{1}{2}(\sin 7x+\sin 3x)=\sin 7x+\sin 3x$.

19 Express $6\sin 6x\cdot\cos 3x$ as a sum or difference. The answer is $3(\sin 9x+\sin 3x)$.

Again, plug in the appropriate product-to-sum identity: $6\left(\frac{1}{2}\left[\sin(6x+3x)+\sin(6x-3x)\right]\right)$. Then simplify using multiplication: $6\cdot\frac{1}{2}(\sin 9x+\sin 3x)=3(\sin 9x+\sin 3x)$.

20 Express $7\sin 8x\cdot\sin 3x$ as a sum or difference. The answer is $\frac{7}{2}(\cos 5x-\cos 11x)$.

You guessed it! Plug in the appropriate product-to-sum identity: $7\left(\frac{1}{2}\left[\cos(8x-3x)-\cos(8x+3x)\right]\right)$.

Simplify using multiplication: $7\cdot\frac{1}{2}(\cos 5x-\cos 11x)=\frac{7}{2}(\cos 5x-\cos 11x)$.

21 Find $\sin 195°-\sin 75°$. The answer is $-\frac{\sqrt{6}}{2}$.

Begin by using the sum-to-product identity to rewrite the expression:

$2\cos\left(\frac{195°+75°}{2}\right)\sin\left(\frac{195°-75°}{2}\right)=2\cos 135°\cdot\sin 60°$. Simplify the result using unit circle values: $2\left(-\frac{\sqrt{2}}{2}\right)\left(\frac{\sqrt{3}}{2}\right)=\cancel{2}\left(-\frac{\sqrt{2}}{\cancel{2}}\right)\left(\frac{\sqrt{3}}{2}\right)=-\frac{\sqrt{6}}{2}$

22 Find $\cos 375° - \cos 75°$. The answer is $\frac{\sqrt{2}}{2}$.

Use the sum-to-product identity to rewrite the expression:
$-2\sin\left(\frac{375°+75°}{2}\right)\sin\left(\frac{375°-75°}{2}\right) = -2\sin 225° \cdot \sin 150°$. Then, plug in the unit circle values and simplify: $-2\left(-\frac{\sqrt{2}}{2}\right)\left(\frac{1}{2}\right) = -\cancel{2}\left(-\frac{\sqrt{2}}{\cancel{2}}\right)\left(\frac{1}{2}\right) = \frac{\sqrt{2}}{2}$

23 Find $\sin\frac{7\pi}{12} + \sin\frac{\pi}{12}$. The answer is $\frac{\sqrt{6}}{2}$.

Rewrite the expression with the appropriate sum-to-product identity:

$$2\sin\left(\frac{\frac{7\pi}{12}+\frac{\pi}{12}}{2}\right)\cos\left(\frac{\frac{7\pi}{12}-\frac{\pi}{12}}{2}\right) = 2\sin\left(\frac{\frac{8\pi}{12}}{2}\right)\cos\left(\frac{\frac{6\pi}{12}}{2}\right) = 2\sin\left(\frac{8\pi}{24}\right)\cos\left(\frac{6\pi}{24}\right) = 2\sin\left(\frac{\pi}{3}\right)\cos\left(\frac{\pi}{4}\right)$$

Use unit circle values to simplify the result: $2\left(\frac{\sqrt{3}}{2}\right)\left(\frac{\sqrt{2}}{2}\right) = \cancel{2}\left(\frac{\sqrt{3}}{\cancel{2}}\right)\left(\frac{\sqrt{2}}{2}\right) = \frac{\sqrt{6}}{2}$

24 Find $\cos\frac{23\pi}{12} + \cos\frac{5\pi}{12}$. The answer is $\frac{\sqrt{6}}{2}$.

Again, start by using the sum-to-product identity to rewrite the expression:

$$2\cos\left(\frac{\frac{23\pi}{12}+\frac{5\pi}{12}}{2}\right)\cos\left(\frac{\frac{23\pi}{12}-\frac{5\pi}{12}}{2}\right) = 2\cos\left(\frac{\frac{28\pi}{12}}{2}\right)\cos\left(\frac{\frac{18\pi}{12}}{2}\right) = 2\cos\left(\frac{28\pi}{24}\right)\cos\left(\frac{18\pi}{24}\right) =$$

$2\cos\left(\frac{7\pi}{6}\right)\cos\left(\frac{3\pi}{4}\right)$. Next, simplify the result using unit circle values:

$$2\left(-\frac{\sqrt{3}}{2}\right)\left(-\frac{\sqrt{2}}{2}\right) = \cancel{2}\left(-\frac{\sqrt{3}}{\cancel{2}}\right)\left(-\frac{\sqrt{2}}{2}\right) = \frac{\sqrt{6}}{2}$$

25 Prove $\cos^2 3x - \sin^2 3x = \cos 6x$.

Working on the left side, use the power-reducing formulas to rewrite both terms:
$\left[\frac{1+\cos 2(3x)}{2}\right] - \left[\frac{1-\cos 2(3x)}{2}\right]$. Combine the fractions into one: $\frac{(1+\cos 6x)-(1-\cos 6x)}{2}$.

Combine like terms: $\frac{2\cos 6x}{2} = \cos 6x$.

26 Prove $(1 + \cos 2x) \cdot \tan^3 x = \tan x \cdot (1 - \cos 2x)$.

Start by rewriting the tangent term so that you can use a power-reducing formula: $(1 + \cos 2x) \cdot \tan x \cdot \tan^2 x$. Now use the power-reducing formula: $(1+\cos 2x)\cdot\left(\frac{1-\cos 2x}{1+\cos 2x}\right)\cdot \tan x$. Cancel terms in the numerator and denominator: $\cancel{(1+\cos 2x)}\cdot\left(\frac{1-\cos 2x}{\cancel{1+\cos 2x}}\right)\cdot \tan x$, and you're there!

27 Express $\sin^4 x - \cos^4 x$ without exponents. The answer is $-\cos 2x$.

For this one, we start by factoring the difference of two squares: $(\sin^2 x + \cos^2 x)(\sin^2 x - \cos^2 x)$. Now use Pythagorean identities (see Chapter 8) to simplify: $(1)(\sin^2 x - \cos^2 x)$. If you replace the cosine term using the same Pythagorean identity, you can combine like terms to have only one squared term remaining: $\sin^2 x - (1 - \sin^2 x) = \sin^2 x - 1 + \sin^2 x = 2\sin^2 x - 1$. From here, you just need to replace the squared term using power-reducing formulas and simplify:
$2\left(\frac{1-\cos 2x}{2}\right) - 1 = \cancel{2}\left(\frac{1-\cos 2x}{\cancel{2}}\right) - 1 = 1 - \cos 2x - 1 = -\cos 2x$.

28 Express $\cot^2(x/2)$ without exponents. The answer is $\dfrac{1+\cos x}{1-\cos x}$

Because you don't have a power-reducing formula for cotangent, you need to start by using ratio identities (see Chapter 8) to rewrite the expression: $\dfrac{1}{\tan^2\left(\frac{x}{2}\right)}$. Now you can use tangent's power-reducing formula: $\dfrac{1}{\dfrac{1-\cos 2\left(\frac{x}{2}\right)}{1+\cos 2\left(\frac{x}{2}\right)}}$. Simplify the complex fraction by multiplying by its reciprocal: $\dfrac{1+\cos 2\left(\frac{x}{2}\right)}{1-\cos 2\left(\frac{x}{2}\right)}$. Finally, make cancellations to simplify: $\dfrac{1+\cos \cancel{2}\left(\frac{x}{\cancel{2}}\right)}{1-\cos \cancel{2}\left(\frac{x}{\cancel{2}}\right)} = \dfrac{1+\cos x}{1-\cos x}$

Chapter 10

Solving Oblique Triangles

In This Chapter

- Using the Law of Sines and the Law of Cosines to solve triangles
- Solving word problems using oblique triangles
- Finding the area of a triangle

The trigonometry functions sine, cosine, and tangent are great for finding missing sides and angles inside right triangles. But what happens when a triangle isn't quite right? This type of triangle is known as an *oblique triangle* — any kind of triangle that isn't a right triangle. As you can see in Chapter 6, the process of finding all the sides and angles in a triangle is known as *solving the triangle.* This chapter helps you figure out that process for oblique triangles.

As long as you know one angle and the side directly across from it, you can use the Law of Sines. The Law of Sines can be used in three different cases: angle-side-angle (ASA), angle-angle-side (AAS), and side-side-angle (SSA). The first two cases have exactly one solution. The third case is known as *the ambiguous case* because it may have one, two, or no solutions. We take a look at each case to show you how to deal with them. In fact, the ambiguous case gets its own section in this chapter.

If you don't know an angle and the side opposite it, you can start off with the Law of Cosines, which can be used for two cases: side-side-side (SSS) and side-angle-side (SAS).

Most books use *standard notation* to label an oblique triangle: Each vertex is labeled with a capital letter, and the side opposite it is the same lowercase letter (across from angle A is side a, and so on). We recommend drawing out the triangle and labeling the information you're given. You'll know right away which case you've got on your hands and, therefore, which formula to use.

Solving a Triangle with the Law of Sines: ASA and AAS

After you draw out your triangle and see that you have two angles and the side in between them (ASA) or two angles and a consecutive side (AAS), proceed to solve the triangle with the Law of Sines:

$$\frac{a}{\sin A} = \frac{b}{\sin B} = \frac{c}{\sin C}$$

In either case, because you know two angles inside the given triangle, you automatically know the third, because the sum of the angles inside any triangle is always 180°. Remember, you'll find exactly one solution in either of these cases of the Law of Sines.

Q. Solve the triangle ABC if A = 54°, B = 28°, and c = 11.2.

A. C = 98°, $a \approx 9.15$, $b \approx 5.31$. First, draw and label the triangle. You can see that you have a side sandwiched between two angles (ASA), so you know you can start with the Law of Sines. Because you know two of the angles in the triangle, you can find the third angle first: 54° + 28° + C = 180°; C = 98°. Now substitute all the given information and the angle you just found into the Law of Sines: $\frac{a}{\sin 54^\circ} = \frac{b}{\sin 28^\circ} = \frac{11.2}{\sin 98^\circ}$.

Using the second two equivalent ratios gives you a proportion you can solve: $\frac{b}{\sin 28^\circ} = \frac{11.2}{\sin 98^\circ}$. Cross multiply to get $b \cdot \sin 98^\circ = 11.2 \cdot \sin 28^\circ$. Divide both sides by sin 98 to get $b = \frac{11.2}{\sin 98^\circ} \cdot \sin 28^\circ$. This goes directly into your calculator to give you b = 5.31.

Note: If you put trig function values into your calculator and round as you go, your final answer will be affected. If, however, you wait until the end like we did in solving for b in this example, the answer will be more precise.

Set the first and third ratios equal to each other to get the proportion that can be solved for a: $\frac{a}{\sin 54^\circ} = \frac{11.2}{\sin 98^\circ}$. Cross-multiply to get $a \cdot \sin 98^\circ = 11.2 \sin 54^\circ$. Divide the sin98 from both sides to get $a = \frac{11.2}{\sin 98^\circ} \cdot \sin 54^\circ$ or a = 9.15.

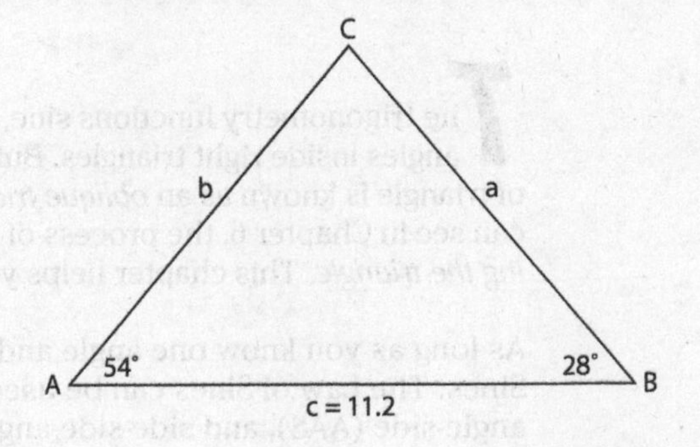

1. Solve the triangle if B = 46°, C = 62°, and a = 21.

Solve It

2. Solve the triangle if A = 19°, C = 100°, and b = 4.4.

Solve It

3. Solve the triangle if A = 49°, B = 21°, and a = 5.

Solve It

4. Solve the triangle if A = 110°, C = 56°, and a = 8.

Solve It

Tackling Triangles in the Ambiguous Case: SSA

A triangle with two sides and a consecutive angle is the *ambiguous case* of the Law of Sines. In this case, you may have one, two, or no solutions. We think it's easiest to assume that two solutions exist. That way, in attempting to find them both, you'll either get an error message on your calculator that tells you that there are no solutions, or you'll find one solution and then, in finding the second one, you'll get an answer that doesn't make sense (meaning there really is one solution). Or, you'll find one solution and then find a second one that works (so there really are two solutions).

When using your calculator to solve these types of problems, anytime you use an inverse trig function (like $\sin^{-1}$, $\cos^{-1}$, or $\tan^{-1}$) to solve for an angle, know that the calculator may give you the reference angle, or the first quadrant answer θ (for more review on this, check out Chapter 6). There can be a second quadrant answer, 180 – θ.

Q. Solve the triangle if a = 25, c = 15, and C = 40°.

A. No solution. Start off by substituting the given information into the Law of Sines: $\frac{25}{\sin A} = \frac{b}{\sin B} = \frac{15}{\sin 40^\circ}$. Notice that the middle ratio has absolutely no information in it at all, so you basically ignore it for now and work with the first and third ratios to get the proportion: $\frac{25}{\sin A} = \frac{15}{\sin 40^\circ}$. Now, cross-multiply to get $15 \cdot \sin A = 25 \cdot \sin 40^\circ$. Solve for the trig function with the variable in it to get $\sin A = \frac{25 \sin 40^\circ}{15}$. Put the expression into your calculator to get that $\sin A \approx 1.07$. Even if you forget that sine has only values between –1 and 1 inclusively and you inverse sine both sides of the equation, you'll get an error message on your calculator that tells you there's no solution.

5. Solve the triangle if $b = 8$, $c = 14$, and $C = 37°$.

Solve It

6. Solve the triangle if $b = 5$, $c = 12$, and $B = 20°$.

Solve It

7. Solve the triangle if $a = 10$, $c = 24$, and $A = 102°$.

Solve It

8. Solve the triangle if $b = 10$, $c = 24$, and $B = 20°$.

Solve It

Conquering a Triangle with the Law of Cosines: SAS and SSS

The Law of Cosines comes in handy when the Law of Sines doesn't work. Specifically, you use the Law of Cosines in two cases:

- You know two sides and the angle in between them (SAS)
- You know all three sides (SSS)

The Law of Cosines is:

$$a^2 = b^2 + c^2 - 2bc\cos A \qquad b^2 = a^2 + c^2 - 2ac\cos B \qquad c^2 = a^2 + b^2 - 2ab\cos C$$

You may also see your textbook present three forms for each of the angles in the triangle:

$$A = \cos^{-1}\left(\frac{b^2 + c^2 - a^2}{2bc}\right) \qquad B = \cos^{-1}\left(\frac{a^2 + c^2 - b^2}{2ac}\right) \qquad C = \cos^{-1}\left(\frac{a^2 + b^2 - c^2}{2ab}\right)$$

You don't have to memorize all these formulas. The ones for the angles are just the first three, each rewritten in terms of the angle. We show the steps for one of them in *Pre-Calculus For Dummies* by Krystle Rose Forseth, Christopher Burger, and Michelle Rose Gilman (Wiley), so if you'd like to see exactly how you arrive at the angle formulas, check it out there.

Q. Solve the triangle if A = 40°, b = 10, and c = 7.

A. a = 6.46, C ≈ 44.1°, B ≈ 95.9°. If you draw the triangle, you notice that this time the angle is between two sides (SAS), and you know that you have to use the Law of Cosines. First, $a^2 = 10^2 + 7^2 - 2(10)(7)\cos 40$. Put this right into your calculator to get $a^2 \approx 41.75$, or $a \approx 6.46$. Next, continue using the Law of Cosines to solve for angles (you'll avoid that pesky ambiguous case!) and get $7^2 = 6.46^2 + 10^2 - 2(6.46)(10)\cos C$, or 49 = 41.75 + 100 – 129.2cosC. Isolate for the trig function next and get 0.718 = cosC (this is why you don't *have* to memorize the formulas for the angles!). Inverse cosine both sides of the equation to get C ≈ 44.1°. Now it's easy to find B because the triangle's angles must total 180°. In this case, B ≈ 95.9°.

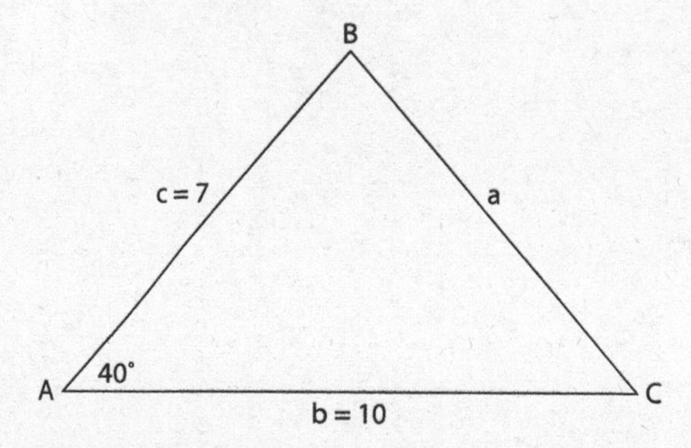

9. Solve the triangle if C = 120°, $a = 6$, and $b = 10$.

Solve It

10. Solve the triangle if A = 70°, $b = 6$, and $c = 7$.

Solve It

11. Solve the triangle if $a = 9$, $b = 5$, and $c = 7$.

Solve It

12. Solve the triangle if $a = 7.3$, $b = 9.9$, and $c = 16$.

Solve It

Using Oblique Triangles to Solve Word Problems

Usually around this time, your textbook or your teacher will present word problems. Argh! Run away, right? Wrong. Draw out a picture and discover that each and every problem at this time is a triangle that isn't right (an *oblique triangle*), where you're looking for one missing piece of information (as opposed to looking for all the angles and sides, like you do in the previous three sections). That means less work for you! Each situation requires you to use the Law of Sines or the Law of Cosines exactly once to solve for the missing information.

Q. A plane flies for 300 miles in a straight line, makes a 45° turn, and continues for 700 miles. How far is it from its starting point?

A. Approximately 936.47 miles. First, draw out a picture like the one that follows. Notice that we use S for the starting point, T for the turning point, and E for the ending point.

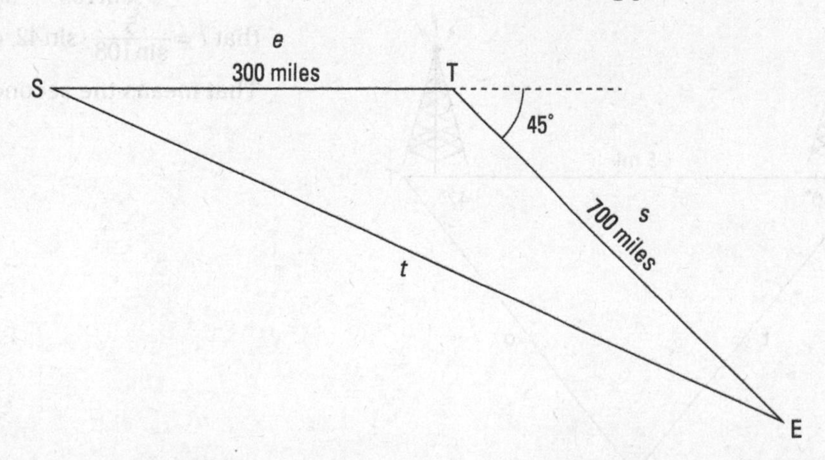

Now that you have the picture, you can figure out whether you need to use the the Law of Sines or the Law of Cosines. Because this is SAS, you start off with the modified Law of Cosines, using the different variables from the picture:

$t^2 = s^2 + e^2 - 2se\cos T$, or $t^2 = 700^2 + 300^2 - 2(700)(300)\cos 135°$. This means that $t^2 = 876{,}984.85$, and $t = 936.47$.

Q. Two fire towers are exactly 5 miles apart in a forest. They both spot a forest fire, one at an angle of 30° and the other at an angle of 42°. Which tower is closer?

A. The second fire tower is closer. Okay, so we lied just a little bit in the introduction to this section. You aren't always looking for only one missing piece of information. In this problem, you have to find how far both towers are from the fire in order to know which one is closer. But you forgive us, right? First, draw out a figure like the one that follows.

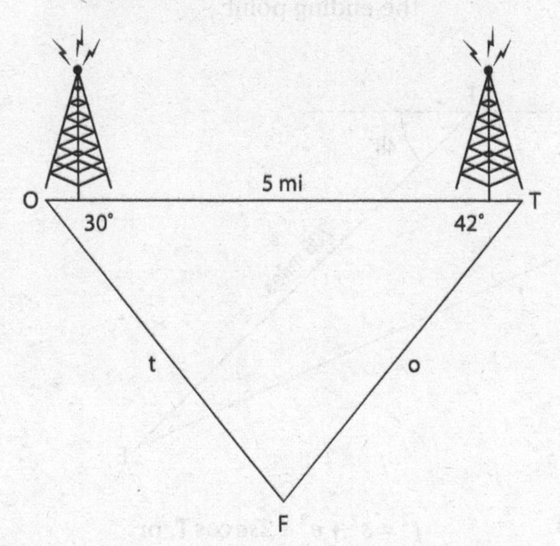

This time, we use O for fire tower one, T for fire tower two, and F for the fire itself. You have a classic case of ASA, so you can use the Law of Sines this time. Knowing two of the angles makes it possible to find the third one easily: F = 108°. Now that you have the third angle, you can use the Law of Sines to set up two proportions: $\frac{5}{\sin 108^\circ} = \frac{o}{\sin 30^\circ}$. Solving for o gets you $o = \frac{5}{\sin 108^\circ} \cdot \sin 30^\circ$, or $o = 2.6$ miles. Now set up another proportion to solve for t: $\frac{5}{\sin 108^\circ} = \frac{t}{\sin 42^\circ}$, which means that $t = \frac{5}{\sin 108} \cdot \sin 42$, or $t = 3.5$ miles.

That means the second tower is closer.

13. Two trains leave a station at the same time on different tracks that have an angle of 100° between them. If the first train is a passenger train that travels 90 miles per hour and the second train is a cargo train that can travel only 50 miles per hour, how far apart are the two trains after three hours?

Solve It

14. A radio tower is built on top of a hill. The hill makes an angle of 15° with the ground. The tower is 200 feet tall and located 150 feet from the bottom of the hill. If a wire is to connect the top of the tower with the bottom of the hill, how long does the wire need to be?

Solve It

15. A mapmaker stands on one side of a river looking at a flagpole on an island at an angle of 85°. She then walks in a straight line for 100 meters, turns, and looks back at the same flagpole at an angle of 40°. Find the distance from her first location to the flagpole.

Solve It

16. Two scientists stand 350 feet apart, both looking at the same tree somewhere in between them. The first scientist measures an angle of 44° from the ground to the top of the tree. The second scientist measures an angle of 63° from the ground to the top of the tree. How tall is the tree?

Solve It

Figuring Area

If you go in a slightly different direction in the proof of the Law of Sines, you discover a handy formula to find the area of an oblique triangle if you know two sides and the angle between them, as shown in Figure 10-1. The area of the triangle formed is $A = \frac{1}{2}ab\sin C$.

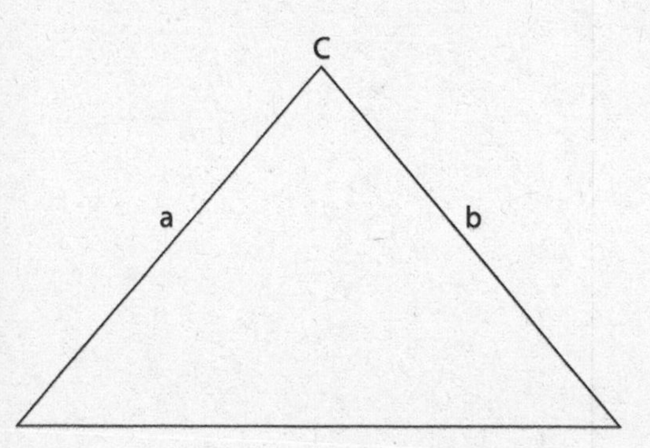

Figure 10-1: If you know two sides of a triangle and the angle between them, you can calculate the area.

Note: The letters in the formula aren't important. The area of the triangle is *always* one-half the product of the two sides and the sine of the angle between them.

You can also find the area when you know all three sides (a, b, and c) by using what's called Heron's Formula. It says that the area of a triangle is

$$\sqrt{s(s-a)(s-b)(s-c)}, \text{ where } s = \frac{1}{2}(a+b+c)$$

The variable s is called the semiperimeter, or half of the triangle's perimeter.

Q. Find the area of the triangle where $b = 4$, $c = 7$, and $A = 36°$.

A. The area is about 8.23 square units. When you have the two sides and the angle between them, you plug the given information into the formula to solve for the area. In this case, $A = \frac{1}{2}(4)(7)\sin 36°$ or 8.23.

17. Find the area of the triangle where $a = 7$, $c = 17$, and $B = 68°$.

Solve It

18. Find the area of the triangle on the coordinate plane with vertices at (–5, 2), (5, 6), and (4, 0).

Solve It

Answers to Problems on Solving Triangles

This section contains the answers for the practice problems presented in this chapter. We suggest you read the following explanations if your answers don't match up with ours (or if you just want a refresher on solving a particular type of problem).

1 Solve the triangle if B = 46°, C = 62°, and a = 21. The answer is A = 72°, b = 15.9, and c = 19.5.

We draw out this first triangle only, then save some trees and leave the rest to you.

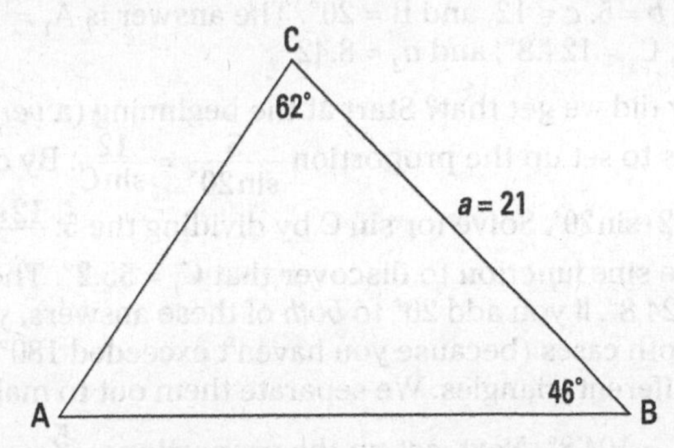

This one is ASA, so you use the Law of Sines to solve it. Because you already know two angles, begin by finding the third: A = 72°. Now set up a proportion from the Law of Sines to solve for the two missing sides. From $\frac{21}{\sin 72^\circ} = \frac{b}{\sin 46^\circ}$, you get that $b = \frac{21}{\sin 72^\circ} \cdot \sin 46^\circ$, or b = 15.9. And from $\frac{21}{\sin 72^\circ} = \frac{c}{\sin 62^\circ}$, you get that c = 19.5.

2 Solve the triangle if A = 19°, C = 100°, and b = 4.4. The answer is B = 61°, a = 1.64, and c = 4.95.

Draw out the figure first. It's ASA again, which means you use the Law of Sines to solve. Find the missing angle first: B = 61°. Now set up the first proportion to solve for a: $\frac{a}{\sin 19^\circ} = \frac{4.4}{\sin 61^\circ}$; $a = \frac{4.4}{\sin 61^\circ} \cdot \sin 19^\circ$, or a = 1.64. Set up another proportion to solve for c: $\frac{4.4}{\sin 61} = \frac{c}{\sin 100}$; $c = \frac{4.4}{\sin 61} \cdot \sin 100$, or c = 4.95.

3 Solve the triangle if A = 49°, B = 21°, and a = 5. The answer is C = 110°, b = 2.37, and c = 6.23.

This AAS case keeps you on your toes, but you still use the Law of Sines. The missing angle is C = 110°. The first proportion is $\frac{5}{\sin 49^\circ} = \frac{b}{\sin 21^\circ}$, which gets you $b = \frac{5}{\sin 49} \cdot \sin 21$, or b = 2.37. The second proportion is $\frac{5}{\sin 49} = \frac{c}{\sin 110}$, which gets you $c = \frac{5}{\sin 49} \cdot \sin 110$, or c = 6.23.

4 Solve the triangle if A = 110°, C = 56°, and a = 8. The answer is B = 14°, b = 2.06, and c = 7.06.

This one is also AAS, so you use the Law of Sines to solve it. First, the missing angle B is 14°. Now set up the proportion, $\frac{8}{\sin 110} = \frac{c}{\sin 56}$, to get that $c = \frac{8}{\sin 110} \cdot \sin 56$, or c = 7.06. Set up another proportion, $\frac{8}{\sin 110} = \frac{b}{\sin 14}$, to get that $b = \frac{8}{\sin 110} \cdot \sin 14$, or b = 2.06.

5 Solve the triangle if b = 8, c = 14, and C = 37°. The answer is A = 122.9°, B = 20.1°, and a = 19.53.

Notice when you draw this one that it's the dreaded SSA, or the ambiguous case. Always assume there are two answers when you're dealing with these types of problems, until you find out otherwise. Set up the proportion $\frac{14}{\sin 37^\circ} = \frac{8}{\sin B}$. This means that $14 \cdot \sin B = 8 \cdot \sin 37^\circ$, or

$\sin B = \frac{8\sin 37^\circ}{14} = 0.344$. Use inverse sine to get that $B_1 = 20.1^\circ$. This is the first quadrant answer. The second quadrant has a second answer: $B_2 = 180^\circ - 20.1^\circ = 159.9^\circ$. However, if you look closely, you notice that we start off with C = 37°. You can't add a 159.9° angle on top of that and still have a triangle, so you throw this second solution away. Only one triangle satisfies the conditions given. Now that you know C and (the one and only) B, it's easy as pi (Get it? Pi!) to find A = 122.9°. Set up another proportion, $\frac{14}{\sin 37^\circ} = \frac{a}{\sin 122.9^\circ}$, which means that $a = \frac{14}{\sin 37^\circ} \cdot \sin 122.9^\circ = 19.53$.

6 Solve the triangle if $b = 5$, $c = 12$, and B = 20°. The answer is $A_1 = 104.8^\circ$, $C_1 = 55.2^\circ$, and $a_1 = 14.13$; or, $A_2 = 35.2^\circ$, $C_2 = 124.8^\circ$, and $a_2 = 8.42$.

Two solutions! How did we get that? Start at the beginning (a *very* good place to start) and use the Law of Sines to set up the proportion $\frac{5}{\sin 20^\circ} = \frac{12}{\sin C}$. By cross-multiplying, you get the equation $5 \cdot \sin C = 12 \cdot \sin 20^\circ$. Solve for sin C by dividing the 5: $\frac{12\sin 20^\circ}{5} \approx 0.821$. If $\sin C \approx 0.821$, then use the inverse sine function to discover that $C_1 = 55.2^\circ$. The second quadrant answer is $C_2 = 180^\circ - 55.2^\circ = 124.8^\circ$. If you add 20° to *both* of these answers, you discover that it's possible to make a triangle in both cases (because you haven't exceeded 180°). This sends you on two different paths for two different triangles. We separate them out to make sure you follow the steps.

If $C_1 = 55.2^\circ$, then $A_1 = 104.8^\circ$. Next, set up the proportion $\frac{5}{\sin 20^\circ} = \frac{a_1}{\sin 104.8^\circ}$. This means that $a_1 = \frac{5}{\sin 20^\circ} \cdot \sin 104.8^\circ = 14.13$.

If $C_2 = 124.8^\circ$, then $A_2 = 35.2^\circ$. Set up another proportion, $\frac{5}{\sin 20^\circ} = \frac{a_2}{\sin 35.2^\circ}$, to then get that $a_2 = \frac{5}{\sin 20^\circ} \cdot \sin 35.2^\circ = 8.42$.

7 Solve the triangle if $a = 10$, $c = 24$, and A = 102°. The answer is no solution.

Here we go again. By the time you're done with this section of pre-calc, you'll be an expert (whether you like it or not) at solving triangles. If you draw this one, you see another ambiguous SSA case. Set up the proportion $\frac{10}{\sin 102^\circ} = \frac{24}{\sin C}$ using the Law of Sines. Cross-multiply to get $10 \cdot \sin C = 24 \cdot \sin 102^\circ$, and then divide the 10 from both sides to get $\sin C = \frac{24\sin 102^\circ}{10} \approx 2.35$. That's when the alarms go off.

Sine can't have a value bigger than 1, so there's no solution.

8 Solve the triangle if $b = 10$, $c = 24$, and B = 20°. The answer is $A_1 = 104.8^\circ$, $C_1 = 55.2^\circ$, and $a_1 = 28.3$; or, $A_2 = 35.2^\circ$, $C_2 = 124.8^\circ$, and $a_2 = 16.8$.

We work it out for you here:

$$\frac{10}{\sin 20^\circ} = \frac{24}{\sin C}$$

$$10 \cdot \sin C = 24 \cdot \sin 20^\circ$$

$$\sin C = \frac{24\sin 20^\circ}{10} \approx 0.821$$

$$C_1 = \sin^{-1}(0.821) = 55.2^\circ$$

$$A_1 = 104.8^\circ$$

$$\frac{10}{\sin 20^\circ} = \frac{a_1}{\sin 104.8^\circ}$$

$$10 \cdot \sin 104.8^\circ = a_1 \cdot \sin 20^\circ$$

$\frac{10}{\sin 20^\circ} \cdot \sin 104.8^\circ = a_1$

$a_1 = 28.3$

$C_2 = 124.8^\circ$ (this solution also works)

$A_2 = 35.2^\circ$

$\frac{10}{\sin 20^\circ} = \frac{a_2}{\sin 35.2^\circ}$

$10 \cdot \sin 35.2^\circ = a_2 \cdot \sin 20^\circ$

$a_2 = 16.8$

9 Solve the triangle if C = 120°, a = 6, and b = 10. The answer is c = 14, A = 21.8°, and B = 38.2°.

This is definitely a Law of Cosines problem when you draw the triangle (SAS). Find c first: $c^2 = a^2 + b^2 = 2ab\cos C$. Plug in what you know: $c^2 = 6^2 + 10^2 - 2(6)(10)\cos 120$. Plug this right into your calculator to get that $c^2 = 196$, or $c = 14$. Now find A: $a^2 = b^2 + c^2 - 2bc\cos A$. Plug in what you know: $6^2 = 10^2 + 14^2 - 2(10)(14)\cos A$. Simplify: $36 = 296 - 280\cos A$. Solve for cosA by first subtracting 296 from both sides: $-260 = -280\cos A$. Now divide the –280 to get $0.929 = \cos A$. Now use inverse cosine to get A = 21.8, and then use the fact that the sum of the angles of a triangle is 180° to figure out that B = 38.2°.

10 Solve the triangle if A = 70°, b = 6, and c = 7. The answer is a = 7.50, B = 48.7°, and C = 61.3°.

By plugging what you know into the Law of Cosines, $a^2 = b^2 + c^2 - 2bc\cos A$, you get $a^2 = 6^2 + 7^2 - 2(6)(7)\cos 70$. This simplifies to $a^2 = 56.27$, or $a = 7.50$.

Now switch the substituting in the second law: $b^2 = a^2 + c^2 - 2ac\cos B$; $6^2 = 7.50^2 + 7^2 - 2(7.50)(7)\cos B$. Simplify: $36 = 105.27 - 105\cos B$. Isolate cosB by first subtracting 105.27: $-69.27 = -105\cos B$, and then dividing the –105: $0.660 = \cos B$. This means that B = 48.7°. From there you can figure out that C = 61.3°.

11 Solve the triangle if a = 9, b = 5, and c = 7. The answer is A = 95.7°, B = 33.6°, and C = 50.7°.

You're solving an SSS triangle using the Law of Cosines, so get crackin' to find A:

$a^2 = b^2 + c^2 - 2bc\cos A$

$9^2 = 5^2 + 7^2 - 2(5)(7)\cos A$

$81 = 74 - 70\cos A$

$7 = -70\cos A$

$95.7^\circ = A$

And again to find B:

$b^2 = a^2 + c^2 - 2ac\cos B$

$5^2 = 9^2 + 7^2 - 2(9)(7)\cos B$

$25 = 130 - 126\cos B$

$-105 = -126\cos B$

$0.833 = \cos B$

$33.6^\circ = B$

Last, but certainly not least, C = 50.7°.

12 Solve the triangle if a = 7.3, b = 9.9, and c = 16. The answer is A = 18.3°, B = 25.3°, and C = 136.4°.

Here we go again to find A:

$$a^2 = b^2 + c^2 - 2bc\cos A$$

$$7.3^2 = 9.9^2 + 16^2 - 2(9.9)(16)\cos A$$

$$53.29 = 98.01 + 256 - 316.8\cos A$$

$$-300.72 = -316.8\cos A$$

$$0.949 = \cos A$$

$$18.3° = A$$

And once more to find B:

$$b^2 = a^2 + c^2 - 2ac\cos B$$

$$9.9^2 = 7.3^2 + 16^2 - 2(7.3)(16)\cos B$$

$$98.01 = 53.29 + 256 - 233.6\cos B$$

$$-211.28 = -233.6\cos B$$

$$0.904 = \cos B$$

$$25.3° = B$$

And finally, C = 136.4°.

13 **Two trains leave a station at the same time on different tracks that have an angle of 100° between them. If the first train is a passenger train that travels 90 miles per hour and the second train is a cargo train that can travel only 50 miles per hour, how far apart are the two trains after three hours? The answer is approximately 330.86 miles apart.**

The two trains depart (F and S) from the same station (T) in a picture like this one:

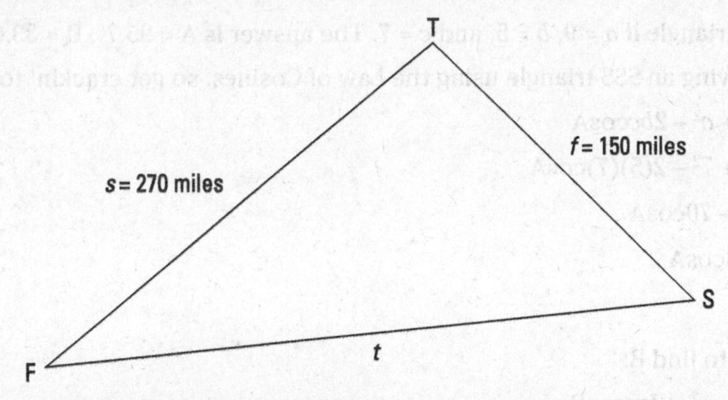

Notice that you have to use the Law of Cosines to solve for how far apart the two trains are, t.

Next, you need to find how far the trains have traveled to know how far apart they are. Using D = rt, for the first train you get D = 90(3) = 270 miles; for the second train, D = 50(3) = 150 miles. Plug these values into the equation: $t^2 = f^2 + s^2 - 2fs\cos T$; $t^2 = 150^2 + 270^2 - 2(150)(270)\cos 100°$. This goes right into your calculator to give you t^2 = 109465.50, or t = 330.86 miles.

14 A radio tower is built on top of a hill. The hill makes an angle of 15° with the ground. The tower is 200 feet tall and located 150 feet from the bottom of the hill. If a wire is to connect the top of the tower with the bottom of the hill, how long does the wire need to be? The answer is about 279.3 feet long.

This time the picture is:

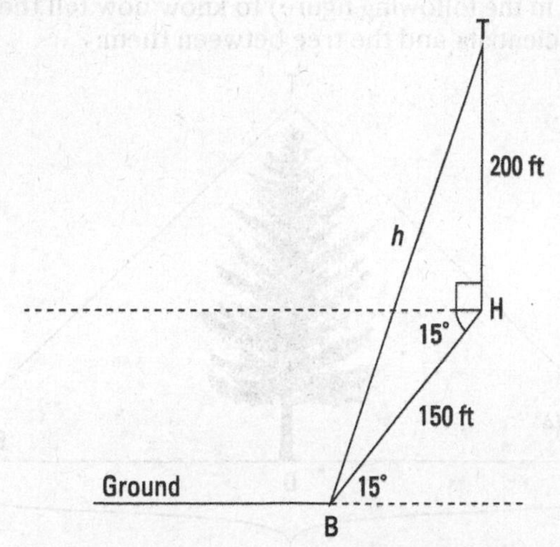

To find ∠H in the picture, you add a horizontal line that's parallel to the ground. Then, using the facts that alternate interior angles are congruent and that the tower has to be completely vertical (or else we have a leaning tower), we know that $H = 15^\circ + 90^\circ = 105^\circ$.

Now, jump in with the Law of Cosines:

$$h^2 = t^2 + b^2 - 2tb\cos H$$

$$h^2 = 150^2 + 200^2 - 2(150)(200)\cos 105^\circ$$

$$h^2 = 78029.14$$

$$h = 279.3 \text{ feet}$$

15 A mapmaker stands on one side of a river looking at a flagpole on an island at an angle of 85°. She then walks in a straight line for 100 meters, turns, and looks back at the same flagpole at an angle of 40°. Find the distance from her first location to the flagpole. The answer is 78.5 meters.

Looking down on the surveyor and the flagpole, here's the picture you use to solve this problem:

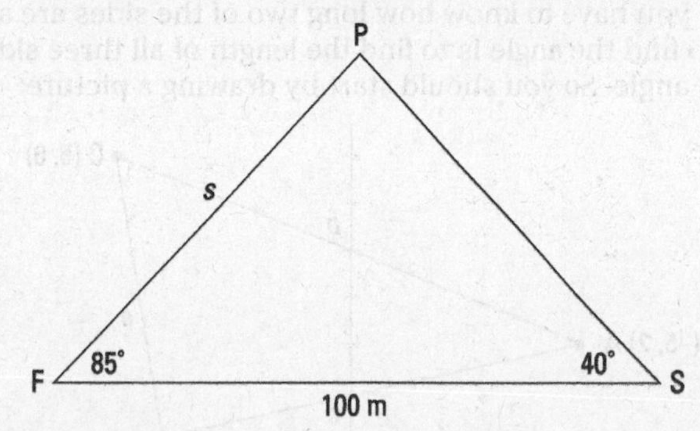

Because you have two angles, you can find that P = 55° and use the Law of Sines. $\frac{100}{\sin 55^\circ} = \frac{s}{\sin 40^\circ}$ tells you that $s = \frac{100}{\sin 55^\circ} \cdot \sin 40^\circ = 78.5$ meters.

16 Two scientists stand 350 feet apart, both looking at the same tree somewhere between them. The first scientist measures an angle of 44° from the ground to the top of the tree. The second scientist measures an angle of 63° from the ground to the top of the tree. How tall is the tree? The answer is 226.53 feet tall.

This problem takes some work. You have to know the distance from either scientist to the top of the tree (FT or TS in the following figure) to know how tall the tree (TB) really is. Here's a drawing of the two scientists and the tree between them:

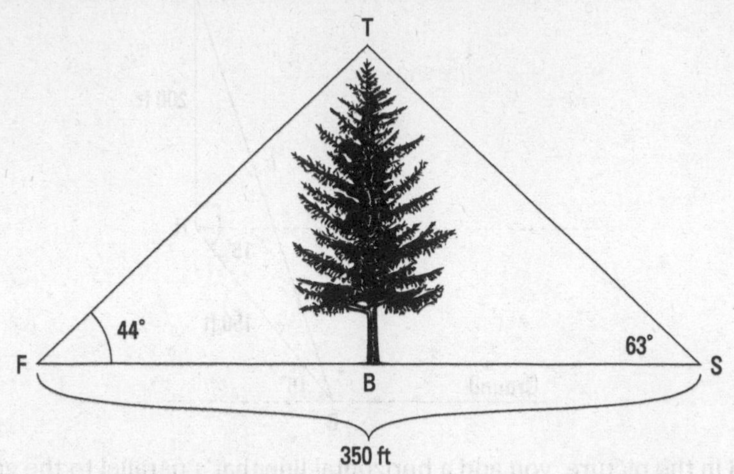

Knowing two angles gets you the third one: T = 73°. Law of Sines it is, then! $\frac{350}{\sin 73^\circ} = \frac{FT}{\sin 63^\circ}$ gives you that $FT = \frac{350}{\sin 73^\circ} \cdot \sin 63^\circ = 326.10$ feet.

Now, because the tree grows straight up, you have a right triangle in which you know one angle and one side. That means you need to go back to SOHCAHTOA (see Chapter 6 for more information) to solve for the missing side, TB. Knowing that you have the hypotenuse and are looking for the opposite side, you can use the sine function and get $\sin 44^\circ = \frac{TB}{326.10}$. This means that TB = 226.53 feet tall. That's one big tree!

17 Find the area of the triangle where $a = 7$, $c = 17$, and B = 68°. The answer is about 55.17 square units.

Knowing two sides and the angle between them makes this an easy one: $A = \frac{1}{2}(7)(17)\sin 68^\circ = 55.17$.

18 Find the area of the triangle on the coordinate plane with vertices at (–5, 2), (5, 6), and (4, 0). The answer is about 28 square units.

To figure this out, you have to know how long two of the sides are and the angle between them. The easiest way to find the angle is to find the length of all three sides and then use the Law of Cosines to find an angle. So you should start by drawing a picture:

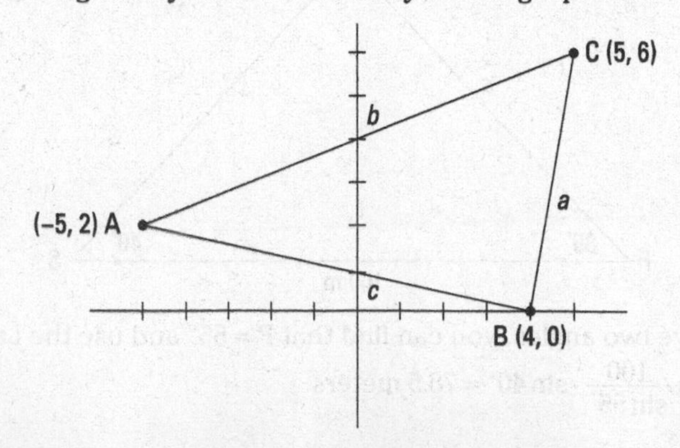

Find all three sides first (for a review of how to find the distance between two points, see Chapter 1):

$AC=\sqrt{(5-(-5))^2+(6-2)^2}=\sqrt{(10)^2+(4)^2}=\sqrt{100+16}=\sqrt{116}$ (this is the same as side a when you use the Law of Cosines)

$AB=\sqrt{(4-(-5))^2+(0-2)^2}=\sqrt{(9)^2+(-2)^2}=\sqrt{81+4}=\sqrt{85}$ (this is side c)

$BC=\sqrt{(4-5)^2+(0-6)^2}=\sqrt{(-1)^2+(-6)^2}=\sqrt{1+36}=\sqrt{37}$ (this is side b)

Now that you've found the length of all three sides, use Heron's Formula to find the area:

$s=\frac{1}{2}\left(\sqrt{116}+\sqrt{85}+\sqrt{37}\right)=13.04$

$A=\sqrt{13.04(13.04-10.77)(13.04-9.22)(13.04-6.08)}=\sqrt{13.04(2.27)(3.82)(6.96)}=\sqrt{787}=28.05$

Part IV
Poles, Cones, Variables, Sequences, and Finding Your Limits

In this part . . .

This part begins with how to perform operations with and graph complex numbers. We also introduce the idea of polar coordinates, a brand new way of graphing equations! We know you've been asking yourself when that was gonna happen . . . well, here you go. Conic sections are also a great thing to graph, so we cover them in detail.

We then move on to the systems of equations and cover solving linear and nonlinear equations, as well as working with matrices. Next, we're on to sequences and series. We discuss how to find any term in a sequence, how to calculate the sum of a sequence, and how to write the formula that determines a given sequence. Lastly, we cover the topics that usually constitute the end of pre-calculus (and the beginning of calculus): limits and continuity.

Chapter 11

Exploring Complex Numbers and Polar Coordinates

In This Chapter

- Working with complex numbers in operations and graphs
- Switching between polar and rectangular coordinates
- Graphing polar coordinates and equations

Once upon a time, mathematicians delved into their imaginations and invented a whole new set of numbers. They decided to call these numbers imaginary numbers, because there was just no way these new numbers would ever pop up in the real world.

Well, they were wrong. These imaginary numbers did eventually appear. Fields like engineering, electricity, and quantum physics all use imaginary numbers in their everyday applications. An *imaginary number* is basically the square root of a negative number. The *imaginary unit,* denoted i, is the solution to the equation $i^2 = -1$.

A *complex number* can be represented in the form $a + bi$, where a and b are real numbers and i denotes the imaginary unit. In the complex number $a + bi$, a is called the real part and b is called the imaginary part. Real numbers can be considered a subset of the complex numbers that have the form $a + 0i$. When a is zero, then $0 + bi$ is written as simply bi and is called a *pure imaginary number.*

This chapter is like the bonus features on your favorite DVD, because it also includes graphing points and equations in a whole new way, called graphing polar coordinates. Hang in there for a bit while we explore complex numbers in depth before moving on to polar coordinates.

Performing Operations with and Graphing Complex Numbers

Complex numbers in the form $a + bi$ can be graphed on a *complex coordinate plane.* Each complex number corresponds to a point (a, b) in the complex plane. The real axis is the line in the complex plane consisting of the numbers that have a zero imaginary part: $a + 0i$. Every real number graphs to a unique point on the real axis. The imaginary axis is the line in the complex plane consisting of the numbers that have a zero real part: $0 + bi$. Figure 11-1 shows several examples of points on the complex plane.

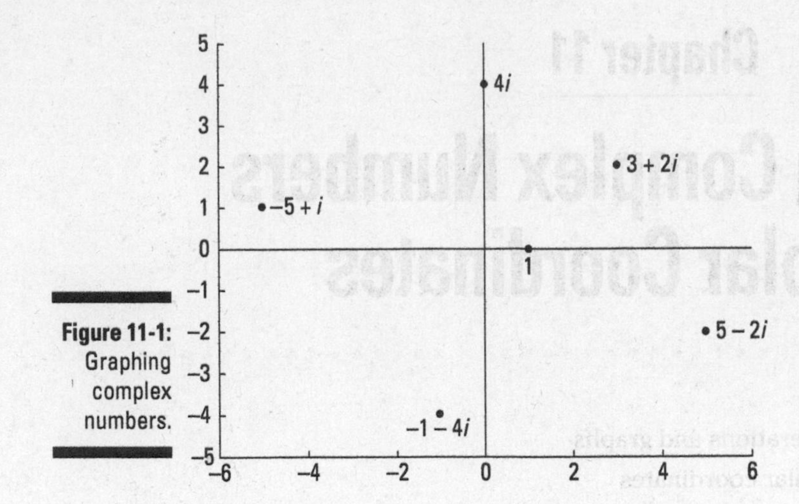

Figure 11-1: Graphing complex numbers.

Adding and subtracting complex numbers is just another example of collecting like terms: You can add or subtract only real numbers, and you can add or subtract only imaginary numbers.

When multiplying complex numbers, you FOIL the two binomials. All you have to do is remember that the imaginary unit is defined such that $i^2 = -1$, so any time you see i^2 in an expression, replace it with –1. When dealing with other powers of *i*, notice the pattern here:

$i = \sqrt{-1}$ $\quad i^5 = i$

$i^2 = -1$ $\quad i^6 = -1$

$i^3 = -i$ $\quad i^7 = -i$

$i^4 = 1$

It continues in this manner forever, repeating in a cycle every fourth power. To find a larger power of *i*, rather than counting forever, realize that the pattern repeats. For example, to find i^{243}, divide 4 into 243 and you get 60 with a remainder of 3. The pattern will repeat 60 times and then you'll have 3 left over, so i^{243} is the same as i^3, which is $-i$.

The *conjugate* of a complex number $a + bi$ is $a - bi$, and vice versa. When you multiply two complex numbers that are conjugates of each other, you always end up with a pure real number:

$(a+bi)(a-bi)$.

FOIL the binomials: $a^2 - abi + abi - b^2i^2$

Cancel the two middle terms: $a^2 - b^2i^2$

Replace i^2 with –1: $a^2 - b^2(-1)$.

Simplify: $a^2 + b^2$.

Remember that absolute value bars enclosing a real number represent distance. In the case of a complex number, $|a + ib|$ represents the distance from the point to the origin. This distance is always the same as the length of the hypotenuse of the right triangle drawn when connecting the point to the *x*- and *y*-axes.

When dividing complex numbers, you end up with a root in the denominator (because if $i^2 = -1$, then $i = \sqrt{-1}$. This means that you have to rationalize the denominator. To do so, you must multiply the complex number in the denominator by its conjugate, and then multiply this same expression in the numerator. (For more information on rationalizing the denominator of any fraction, see Chapter 2.)

We show you how to do this with variables, but you may want to go to the forthcoming example to see how to rationalize a fraction with complex numbers involved.

$\frac{a+bi}{c+di}$

Multiply both the numerator and the denominator by the conjugate of the denominator: $\frac{a+bi}{c+di} \cdot \frac{c-di}{c-di}$

FOIL the numerators and the denominators: $\frac{ac-adi+bci-bdi^2}{c^2-cdi+cdi-d^2i^2}$

Cancel like terms in the denominator and replace the i^2 with –1 in the numerator and denominator: $\frac{ac-adi+bci+bd}{c^2+d^2}$

Q. Perform the indicated operation: $(3-4i)+(-2+5i)^2$

A. $-18 - 24i$. Follow your order of operations (PEMDAS — Parentheses Exponents Multiplication Division Addition Subtraction) and square the second binomial by FOILing it times itself: $(3 - 4i) + (-2 + 5i)(-2 + 5i)$; $(3-4i)+(4-10i-10i+25i^2)$. Combine the like terms: $(3-4i)+(4-20i+25i^2)$. Substitute –1 for i^2: $(3 - 4i) + (4 - 20i - 25)$. Combine like terms: $(3 - 4i) + (-21 - 20i)$. Combine like terms one more time — reals with reals and imaginary with imaginary: $-18 - 24i$.

Q. Perform the indicated operation: $\frac{6}{2-9i}$

A. $\frac{12+54i}{85}$. The conjugate of the denominator is $2 + 9i$. Multiply this on the top and bottom of the fraction: $\frac{6}{2-9i} \cdot \frac{2+9i}{2+9i}$. Distribute the numerator and FOIL the denominator: $\frac{12+54i}{4+18i-18i-81i^2}$. Cancel like terms in the denominator and replace i^2 with –1: $\frac{12+54i}{4+81}$. This simplifies to the answer $\frac{12+54i}{85}$.

Q. Graph $4 - 6i$ on the complex coordinate plane.

A. See the following graph. Go to the right 4 units on the real axis and 6 units down on the imaginary axis and place a point. Easy!

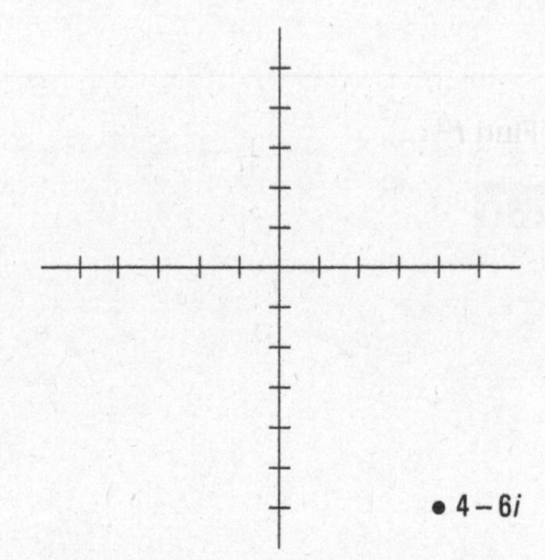

Q. Find real values x and y such that $3x + 4yi = 6 - 2i$.

A. **$x = 2$, $y = -\frac{1}{2}$. The real parts of both sides of this equation must equal each other: $3x = 6$. Divide both sides by 3 to get the solution $x = 2$. Furthermore, the imaginary parts must also equal each other: $4y = -2$, so that when you divide both sides by 4 and reduce, you get $y = -\frac{1}{2}$.**

1. Plot the point $4 - 3i$ on the complex plane.

Solve It

2. Find $|3 - 4i|$

Solve It

3. Find i^{22}

Solve It

4. Solve the equation $5x^2 - 2x + 3 = 0$.

Solve It

Round a Pole: Graphing Polar Coordinates

Up until now in your math career, you've been graphing everything based on the *rectangular coordinate system.* It's called that because it's based on two number lines perpendicular to each other. Pre-calc takes that concept further when it introduces *polar coordinates.*

In polar coordinates, every point is located around a central point, called the *pole,* and is named (r, θ). r is the radius, and θ is the angle formed between the polar axis (think of it as what *used* to be the positive x-axis) and the segment connecting the point to the pole (what *used* to be the origin).

Most books use radians when measuring angles in polar coordinates, so we do the same thing. However, some figures in this chapter may be labeled in degrees or in both degrees and radians. If you need a recap on radians, see Chapter 6. Figure 11-2 shows the polar coordinate plane.

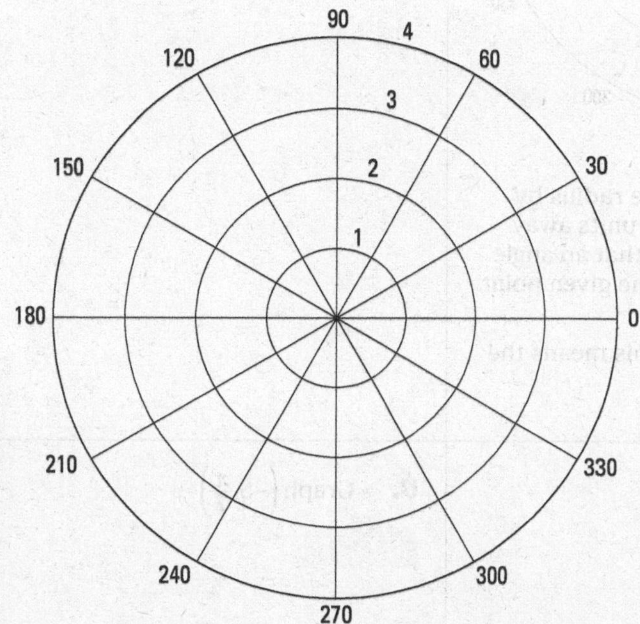

Figure 11-2: Graphing round and round on the polar coordinate plane.

Notice that a point on the polar coordinate plane has more than one name. Because you're moving in a circle, you can always add or subtract 2π to any angle and end up at the same point. This concept is so important in graphing equations in polar forms that we dedicate this entire section to making sure that you understand it.

When both the radius and the angle are positive, most of our students have no difficulties finding the point on the polar plane. If the radius is positive and the angle is negative, the point moves in a clockwise direction, just like radians do. If the radius is negative and the angle is positive, find the point where both are positive first and then reflect that point across the pole. If both the radius and the angle are negative, find the point where the radius is positive and the angle is negative and then reflect that across the pole.

Q. What's the polar coordinate of Point P in the following figure?

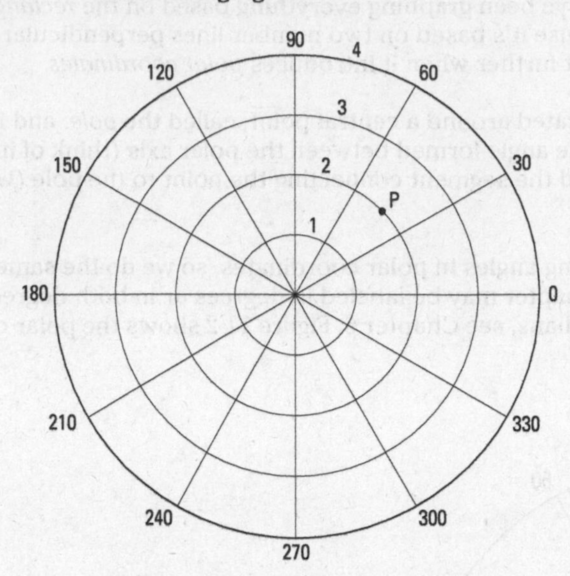

A. (2, π/4). First determine the radius by noticing that the point is 2 units away from the pole. Also notice that an angle forms when you connect the given point to the pole and that the polar axis is a π/4 angle. This means the point is $\left(2, \frac{\pi}{4}\right)$.

Q. Name two other points that determine the same point in the preceding question.

A. Possible answers include $\left(2, \frac{9\pi}{4}\right)$, $\left(2, \frac{17\pi}{4}\right)$ and $\left(2, -\frac{7\pi}{4}\right)$. You add 2π to the angle twice and subtract it once to get these three angles. It's all about finding the common denominator at that point: $\frac{\pi}{4} + 2\pi = \frac{\pi}{4} + \frac{8\pi}{4} = \frac{9\pi}{4}$, our first answer. Then take that answer and add 2π to it to get the next one. You can do this for the rest of your life and still not list all the possibilities.

5. Graph $\left(4, -\frac{5\pi}{3}\right)$

Solve It

6. Graph $\left(-5, \frac{\pi}{2}\right)$

Solve It

7. Name two other polar coordinates for the point in Question 5: one with a negative angle and one with a positive angle.

Solve It

8. Name two other polar coordinates for the point in Question 6: one with a negative angle and one with a positive radius.

Solve It

Changing to and from Polar

You can use both polar and rectangular coordinates to name the same point on the coordinate plane. Sometimes it's easier to write an equation in one form than the other, but usually, pre-calc books begin by having you switch between the two just to get used to them. Figure 11-3 shows how to determine the relationship between these two not-so-different methods.

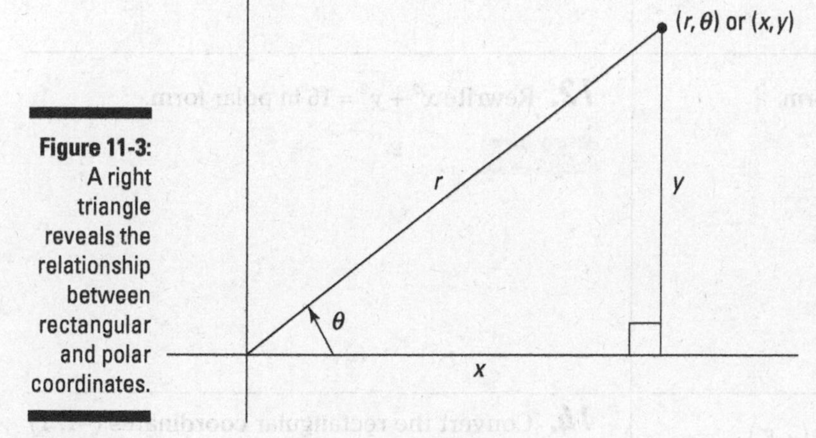

Figure 11-3: A right triangle reveals the relationship between rectangular and polar coordinates.

Some simple right triangle trig and a little Pythagorean Theorem determine the relationship:

- $\sin\theta = \frac{y}{r}$ or $y = r\sin\theta$
- $\cos\theta = \frac{x}{r}$ or $x = r\cos\theta$
- $x^2 + y^2 = r^2$
- $\tan\theta = \frac{y}{x}$ or $\theta = \tan^{-1}\left(\frac{y}{x}\right)$

Q. Rewrite the equation in rectangular form: $r = 2\sin\theta$.

A. $x^2 + y^2 = 2y$. First, convert all trig functions to x, y, and r. If $r = 2\sin\theta$, then $r = 2 \cdot \frac{y}{r}$. Then multiply the r to the other side to get $r^2 = 2y$. Then you can replace r^2 from the Pythagorean Theorem and get $x^2 + y^2 = 2y$.

Q. Rewrite the equation $r = 2\csc\theta$ in rectangular form.

A. $y = 2$. First, realize that cosecant is the reciprocal of sine (see Chapter 6 for a refresher). If $\sin\theta = \frac{y}{r}$, then $\csc\theta = \frac{r}{y}$. Substitute this into the equation to get $r = 2\frac{r}{y}$.

Multiply the y to the other side and get $ry = 2r$. Observe that the original equation implies that r is not zero, which allows you to cancel the r's here. This means that $y = 2$.

9. Rewrite $r = 3\sin\theta + 4\cos\theta$ in rectangular form.

Solve It

10. Rewrite $y = 2x - 1$ in polar form.

Solve It

11. Rewrite $3x - 5y = 10$ in polar form.

Solve It

12. Rewrite $x^2 + y^2 = 16$ in polar form.

Solve It

13. Convert the polar coordinates $\left(4, \frac{\pi}{6}\right)$ to rectangular coordinates.

Solve It

14. Convert the rectangular coordinates $(-1, 1)$ to polar coordinates.

Solve It

Graphing Polar Equations

When given an equation in polar coordinates and asked to graph it, most students go with the plug-and-chug method: Pick values for θ from the unit circle that you know so well and find the value of r. Polar equations have various types of graphs, and we take a closer look at each one. Be sure to also see Chapter 12 (conic sections) for information about how to graph conic sections in polar coordinates.

Archimedean spiral

$r = a\theta$ gives a graph that forms a spiral. a is a constant that's multiplying the angle, and then the radius is the same. If a is positive, the spiral moves in a counterclockwise direction, just like positive angles do. If a is negative, the spiral moves in a clockwise direction.

Cardioid

You may recognize the word *cardioid* if you've ever worked out and done your cardio. The word relates to the heart, and when you graph a cardioid, it does look like a heart, of sorts. Cardioids are written in the form $r = a(1 \pm \sin\theta)$ or $r = a(1 \pm \cos\theta)$. The cosine equations are hearts that point to the left or right, and the sine equations open up or down.

Rose

A rose by any other name is . . . a polar equation. If $r = a\sin n\theta$ or $r = a\cos n\theta$, the graphs look like flowers with petals. The number of petals is determined by n. If n is odd, then there are n (the same number of) petals. If n is even, there are $2n$ petals.

Circle

When $r = a\sin\theta$ or $r = a\cos\theta$, you end up with a circle with a diameter of a. Circles with cosine in them are centered on the x-axis, and circles with sine in them are centered on the y-axis. These are particular types of circles passing through the origin.

Lemniscate

A lemniscate makes a figure eight; that's the best way to remember it. $r^2 = \pm a^2 \sin 2\theta$ forms a figure eight between the axes, and $r^2 = \pm a^2 \cos 2\theta$ forms a figure eight that lies on one of the axes as a line of symmetry.

Limaçon

A cardioid is really a special type of limaçon, which is why they look similar to each other when you graph them. The familiar forms of limaçons are

$$r = a \pm b\sin\theta$$

$$r = a \pm b\cos\theta$$

Q. Sketch the graph of $r = 1 + \sin\theta$.

A. See the following graph. This is a cardioid. If you're lucky enough to have a teacher who lets you use a graphing calculator to graph polar equations, make sure your calculator is set to radians, input the equation into the grapher, and presto, you have yourself a graph. If not, just plug and chug the equation to get the graph. For instance, if $\theta = 0$, then $r = 1 + \sin 0 = 1 + 0 = 1$. If $r = \frac{\pi}{2}$, then $r = 1 + \sin\frac{\pi}{2} = 1 + 1 = 2$. Keep going in this manner until you end up with the graph.

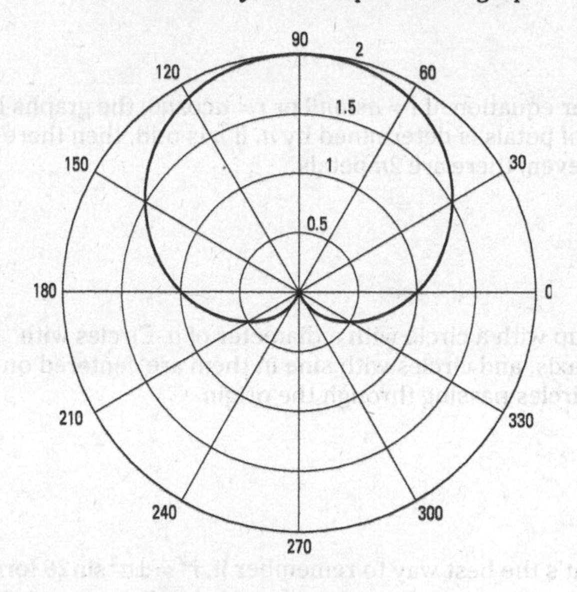

Q. Sketch the graph of $r = \cos 3\theta$.

A. See the following graph. This is a rose with three petals because the coefficient on the inside is odd. Plug and chug this one as well, as shown in the following chart (we show the first quadrant values only):

θ	$\cos 3\theta$	r
0	$\cos 3(0)$	1
$\pi/6$	$\cos \pi/2$	0
$\pi/4$	$\cos 3\pi/4$	$-\frac{\sqrt{2}}{2}$
$\pi/3$	$\cos \pi$	−1
$\pi/2$	$\cos 3\pi/2$	0

15. Sketch the graph of $r^2 = 9\cos 2\theta$.

Solve It

16. Sketch the graph of $r = 2\theta$.

Solve It

17. Sketch the graph of $r = 1 - 3\sin\theta$.

Solve It

18. Sketch the graph of $r = 3 + 2\sin\theta$.

Solve It

Answers to Problems on Complex Numbers and Polar Coordinates

Following are the answers to problems dealing with trig functions. We also provide guidance on getting the answers if you need to review where you went wrong.

1 Plot the point $4-3i$ on the complex plane. See the following graph for the answer.

The real unit is 4 to the right and the imaginary unit is 3 down. This lands you at the point in the figure.

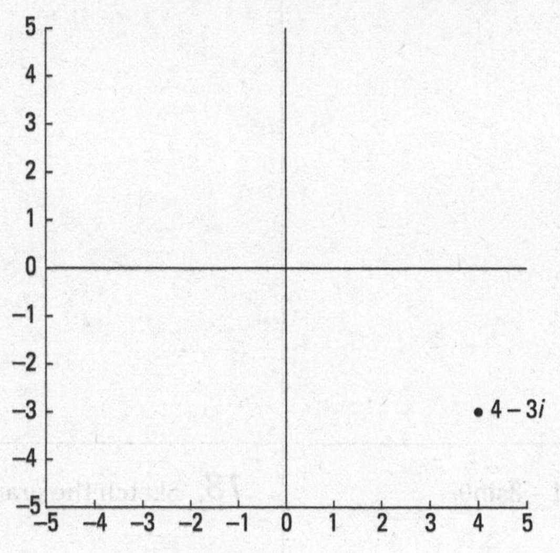

2 Find $|3-4i|$. The answer is 5.

Remember that absolute value bars represent distance. In the case of complex numbers, they represent the distance from the point to the origin. This distance is always the same as the length of the hypotenuse of the right triangle drawn when connecting the point to the *x*- and *y*-axes.

$$d=\sqrt{3^2+(-4)^2}=\sqrt{9+16}=\sqrt{25}=5$$

3 Find i^{22}. The answer is –1.

The pattern for the powers of i repeats every four times when its exponent increases consecutively. Divide 22 by 4 and get 5 with a remainder of 2. This means that $i^{22}=i^2=-1$.

4 Solve the equation $5x^2-2x+3=0$. The answer is $\frac{1\pm\sqrt{14}i}{5}$

This quadratic equation can't be factored, so you have to resort to the quadratic formula to solve it: $x=\frac{2\pm\sqrt{(-2)^2-4(5)(3)}}{2(5)}=\frac{2\pm\sqrt{4-60}}{10}=\frac{2\pm\sqrt{-56}}{10}=\frac{2\pm2\sqrt{14}i}{10}$, which simplifies to $\frac{1\pm\sqrt{14}i}{5}$. If you've forgotten how to deal with the quadratic formula and/or how to simplify roots, see Chapter 4.

5 Graph $\left(4,-\frac{5\pi}{3}\right)$. See the following graph for the answer.

The radius is 4 and the angle is negative, which moves in a clockwise direction and ends up in the first quadrant.

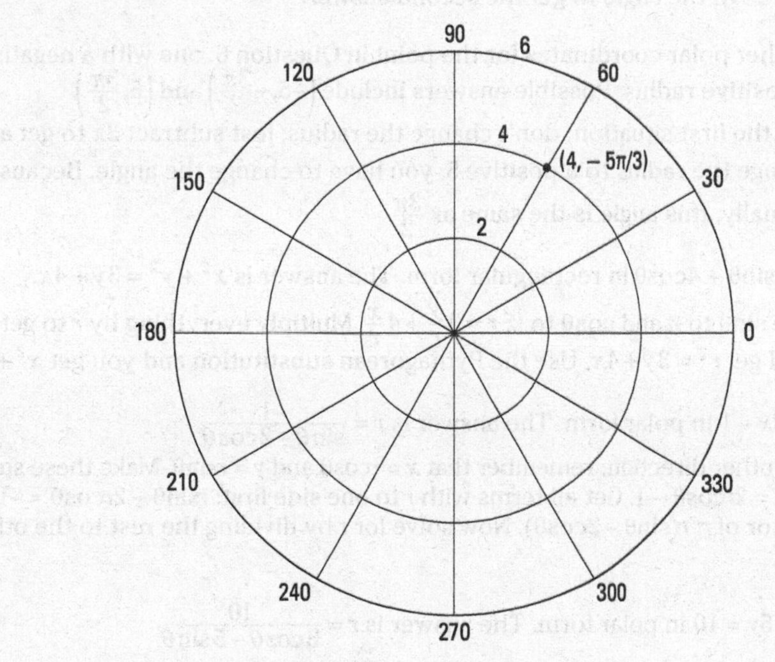

6 Graph $\left(-5,\frac{\pi}{2}\right)$. See the following graph for the answer.

The radius for this one is negative, which reflects the point $\left(5,\ \frac{\pi}{2}\right)$ over the pole.

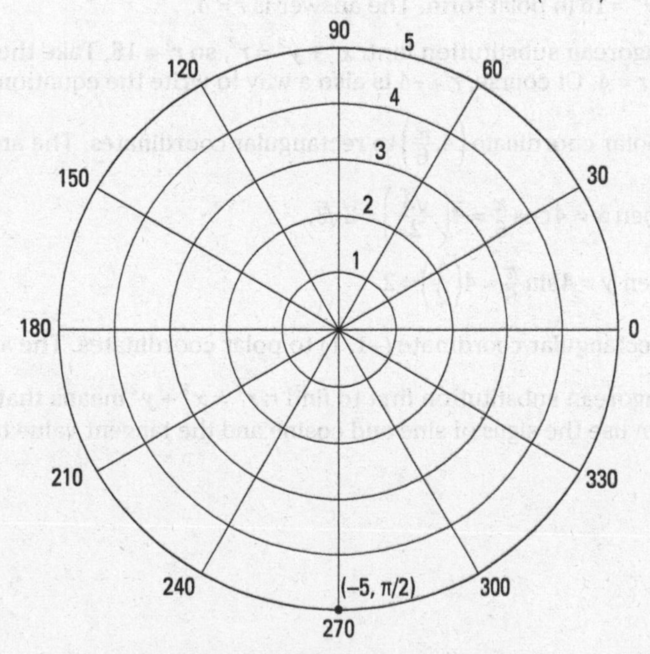

7 Name two other polar coordinates for the point in Question 5: one with a negative angle and one with a positive angle. Possible answers include $\left(4, -\frac{11\pi}{3}\right)$ and $\left(4, \frac{\pi}{3}\right)$.

The radius isn't changing, so don't do anything to it. Subtract 2π from the angle to get the first answer. Add 2π to the angle to get the second answer.

8 Name two other polar coordinates for the point in Question 6: one with a negative angle and one with a positive radius. Possible answers include $\left(-5, -\frac{3\pi}{2}\right)$ and $\left(5, \frac{3\pi}{2}\right)$.

To deal with the first situation, don't change the radius; just subtract 2π to get a negative angle: $-\frac{3\pi}{2}$. To change the radius to a positive 5, you have to change the angle. Because the point was down 5 originally, this angle is the same as $\frac{3\pi}{2}$

9 Rewrite $r = 3\sin\theta + 4\cos\theta$ in rectangular form. The answer is $x^2 + y^2 = 3y + 4x$.

First, change $\sin\theta$ to y/r and $\cos\theta$ to x/r: $r = 3\frac{y}{r} + 4\frac{x}{r}$. Multiply everything by r to get rid of the fractions and get $r^2 = 3y + 4x$. Use the Pythagorean substitution and you get $x^2 + y^2 = 3y + 4x$.

10 Rewrite $y = 2x - 1$ in polar form. The answer is $r = \frac{-1}{\sin\theta - 2\cos\theta}$

To move the other direction, remember that $x = r\cos\theta$ and $y = r\sin\theta$. Make these substitutions first and get $r\sin\theta = 2r\cos\theta - 1$. Get all terms with r to one side first: $r\sin\theta - 2r\cos\theta = -1$. Factor out the common factor of r: $r(\sin\theta - 2\cos\theta)$. Now solve for r by dividing the rest to the other side: $\frac{-1}{\sin\theta - 2\cos\theta}$

11 Rewrite $3x - 5y = 10$ in polar form. The answer is $r = \frac{10}{3\cos\theta - 5\sin\theta}$

Use the same substitutions as you did in the preceding question to get $3r\cos\theta - 5r\sin\theta = 10$. Factor: $r(3\cos\theta - 5\sin\theta) = 10$. Divide: $r = \frac{10}{3\cos\theta - 5\sin\theta}$

12 Rewrite $x^2 + y^2 = 16$ in polar form. The answer is $r = 4$.

Use the Pythagorean substitution first: $x^2 + y^2 = r^2$, so $r^2 = 16$. Take the square root of both sides and get $r = 4$. Of course, $r = -4$ is also a way to write the equation for this circle.

13 Convert the polar coordinate $\left(4, \frac{\pi}{6}\right)$ to rectangular coordinates. The answer is $\left(2\sqrt{3}, 2\right)$.

If $x = r\cos\theta$, then $x = 4\cos\frac{\pi}{6} = 4\left(\frac{\sqrt{3}}{2}\right) = 2\sqrt{3}$.

If $y = r\sin\theta$, then $y = 4\sin\frac{\pi}{6} = 4\left(\frac{1}{2}\right) = 2$.

14 Convert the rectangular coordinate $(-1, 1)$ to polar coordinates. The answer is $\left(\sqrt{2}, \frac{-\pi}{4}\right)$

Use the Pythagorean substitution first to find r: $r^2 = x^2 + y^2$ means that $r^2 = (-1)^2 + (1)^2 = 1 + 1 = 2$, or $r = \sqrt{2}$. Then use the signs of sine and cosine and the tangent value to find the angle: $\tan\theta = -1$, or $\theta = \frac{3}{4}\pi$.

15 Sketch the graph of $r^2 = 9\cos 2\theta$. See the following graph for the answer.

This graph is a lemniscate. If $\theta = 0$, $r = 3$ or $r = -3$. If $\theta = \frac{\pi}{6}$, then $r \approx 2.12$ or $r \approx -2.12$, and so on. Plug and chug with plenty of points to wind up with the graph.

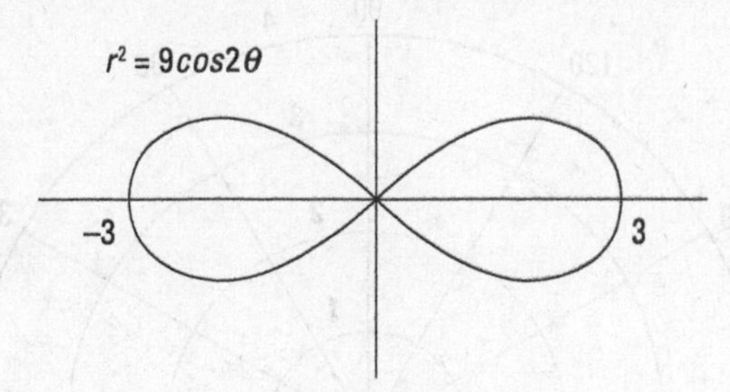

16 Sketch the graph of $r = 2\theta$. See the following graph for the answer.

This is a spiral. If $\theta = 0$, $r = 0$; if $\theta = \frac{\pi}{6}$, $r \approx 1.047$, and so on.

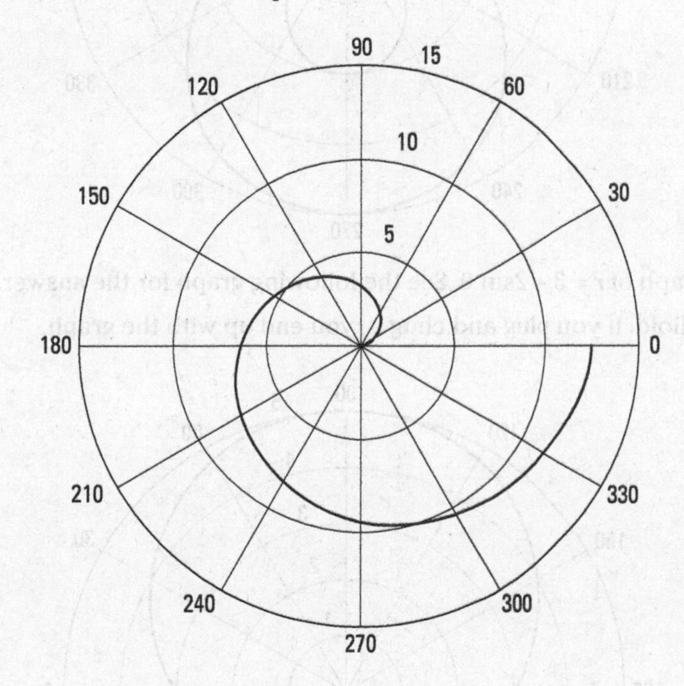

17 Sketch the graph of $r = 1 - 3\sin\theta$. See the following graph for the answer.

This is a limaçon. Plug and chug it with enough points to get the graph.

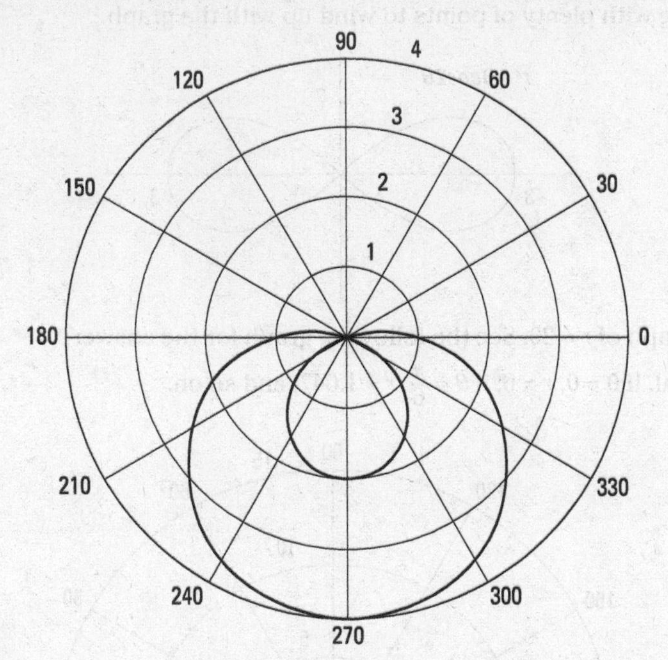

18 Sketch the graph of $r = 3 + 2\sin\theta$. See the following graph for the answer.

This is a cardioid. If you plug and chug it, you end up with the graph.

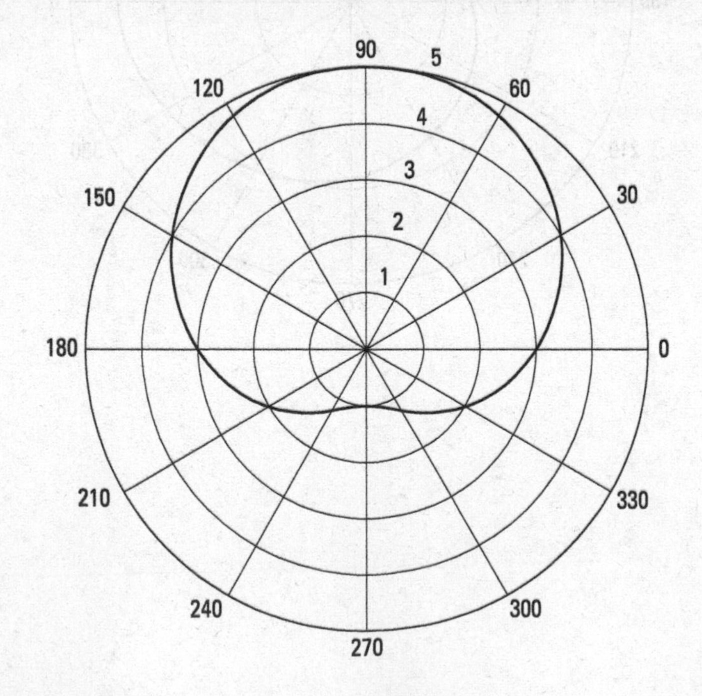

Chapter 12

Conquering Conic Sections

In This Chapter

- Closing in on circles
- Graphing parabolas, ellipses, and hyperbolas
- Recognizing the different conic sections
- Working with parametric form and polar coordinates

Who doesn't love a good cone? We're big fans of the ice cream cone ourselves, but maybe you're partial to the traffic cone because it keeps you safe on the road. Whatever type of cone is your favorite, for mathematicians the cone is the creative fuel for the fire of a whole bunch of ideas.

You see, about 2,200 years ago, some smart mathematician named Apollonius of Perga decided one day to stack two cones point to point. He sliced them in different directions and came up with four different *conic sections:* the *circle,* the *ellipse,* the *parabola,* and the *hyperbola.* Each conic section has its own equation and its own parts, which you need to determine in order to graph it. Typically, a textbook question will ask you to graph a conic section, identify certain parts, or write its equation. To do *any* of these tasks, you must be able to recognize what kind of conic section it is (more on that later) and write it in its own equation form.

When a conic section is already written in its standard form, you know most of the information about the parts of that particular conic. When a conic section isn't written in its standard form, well, you just have to write it that way. How do you do that? We're so glad you asked. It's the process called *completing the square.* That should sound familiar to you from previous math classes. If not, review Chapter 4.

A Quick Conic Review

To save you some time, we quickly review here how to complete the square for conic sections. When the equation of a conic section isn't written in its standard form, completing the square is the only way to convert the equation to its standard form. The steps of the process are as follows:

1. **Add/subtract any constant to the opposite side of the given equation, away from all the variables.**
2. **Factor the leading coefficient out of all terms in front of the set of parentheses.**
3. **Divide the remaining linear coefficient by two, but only in your head.**

4. **Square the answer from Step 3 and add that inside the parentheses.**

   Don't forget that if you have a coefficient from Step 2, you must multiply the coefficient by the number you get in this step and add *that* to both sides.

5. **Factor the quadratic polynomial as a perfect square trinomial.**

This chapter is dedicated to each mathematical conic superstar. One at a time, we tell you how to write each one in its form and how to graph it.

Going Round and Round with Circles

A circle is simple, but not so plain. You can do so much with a circle. The very tires you drive on are, after all, circular in shape. A circle has one point in the very middle called the *center.* All the points on the circle are the same distance, called the *radius,* from the center.

When a circle is drawn centered at the origin of the coordinate plane, the equation that describes it is simple as well:

$$x^2 + y^2 = r^2$$

r is the parameter that represents the circle's radius.

When the circle is moved around the coordinate plane with horizontal and/or vertical shifts, its equation looks like this:

$$(x-h)^2 + (y-v)^2 = r^2$$

The circle is centered at (h, v), h is the horizontal shift from the origin, v is the vertical shift from the origin, and r is still the radius. If a circle isn't written in this form, you'll still recognize the equation as one that may define a circle because the equation will have *both* an x^2 and a y^2, and the coefficients on both will be equal.

Don't be intimidated if you see this written another way in your pre-calc textbook. Different books may write this equation differently. Just know that it's always written as $(x - \text{horizontal})^2 + (y - \text{vertical})^2 = r^2$.

Q. What's the center and the radius of the circle $(x+3)^2 + y^2 = 16$?

A. The center is $(-3, 0)$; radius $r = 4$. Because we've been talking about circles, we thought we'd start you off with training wheels. This equation is already in the proper form of a circle, so finding its information is easy. The horizontal value with x is 3, so $h = -3$; the vertical shift is missing, so $v = 0$. This means that the center is located at the point $(-3, 0)$. Meanwhile, the other side of the equation gives you the radius squared. Setting $r^2 = 16$ gives you the solution $r = 4$.

Q. Graph $x^2 + y^2 - 6y + 2 = 0$.

A. **See the following graph. Now the training wheels come off! First subtract the constant to move it to the other side: $x^2 + y^2 - 6y = -2$. Because the leading coefficients are 1, you don't have to factor anything out and can move onto the next step. Then, because the *x* variable doesn't have a linear term, you don't have to complete the square there. Notice, however, that the *y* variable does have one, so you follow the process: $\left(-\frac{6}{2}\right)^2 = 9$. This value gets added to both sides: $x^2 + y^2 - 6y + 9 = 7$. Factor the *y*'s as a perfect square trinomial to get $x^2 + (y-3)^2 = 7$. *Voilá!* You have a circle. Its center is (0, 3), and its radius is $\sqrt{7}$.**

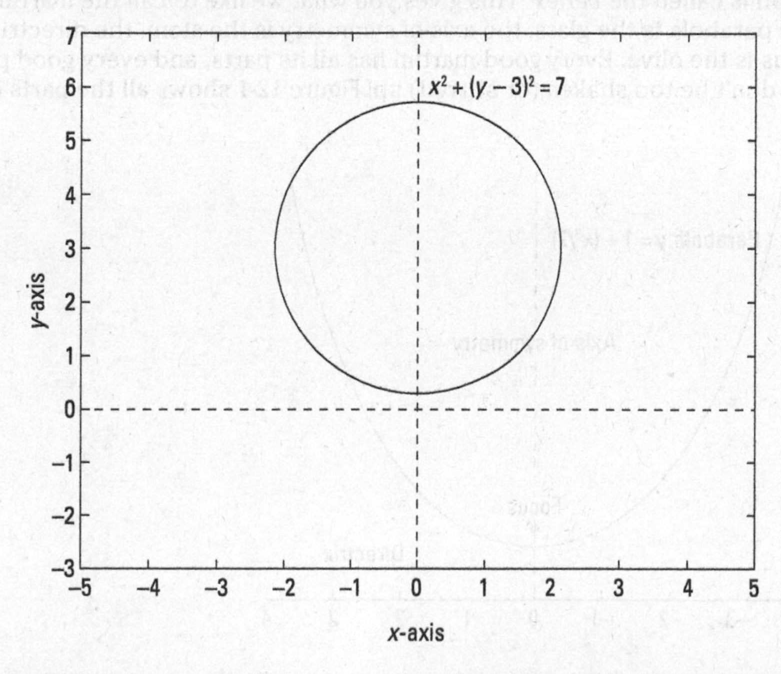

1. Find the center and radius of the circle $2x^2 + 2y^2 - 4x = 15$. Then graph the circle.

Solve It

2. Write the equation of the circle with the center (–1, 4) if the circle passes through the point (3, 1).

Solve It

The Ups and Downs: Graphing Parabolas

When you graph a *quadratic polynomial* (see Chapter 4 for more information on this type of polynomial), you always get a parabola. Typically, up until this point in pre-calc, the graphs of parabolas have been vertical: they open up or down. In conic sections, however, they can also open horizontally: to the left or to the right. We take a look at each situation in the following sections.

Officially, a *parabola* is the set of all points on a plane that are the same distance from a given point (*focus*) and a given line (*directrix*). Each parabola can be folded exactly in half over a line called the *axis of symmetry*. The point where the axis of symmetry intersects the graph is called the *vertex*. This gives you what we like to call the martini of conic sections: The parabola is the glass, the axis of symmetry is the stem, the directrix is the base, and the focus is the olive. Every good martini has all its parts, and every good parabola does, too. But don't be too shaken (or stirred) up! Figure 12-1 shows all the parts of a parabola.

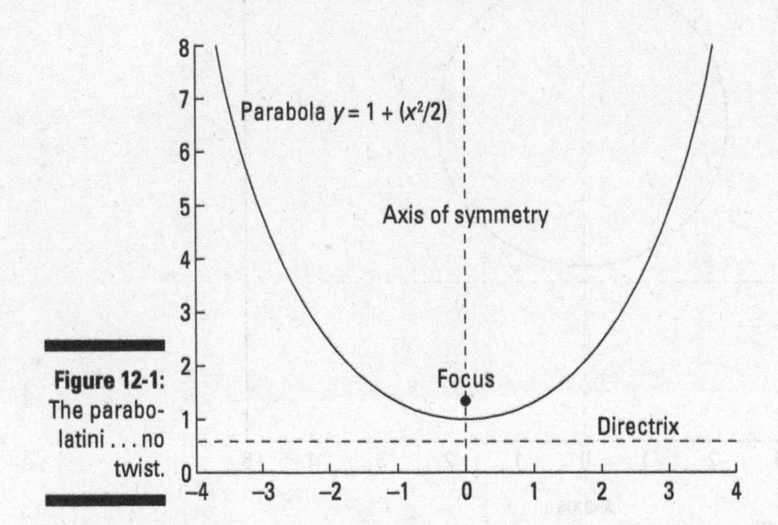

Figure 12-1: The parabo-latini . . . no twist.

Standing tall: Vertical parabolas

The equation of a vertical parabola is

$$y = a(x-h)^2 + v$$

where h is the horizontal shift from the origin, v is the vertical shift from the origin, and a is the vertical stretch or shrink. You know you're dealing with a vertical parabola when x is squared but y isn't. We discuss these types of transformations ad nauseum in Chapter 3, and it shouldn't surprise you that nothing changes just because we're calling them conic sections. Each vertical parabola still has the following parts:

- **Vertex:** (h, v)
- **Axis of symmetry:** $x = h$
- **Focus:** $\left(h, v + \frac{1}{4a}\right)$
- **Directrix:** $y = v - \frac{1}{4a}$

Don't memorize these as formulas to find the parts. Instead, use what you know about a parabola to get the job done. The vertex is always the first point that you graph on a parabola. From there, know that the focus and the directrix are $\frac{1}{4a}$ away from it. One is above and one is below, depending on the value of a. If $a < 0$, the parabola opens down and the focus moves down while the directrix moves up. If $a > 0$, the parabola opens up, the focus moves up, and the directrix moves down.

Q. Graph the parabola $y = -2(x - 1)^2 + 5$.

A. See the following graph. Because this equation is already in the proper parabola form, you should be able to go right to graphing. The vertex is at (1, 5). The vertical transformation is 2, and the graph is turned upside down. If you don't know where this information comes from, we strongly recommend you go back and read Chapter 3 now, which is chock-full of information on transforming any function.

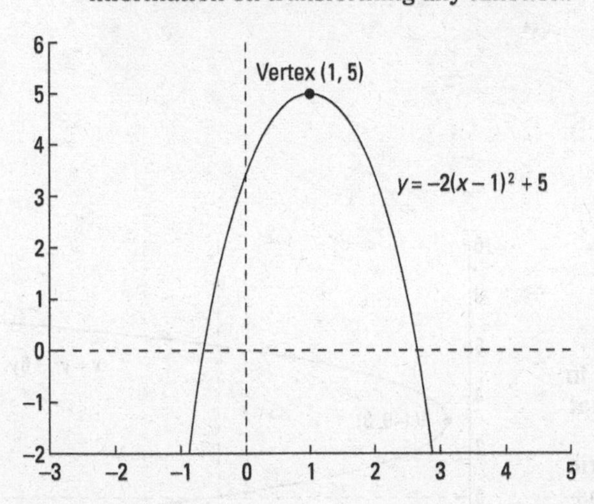

Q. State the vertex, axis of symmetry, focus, and directrix of $y = 3x^2 - 4x + 1$.

A. Vertex: $(\frac{2}{3}, -\frac{1}{3})$; axis of symmetry: $x = \frac{2}{3}$; focus: $(\frac{2}{3}, -\frac{1}{4})$; directrix: $y = -\frac{5}{12}$. That's an awful lot of fractions, but the process doesn't change. Start by subtracting 1 from both sides: $y - 1 = 3x^2 - 4x$. Then factor out the 3: $y - 1 = 3(x^2 - \frac{4x}{3})$. Now complete the square and be sure to keep the equation balanced: $y - 1 + \frac{4}{3} = 3(x^2 - \frac{4x}{3} + \frac{4}{9})$. Simplify and factor: $y + \frac{1}{3} = 3\left(x - \frac{2}{3}\right)^2$. Then solve for y to put the parabola in its proper form: $y = 3(x - \frac{2}{3})^2 - \frac{1}{3}$. Next, use the formulas to figure out all the parts — the vertex is $(\frac{2}{3}, -\frac{1}{3})$, the axis of symmetry is $x = \frac{2}{3}$, the focus is $(\frac{2}{3}, -\frac{1}{4})$, and the directrix is $y = -\frac{1}{3} - \frac{1}{12} = -\frac{5}{12}$.

3. What's the vertex of the parabola $y = -x^2 + 4x - 6$? Sketch the graph of this parabola.

Solve It

4. Find the focus and the directrix of the parabola $y = 4x^2$.

Solve It

Lying down on the job: Horizontal parabolas

The equation of a horizontal parabola is very similar to the vertical one we discuss in the preceding section. Here it is:

$$x = a(y - v)^2 + h$$

This is a horizontal parabola because y is squared but x isn't. Notice, also, that h and v are still there for the horizontal and vertical shifts, respectively, but that they've switched places. Because this parabola is horizontal, a also switches to become the horizontal transformation. This is the first time we've talked about a horizontal transformation, so we'll take a few moments to explain the idea. A *horizontal transformation* does the same thing as a vertical transformation, but it affects what the function does from left to right. A horizontal transformation where a is a fraction between 0 and 1 is a *horizontal shrink,* and a horizontal transformation where $a > 1$ is called a *horizontal stretch.* All of these numbers are positive, so the parabola opens to the right. When a is negative, the parabola does the same thing, but the graph is reflected in the opposite direction (so the parabola opens to the left).

Here are the parts of a horizontal parabola:

- **Vertex:** (h, v)
- **Axis of symmetry:** $y = v$
- **Focus:** $\left(h + \frac{1}{4a}, v\right)$
- **Directrix:** $x = h - \frac{1}{4a}$

Q. Graph the parabola if its equation is $x = y^2 - 6y$.

A. See the following figure. Complete the square to get this horizontal parabola in its form: $x = (y - 3)^2 - 9$. This means that the parabola's vertex is located at the point (–9, 3) and the graph is symmetric about $y = 3$. There's no horizontal transformation ($a = 1$), so from the vertex the graph moves up 1, over 1; up 1, over 3; up 1, over 5. Using these points and the symmetry of the parabola gives you the graph in the figure.

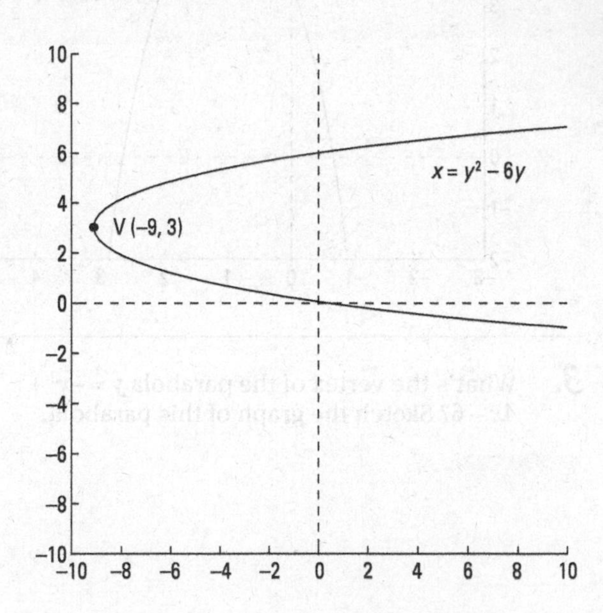

Q. Write the equation of the parabola whose vertex is (–2, 1) if the focus is at the point (–4, 1).

A. $x = \frac{-1}{8}(y-1)^2 - 2$. This one takes some brainpower to work out. Because the focus is to the left of the vertex, you know that the parabola is a horizontal one (forgetting, of course, that you're in the section of the book on horizontal parabolas). Start with the equation of any horizontal parabola: $x = a(y - v)^2 + h$. Then plug in the vertex values (–2, 1): $x = a(y - 1)^2 - 2$. Now, all you have to do is figure out what that pesky value of a is. You know that the equation to find the focus is $\left(h + \frac{1}{4a}, v\right)$. This tells you that $h + \frac{1}{4a} = -4$. You also know that h is –2, so substitute and get $-2 + \frac{1}{4a} = -4$. Solve to get $a = \frac{-1}{8}$. Finally, write the equation of the parabola: $x = \frac{-1}{8}(y-1)^2 - 2$.

5. Sketch the graph of $x = 2(y - 4)^2$.

Solve It

6. Determine whether this parabola opens left or right: $x = -y^2 - 7y + 3$.

Solve It

The Fat and the Skinny: Graphing Ellipses

An *ellipse* is defined as the set of all points on a plane, such that the sum of the distances from any point on the curve to two fixed points, the foci, is a constant. Think of an ellipse as a circle that has gone flat, like a soda left out overnight. All the fizz has left. That doesn't mean that the ellipse has no flavor! It has its own unique parts and equations, depending on whether (you guessed it) it's horizontal or vertical. Here's what you need to know about any ellipse, whether it's horizontal or vertical:

- The center is at the point (h, v).
- The longer axis of symmetry is called the *major axis,* and the distance from the center to a point on the ellipse along the major axis, often called the *semi-major axis,* is represented by a. The points where the ellipse intersects this axis are called the *vertices.*
- The shorter axis of symmetry is called the *minor axis,* and the distance from the center to a point on the ellipse along the minor axis, often called the *semi-minor axis,* is represented by b. The points where the ellipse intersects this axis are called the *co-vertices.*
- This means that a is always greater than b in an ellipse.
- You can find the foci of the ellipse along its major axis by using the equation $f^2 = a^2 - b^2$.

Figure 12-2 shows how all these pieces fall into place for a horizontal and a vertical ellipse.

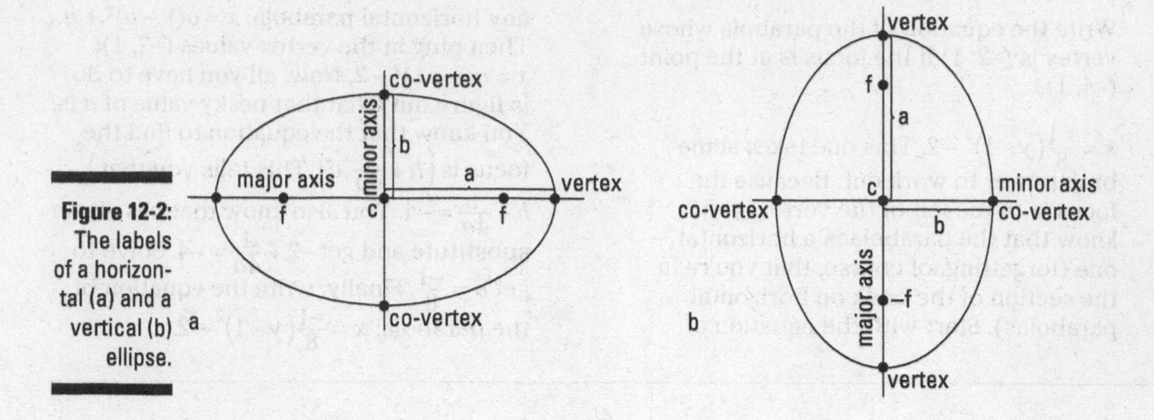

Figure 12-2: The labels of a horizontal (a) and a vertical (b) ellipse.

Short and fat: The horizontal ellipse

The equation of a horizontal ellipse is

$$\frac{(x-h)^2}{a^2}+\frac{(y-v)^2}{b^2}=1$$

Notice that all the variables (except for *f*, the focus) we mention make their appearance in the equation. Notice also that because $a > b$, for any ellipse, $a^2 > b^2$. The fact that the bigger number, a^2, is in the denominator of the *x* fraction tells you that the ellipse is horizontal. To graph any horizontal ellipse after it's written in this form, mark the center first. Then count out *a* units to the left and right and *b* units up and down. These four points determine the ellipse's shape. The vertices are points found at $(h \pm a, v)$. The co-vertices are $(h, v \pm b)$. The two foci are $2f$ units apart along the major axis. As points, the foci are $(h \pm f, v)$.

Q. State the center, vertices, and foci of the ellipse $3x^2 - 18x + 5y^2 = 3$.

A. Center: (3, 0); vertices: $(3 \pm \sqrt{10}, 0)$; foci: (5, 0) and (1, 0). As usual, you need to complete the square to write this ellipse in its standard form. The constant is already on the opposite side, so begin by factoring: $3(x^2 - 6x) + 5y^2 = 3$. Next, complete the square and balance the equation: $3(x^2 - 6x + 9) + 5y^2 = 3 + 27$. Factor the perfect square: $3(x - 3)^2 + 5y^2 = 30$.

Divide everything by 30: $\frac{(x-3)^2}{10}+\frac{y^2}{6}=1$. This tells you the center is (3, 0). Then, if $a^2 = 10$, $a = \sqrt{10}$, and if $b^2 = 6$, $b = \sqrt{6}$. This gives you the vertices at $(3 \pm \sqrt{10}, 0)$. It also tells you the co-vertices are at $(3, \pm\sqrt{6})$. Lastly, $f^2 = 10 - 6$, so $f^2 = 4$, which tells you that $f = 2$ and the foci are at (5, 0) and (1, 0).

Q. Sketch the graph of the ellipse $\frac{(x+2)^2}{25}+\frac{(y-1)^2}{16}=1$.

A. See the following figure. This ellipse is written in the proper form, so to graph it, all you have to do is identify its parts. The center is (–2, 1). If $a^2 = 25$, then $a = 5$. This means your vertices are 5 units to the left and the right from the center, at (–7, 1) and (3, 1). If $b^2 = 16$, then $b = 4$. This means your co-vertices are 4 units above and below the center at (–2, 5) and (–2, –3).

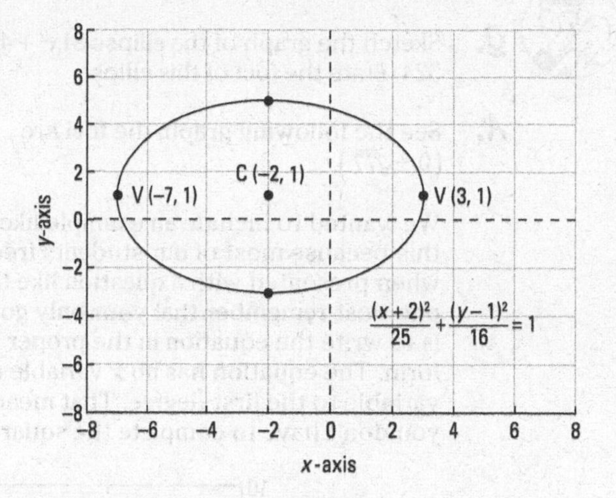

7. Sketch the graph of the ellipse $4x^2 + 12y^2 - 8x - 24y = 0$.

Solve It

8. Write the equation of the ellipse with vertices at (–1, 1) and (9, 1) and foci at $(4\pm\sqrt{21},1)$.

Solve It

Tall and skinny: The vertical ellipse

The equation of a vertical ellipse is

$$\frac{(x-h)^2}{b^2}+\frac{(y-v)^2}{a^2}=1.$$

This equation looks awfully familiar, doesn't it? The only difference between a horizontal ellipse and a vertical one is the location of *a*. When the bigger number is under *x*, it's a horizontal ellipse. When the bigger number is under *y*, it's a vertical ellipse. You graph this ellipse by marking the center, counting up and down *a* units to find the vertices, and then counting left and right *b* units to find the co-vertices. This means your vertices are at $(h,v\pm a)$ and your co-vertices are at $(h\pm b,v)$. Your foci move in the same direction as your vertices, so they're at $(h,v\pm f)$.

Q. Sketch the graph of the ellipse $81x^2 + 4y^2 = 324$. State the foci of this ellipse.

A. See the following graph; the foci are $\left(0, \pm\sqrt{77}\right)$.

We wanted to include an example like this because most of our students freeze when presented with a question like this one. Just remember that your only goal is to write the equation in the proper form. The equation has no *x* variable or *y* variable to the first degree. That means you don't have to complete the square!

Say what? All you need to do is get 1 on the right side of the equation by dividing everything by 324.

When you do, it reduces to $\frac{x^2}{4} + \frac{y^2}{81} = 1$, which also conveniently puts it in the form you want. This ellipse has its center at the origin (0, 0). Then, $a^2 = 81$, so *a* moves up and down 9 units, while $b^2 = 4$, so *b* moves left and right 2 units and gives you the following graph. Lastly, $f^2 = 81 - 4 = 77$, so $f = \sqrt{77}$. This means your foci are at $\left(0, \pm\sqrt{77}\right)$.

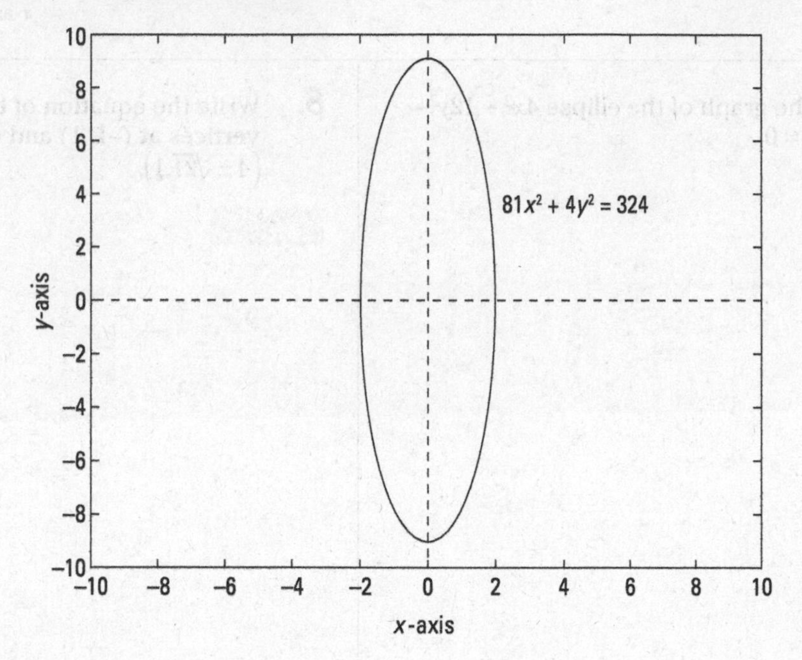

Q. Write the equation of the vertical ellipse with its center at (–4, 1) if its major axis has a length of 10 and its minor axis has a length of 8.

A. $\frac{(x+4)^2}{16} + \frac{(y-1)^2}{25} = 1$. You have all the information you need to write the equation. The center is given to you as (–4, 1).

If the major axis has a length of 10, then $2a = 10$, or $a = 5$. Also, the minor axis has a length of 8, so $2b = 8$, or $b = 4$. Knowing that the ellipse is vertical tells you to put a^2 under the *y* term and b^2 under the *x* term. This gives you the equation $\frac{(x+4)^2}{16} + \frac{(y-1)^2}{25} = 1$.

9. Sketch the graph of the ellipse $\frac{(x-1)^2}{8}+\frac{(y+2)^2}{6}=1$

Solve It

10. State the ordered pair for the vertices, the co-vertices, and the foci of the ellipse $(x+1)^2+\frac{(y-4)^2}{16}=1$

Solve It

No Caffeine Required: Graphing Hyperbolas

A *hyperbola* is the set of all points in the plane such that the difference of the distances from two fixed points (the foci) is a positive constant. Hyperbolas always come in two parts. Each one is a perfect mirror reflection of the other. Maybe they're the narcissists of the math world, always checking themselves out in the mirror that is their axis of symmetry. There are horizontal and vertical hyperbolas. Regardless of how the hyperbola opens, you always find the following parts:

- The center is at the point (h, v).
- The graph on both sides gets closer and closer to two diagonal lines known as *asymptotes.* The equation of the hyperbola, regardless of whether it's horizontal or vertical, gives you two values: a and b. These help you draw a box, and when you draw the diagonals of this box, you find the asymptotes.
- There are two axes of symmetry:
  - The one passing through the *vertices* is called the *transverse axis.* The distance from the center along the transverse axis to the vertex is represented by a.
  - The one perpendicular to the transverse axis through the center is called the *conjugate axis.* The distance along the conjugate axis from the center to the edge of the box that determines the asymptotes is represented by b.
  - a and b have no relationship; a can be less than, greater than, or equal to b.
- You can find the foci by using the equation $f^2 = a^2 + b^2$.

Figure 12-3 shows the parts of a vertical hyperbola.

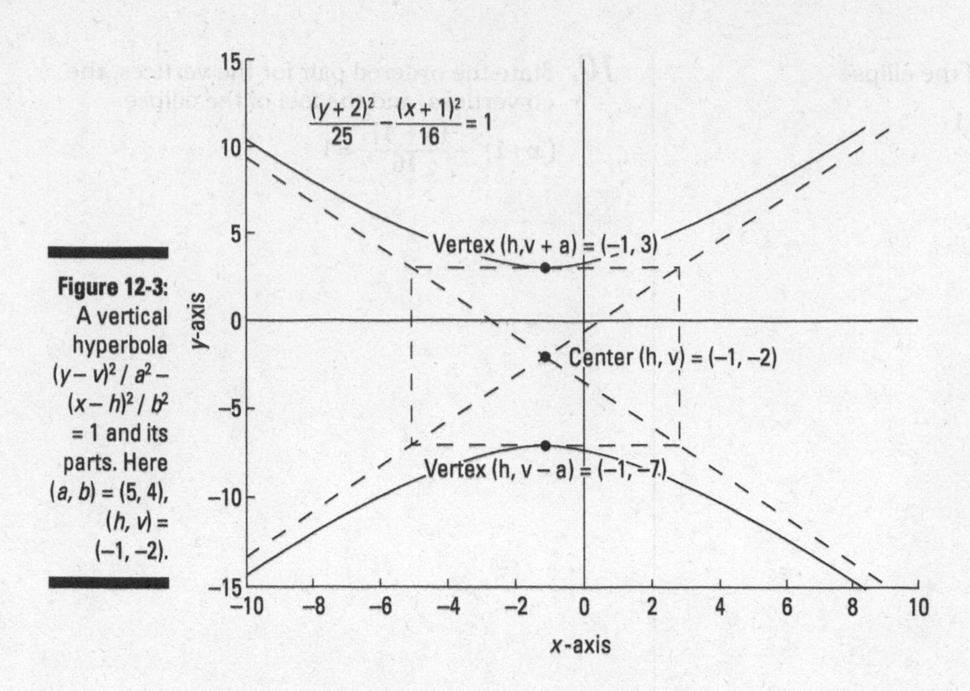

Figure 12-3: A vertical hyperbola $(y-v)^2/a^2 - (x-h)^2/b^2 = 1$ and its parts. Here $(a, b) = (5, 4)$, $(h, v) = (-1, -2)$.

Hip horizontal hyperbolas

How hip is the horizontal hyperbola? It's *so* hip that it doesn't even have to stand up — it just lies around all day. The equation of a horizontal hyperbola is

$$\frac{(x-h)^2}{a^2} - \frac{(y-v)^2}{b^2} = 1$$

This one looks really similar to the equation of the horizontal ellipse. But if you look closely, you notice the subtraction sign between the two fractions. To begin graphing, identify a, which helps determine one edge of a box that you can use to find the hyperbola's asymptotes. The corners of this imaginary box are two points of the asymptotes, so they can be used to draw those lines. The value of a is in the denominator of the x fraction, so it will be left and right from the center. The vertices are at $(h \pm a, v)$. The other edge of the box is found from b, under the y fraction. It moves up and down. The foci move in the same direction as a and can be found at $(h \pm f, v)$. Equations of the asymptotes of a horizontal hyperbola are given by $y = \pm\frac{b}{a}(x-h)+v$.

Q. Sketch the graph of $2x^2 - 3y^2 + 10x + 6y = \frac{41}{2}$

A. See the following figure. Put it in its standard form by completing the square. (Did we mention you'd be an expert at this by the end of this chapter?) Write the equation in order first: $2x^2 + 10x - 3y^2 + 6y = \frac{41}{2}$. Then factor out the coefficients: $2(x^2 + 5x) - 3(y^2 - 2y) = \frac{41}{2}$.

Watch out for the negative sign when factoring. Complete the square and keep the equation balanced: $2\left(x^2 + 5x + \frac{25}{4}\right) - 3(y^2 - 2y + 1) = \frac{41}{2} + \frac{25}{2} - 3$

Factor the difference of squares and simplify $2\left(x + \frac{5}{2}\right)^2 - 3(y-1)^2 = 30$. Last, divide everything by 30: $\frac{\left(x+\frac{5}{2}\right)^2}{15} - \frac{(y-1)^2}{10} = 1$. From the equation you know that the center is $(-5/2, 1)$, $a = \sqrt{15}$, and $b = \sqrt{10}$.

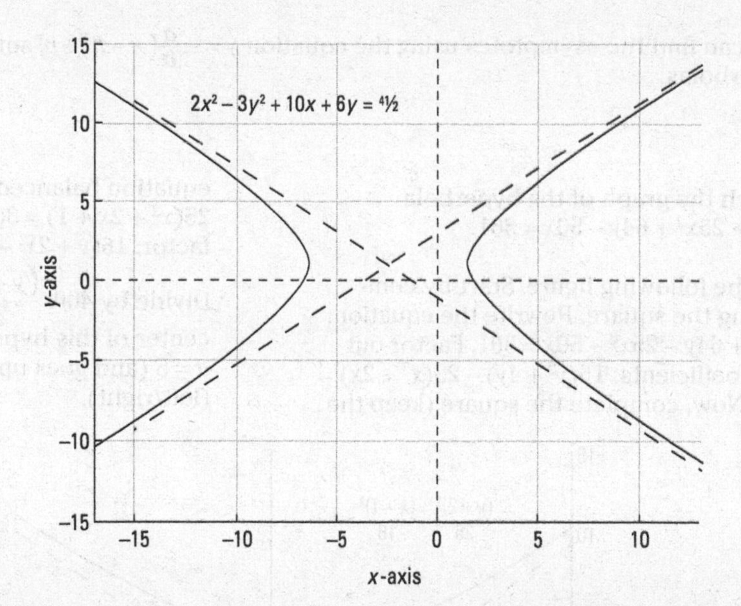

Q. Find the equations of the asymptotes for the hyperbola $\frac{x^2}{25} - \frac{(y-6)^2}{9} = 1$.

A. $y = \pm\frac{3}{5}(x-0)+6 = \pm\frac{3}{5}x+6$. Because this equation is in its form, the information is easy to find. $h = 0$, $v = 6$, $a = 5$, and $b = 3$. Put those into the equation to find the asymptotes for a horizontal hyperbola, and you get $y = \pm\frac{3}{5}(x-0)+6 = \pm\frac{3}{5}x+6$.

11. Find the equation of the hyperbola that meets the following criteria: It has its center at (4, 1), one of its vertices is at (7, 1), and one of its asymptotes is $3y = 2x - 5$.

Solve It

12. Sketch the graph of the equation $(2x - y)(x + 5y) - 9xy = 10$.

Solve It

Vexing vertical hyperbolas

The equation of a vertical hyperbola is

$$\frac{(y-v)^2}{a^2} - \frac{(x-h)^2}{b^2} = 1$$

Do you see the differences between the horizontal and vertical hyperbolas? The x and y switch places (along with the h and v). The a stays on the left, and the b stays on the right. When you write a hyperbola in its standard form, you need to make sure that the positive squared term is always first. The vertices are at $(h, v \pm a)$ and the foci are at $(h, v \pm f)$.

You can find the asymptotes using the equation $y = \pm\frac{a}{b}(x-h)+v$, sort of like the horizontal hyperbolas.

Q. Sketch the graph of the hyperbola $16y^2 - 25x^2 + 64y - 50x = 361$.

A. See the following figure. Start by completing the square. Rewrite the equation: $16y^2 + 64y - 25x^2 - 50x = 361$. Factor out the coefficients: $16(y^2 + 4y) - 25(x^2 + 2x) = 361$. Now, complete the square (keep the equation balanced, too): $16(y^2 + 4y + 4) - 25(x^2 + 2x + 1) = 361 + 64 - 25$. Factor, factor: $16(y + 2)^2 - 25(x + 1)^2 = 400$. Divide by 400: $\frac{(y+2)^2}{25} - \frac{(x+1)^2}{16} = 1$. The center of this hyperbola is (–1,–2), where $a = 5$ (and goes up and down) and $b = 4$ (left/right).

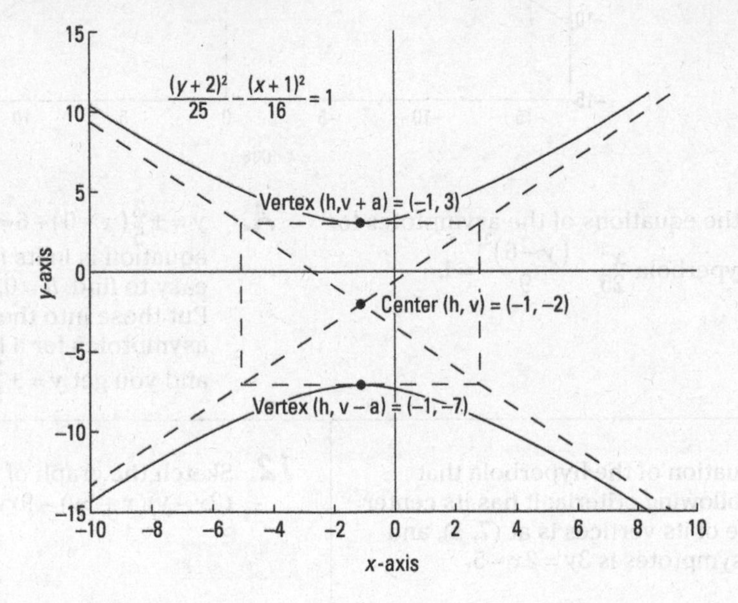

Q. Two science stations are two miles apart. They record an explosion, one station two seconds after the other. Write the equation of the hyperbola that describes the locus of explosion points having the above property by placing both stations on the y-axis with the origin at the center. (***Note:*** Sound travels at 1,100 feet per second.)

A. $\frac{y^2}{1210000} - \frac{x^2}{26668400} = 1$. If the two stations are two miles apart, that means one is a mile below the origin (0, –1) and the other is a mile above it (0, 1). Right away, you should notice that all other measurements in this problem are given in feet, so you have to convert 1 mile to 5280 feet. This means the microphones of the two stations are at (0, –5280) and (0, 5280). This gives you the foci of the hyperbola: $f = 5280$. To find the vertices, you can set up two different equations based on the distance ($d = rt$) the explosion travels to get to the two stations: $d_1 = 1100t$ and $d_2 = 1100(t + 2)$. The absolute value of the difference of the distances gives you the vertices of the hyperbola: $|d_1 - d_2| = |1100t - 1100(t+2)| = |1100t - 1100t - 2200| = |-2200| = 2200$. a for your vertex is half of this: $2200/2 = 1100$, so $a^2 = 1210000$. Now that you know f and a, you can find b^2 from the equation $f^2 = a^2 + b^2$: $b^2 = 26668400$. Lastly, because the hyperbola has its center at the origin, you can write the equation of the hyperbola as $\frac{y^2}{1210000} - \frac{x^2}{26668400} = 1$.

13. Sketch the graph of the equation $x^2 + 2x - 4y^2 + 32y = 59$.

Solve It

14. Write the equation of the hyperbola that has its center at (–3, 5), has one vertex at (–3, 1), and passes through the point $\left(1,5-4\sqrt{2}\right)$.

Solve It

Identifying Conic Sections

Often, you'll be presented with an equation and asked to graph it, but you won't be told what type of conic section it is. You have to be able to identify what type it is before doing any work. That's easier than it sounds, because there are only the four conics, and they have distinct differences:

- **Circles** have x^2 and y^2 with equal coefficients on both.
- **Parabolas** have x^2 or y^2, but not both.
- **Ellipses** have x^2 and y^2 with different (not equal but of the same sign) coefficients on each.
- **Hyperbolas** have x^2 and y^2 where exactly one coefficient is negative.

Table 12-1 has all the information you need to know about the four conics in one handy-dandy chart.

Table 12-1 Types of Conic Sections and Their Parts

Type of Conic	*Parts of the Conic*
Circle	Center (h, v)
$(x-h)^2 + (y-v)^2 = r^2$	Radius r
Horizontal parabola	Vertex (h, v)
$x = a(y-v)^2 + h$	Focus $\left(h+\frac{1}{4a}, v\right)$
	Directrix $x = h-\frac{1}{4a}$
	Axis of symmetry $y = v$

(continued)

Table 12-1 *(continued)*

Type of Conic	***Parts of the Conic***
Vertical parabola	Vertex (h, v)
$y = a(x-h)^2 + v$	Focus $\left(h, v + \frac{1}{4a}\right)$
	Directrix $y = v - \frac{1}{4a}$
	Axis of symmetry $x = h$
Horizontal ellipse	Center (h, v)
$(x-h)^2 / a^2 + (y-v)^2 / b^2 = 1$	Vertices $(h \pm a, v)$
$f^2 = a^2 - b^2,\ a > b$	Co-vertices $(h, v \pm b)$
	Foci $(h \pm f, v)$
Vertical ellipse	Center (h, v)
$(x-h)^2 / b^2 + (y-v)^2 / a^2 = 1$	Vertices $(h, v \pm a)$
$f^2 = a^2 - b^2,\ a > b$	Co-vertices $(h \pm b, v)$
	Foci $(h, v \pm f)$
Horizontal hyperbola	Center (h, v)
$(x-h)^2 / a^2 - (y-v)^2 / b^2 = 1$	Vertices $(h \pm a, v)$
$f^2 = a^2 + b^2$	Foci $(h \pm f, v)$
	Asymptotes $y = \pm\frac{b}{a}(x-h)+v$
Vertical hyperbola	Center (h, v)
$(y-v)^2 / a^2 - (x-h)^2 / b^2 = 1$	Vertices $(h, v \pm a)$
$f^2 = a^2 + b^2$	Foci $(h, v \pm f)$
	Asymptotes $y = \pm\frac{a}{b}(x-h)+v$

15. Sketch the graph of $3x^2 + 4y^2 - 6x + 16y - 5 = 0$.

Solve It

16. Sketch the graph of $4x^2 - 8x - 1 = 4y^2 - 4y$.

Solve It

17. Sketch the graph of $4(x - 2) = 2y^2 + 6y$.

Solve It

18. Sketch the graph of $2y^2 - 4x^2 + 8x - 8 = 0$.

Solve It

19. Sketch the graph of $4x^2 + 4y^2 - 8y + 16x - 4 = 0$.

Solve It

20. Sketch the graph of $3x^2 - 4y^2 + 3x - 2y - \frac{23}{2} = 0$.

Solve It

Conic Sections in Parametric Form and Polar Coordinates

So far, you've graphed all the conics in *rectangular form* (x, y). However, you can graph a conic section in two other ways:

- **Parametric form:** This form is for conics that can't be easily written as a function $y = f(x)$. Indeed, unless you have a vertical parabola, a conic will never be expressible as $y = f(x)$. Both x and y are written in two different equations as being dependent on one other variable (usually t).
- **Polar form:** You recognize this from Chapter 11, where every point is expressed as (r, θ).

We show you how to deal with both of these forms in the following sections.

Parametric form for conic sections

Parametric form defines both x and y in terms of another arbitrary value called the *parameter*. Most often, this is represented by t, as many real-world applications set the definitions based on time. You can find x and y by picking values for t. Why change? In parametric form you can find how far an object has moved over time (the x equation) and the object's height over time (the y equation).

Q. Sketch the curve given by the parametric equations $x = 2t + 1$, $y = t^2 - 3t + 1$, and $1 < t \leq 5$.

A. See the following figure. Create a table for t, x, and y. Pick values of t between the interval values given to you, and then figure out what the x and y values are for each t value. Table 12-2 shows these values, and following is the graph of this parametric function.

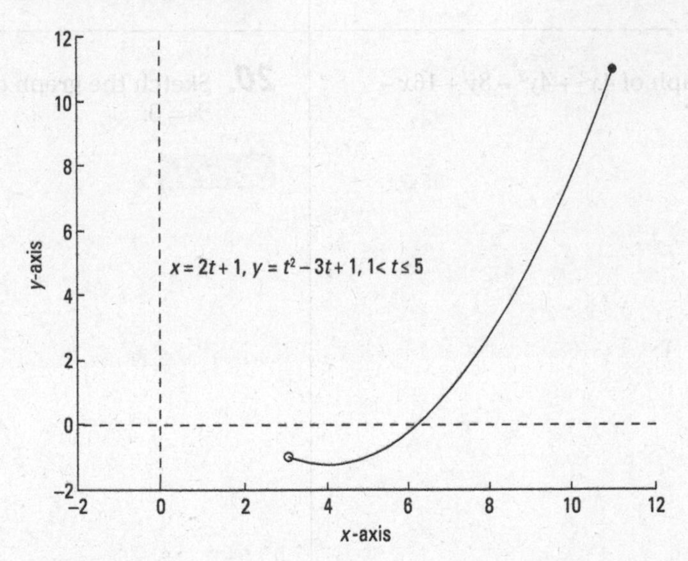

Even though $t > 1$ in the given interval, you need to start your table off with this value to see what the function would have been. Your graph has an open circle on it at this point to indicate that the value isn't included in the graph or the interval.

Table 12-2 Plug and Chug a Parametric Equation

t value	x value	y value
1	3	−1
2	5	−1
3	7	1
4	9	5
5	11	11

21. Sketch the the curve defined by the parametric equations $x = t^2 - 1$, $y = 2t$, and $-2 \le t \le 3$.

Solve It

22. Sketch the graph of the parametric equations $x = \frac{1}{\sqrt{t-1}}$, $y = \frac{1}{t-1}$, and $t > 1$.

Solve It

Changing from parametric form to rectangular form

The only other way to graph a parametric curve is to write it in rectangular form. To do this, you must solve one equation for the parameter and then substitute that value into the other equation. It's easiest if you pick the equation you *can* solve for the parameter (choose the equation that's linear if possible). To show you how it works, we use the example from the last section.

Q. Write the parametric equations $x = 2t + 1$, $y = t^2 - 3t + 1$, and $1 < t \le 5$ in rectangular form.

A. $y = \frac{1}{4}x^2 - 2x + \frac{11}{4}$, x in (3, 11]. First, solve the equation that's linear for t: $t = \frac{x-1}{2}$. Then substitute this value into the other equation for t: $y = \left(\frac{x-1}{2}\right)^2 - 3\left(\frac{x-1}{2}\right) + 1$. Simplify this equation to get $y = \frac{1}{4}x^2 - 2x + \frac{11}{4}$

23. Eliminate the parameter and find an equation in x and y whose graph contains the curve defined by the parametric equations $x = t^2$, $y = 1 - t$, and $t \ge 0$.

Solve It

24. Eliminate the parameter of the parametric equations $x = t - 5$ and $y = \sqrt{t}$.

Solve It

Conic sections on the polar coordinate plane

Conic sections on the polar coordinate plane are all based on a special value known as *eccentricity,* or *e.* This value describes what kind of conic section it is, as well as the conic's shape. Knowing what kind of conic section you're dealing with is difficult until you know what the eccentricity is:

- If $e = 0$, the conic is a circle.
- If $0 < e < 1$, the conic is an ellipse.
- If $e = 1$, the conic is a parabola.
- If $e > 1$, the conic is a hyperbola.

When you know *e*, all conics are expressed in polar form based on (r, θ), where *r* is the radius and θ is the angle. See Chapter 11 for more information on polar equations.

All conics in polar form are written based on four different equations:

$$r = \frac{ke}{1 - e\cos\theta} \text{ or } r = \frac{ke}{1 - e\sin\theta}$$

$$r = \frac{ke}{1 + e\cos\theta} \text{ or } r = \frac{ke}{1 + e\sin\theta}$$

where *e* is eccentricity and *k* is a constant value. To graph any conic section in polar form, substitute sufficiently many evenly spaced values of θ and plug and chug away until you get a picture!

Q. Graph the equation $r = \dfrac{2}{4 - \cos\theta}$

A. See the following figure. First, notice that the equation as shown doesn't fit exactly into any of the equations we just gave you. All those denominators begin with 1, and this equation begins with 4! To deal with this, factor out the 4 from the denominator to get $\dfrac{2}{4\left(1 - \frac{1}{4}\cos\theta\right)}$, which is the same as $\dfrac{2 \cdot \frac{1}{4}}{1 - \frac{1}{4}\cos\theta}$.

Notice that this makes *e* the same in the numerator and denominator (¼) and that *k* is 2. Now that you know *e* is ¼, that tells you the equation is an ellipse. Plugging in values gives you points $\theta = 0$, $r = \frac{2}{3}$; $\theta = \frac{\pi}{2}$, $r = \frac{1}{2}$; $\theta = \pi$, $r = \frac{2}{5}$; $\theta = \frac{3\pi}{2}$, $r = \frac{1}{2}$.

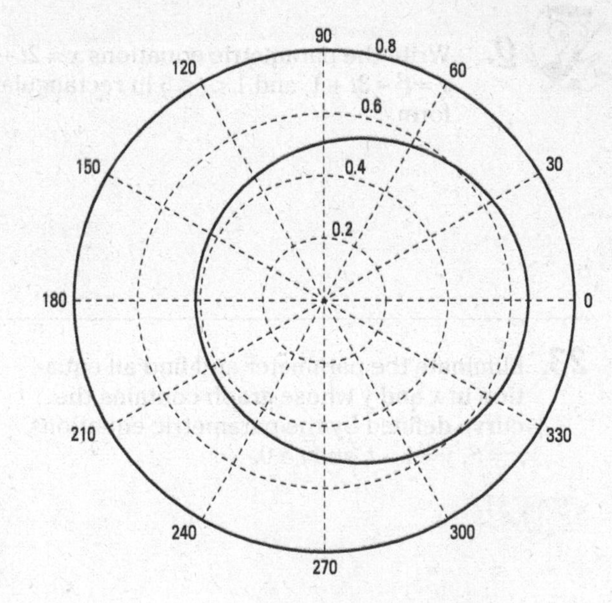

25. Graph the equation of $r = \frac{8}{1+\cos\theta}$ and label any vertices.

Solve It

26. Identify the conic section whose equation is $r = \frac{12}{3-4\sin\theta}$ by stating its eccentricity.

Solve It

Answers to Problems on Conic Sections

This section contains the answers for the practice problems presented in this chapter. We suggest you read the following explanations if your answers don't match up with ours (or if you just want a refresher on solving a particular type of problem).

1 Find the center and radius of the circle $2x^2 + 2y^2 - 4x = 15$. Then graph the circle. The center is (1, 0) and the radius is approximately 2.92.

Rewrite the equation so the x and y variables are together to get $2x^2 - 4x + 2y^2 = 15$. Factor out the coefficient: $2(x^2 - 2x) + 2y^2 = 15$. Complete the square: $2(x^2 - 2x + 1) + 2y^2 = 15 + 2$. Factor and get $2(x - 1)^2 + 2y^2 = 17$. Divide everything by 2 to write the circle in its standard form: $(x - 1)^2 + y^2 = 8.5$.

This means the center is (1, 0) and the radius is $\sqrt{8.5}$, or about 2.92.

2 Write the equation of the circle with the center (–1, 4) if the circle passes through the point (3, 1). The answer is $(x+1)^2+(y-4)^2=25$.

If you're given the center and a point, you can find the distance between the two points using the distance formula from Chapter 1: $d=\sqrt{(x_2-x_1)^2+(y_2-y_1)^2}=\sqrt{(3+1)^2+(1-4)^2}=\sqrt{4^2+3^2}=\sqrt{25}=5$.

This tells you the radius of the circle. Now that you know both the radius and the center, you can write the equation: $(x + 1)^2 + (y - 4)^2 = 25$.

3 What's the vertex of the parabola $y = -x^2 + 4x - 6$? Sketch the graph of this parabola. The answer is (2, –2); see the following graph.

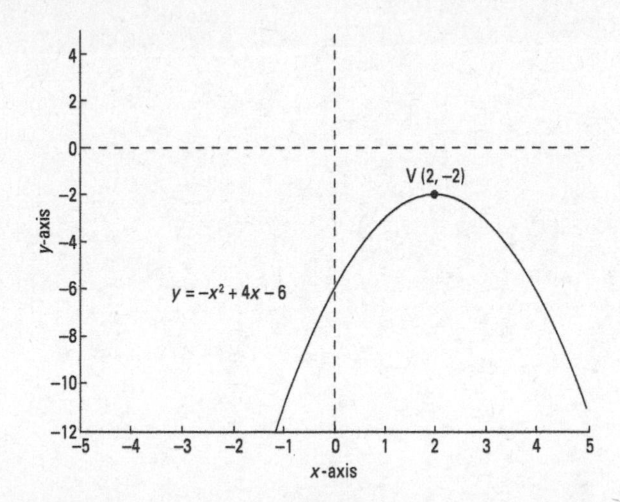

Add the 6 to both sides: $y + 6 = -x^2 + 4x$. Now, factor out the coefficient: $y + 6 = -1(x^2 - 4x)$. Complete the square and balance the equation: $y + 6 - 4 = -1(x^2 - 4x + 4)$. Simplify and factor: $y + 2 = -1(x - 2)^2$. Lastly, subtract 2 from both sides to write the equation in its proper form: $y = -1(x - 2)^2 - 2$. This means the vertex is located at the point (2, –2).

4 Find the focus and the directrix of the parabola $y = 4x^2$. The focus is $(0, \frac{1}{16})$ and the directrix is $y = -\frac{1}{16}$.

There's no square to complete, so if it helps you to fill in the missing information with zeros, then rewrite the equation as $y = 4(x - 0)^2 + 0$. This puts the vertex at the origin (0, 0). Because $a = 4$, the focus is $\frac{1}{4a}$ units above this point at $(0, \frac{1}{16})$ and the directrix is the line that runs $\frac{1}{4a}$ units below the vertex, perpendicular to the axis of symmetry at $y = -\frac{1}{16}$.

5 Sketch the graph of $x = 2(y - 4)^2$. See the following graph for the answer.

This equation is written in the proper form, unless you'd like to rewrite it as $x = 2(y - 4)^2 + 0$ because the h is missing. This is a horizontal parabola with its vertex at (0, 4). It opens to the right, with a horizontal transformation of 2.

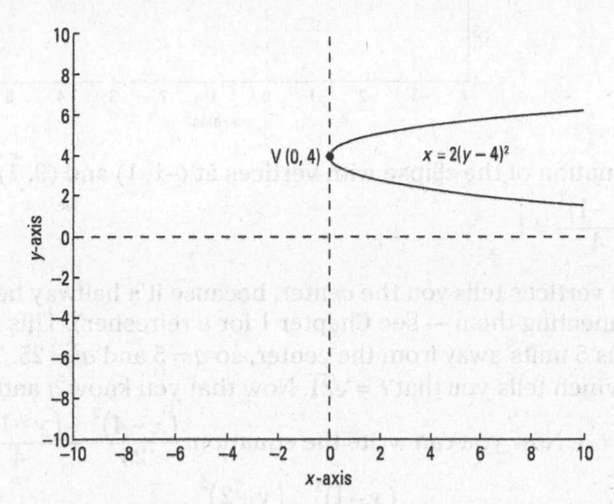

6 Determine whether this parabola opens left or right: $x = -y^2 - 7y + 3$. The answer is the parabola opens to the left.

Did we get you on this one? Did you start completing the square? We know you're used to doing it by now, but make sure you look at the directions of the problem. This one just asks you whether the parabola opens to the left or right, and you can tell from the leading coefficient of –1 (without doing any work at all) that the parabola opens to the left. If you do actually want to graph this equation, the standard form is $x = -1(y + \frac{7}{2})^2 + \frac{61}{4}$.

7 Sketch the graph of the ellipse $4x^2 + 12y^2 - 8x - 24y = 0$. See the following graph for the answer.

You have to complete the square twice for this one, so maybe we make up for that last question here. There's no constant to move, so rewrite the equation and factor out the coefficients: $4(x^2 - 2x) + 12(y^2 - 2y) = 0$. Now complete the square and balance the equation: $4(x^2 - 2x + 1) + 12(y^2 - 2y + 1) = 0 + 4 + 12$. Factor: $4(x - 1)^2 + 12(y - 1)^2 = 16$. Divide everything by 16: $\frac{(x-1)^2}{4} + \frac{3(y-1)^2}{4} = 1$. But don't start to graph it yet, because each ellipse has a coefficient of 1 on the numerator, and you have to divide that 3 in the second fraction to get $\frac{(x-1)^2}{4} + \frac{(y-1)^2}{\frac{4}{3}} = 1$.

Now that the equation is written in standard form, you can graph it. The center is (1, 1), the vertices are (3, 1) and (–1, 1), and the co-vertices are (1, 2.2) and (1, –0.2).

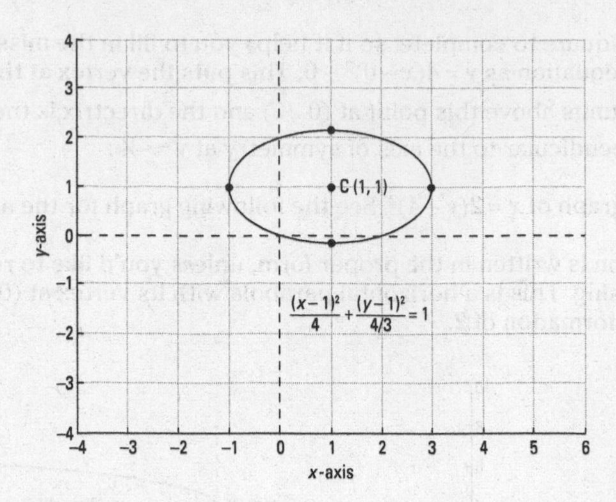

8 Write the equation of the ellipse with vertices at (–1, 1) and (9, 1) and foci at $\left(4 \pm \sqrt{21}, 1\right)$. The answer is $\frac{(x-4)^2}{25}+\frac{(y-1)^2}{4}=1$.

Knowing the vertices tells you the center, because it's halfway between them (the midpoint of the segment connecting them — see Chapter 1 for a refresher). This means the center is at (4, 1) and that each vertex is 5 units away from the center, so $a = 5$ and $a^2 = 25$. The foci are $\pm\sqrt{21}$ units away from the center, which tells you that $f = \sqrt{21}$. Now that you know a and f, you can find b^2 using $f^2 = a^2 - b^2$. In this case, $b^2 = 4$. Now you can write the equation: $\frac{(x-4)^2}{25}+\frac{(y-1)^2}{4}=1$.

9 Sketch the graph of the ellipse $\frac{(x-1)^2}{8}+\frac{(y+2)^2}{6}=1$. See the following graph for the answer.

How convenient! This equation is written in the proper form, so you don't have to complete the square. The center is (1, –2), a is approximately 2.83, and b is approximately 2.45. That gives you the ellipse shown in the graph.

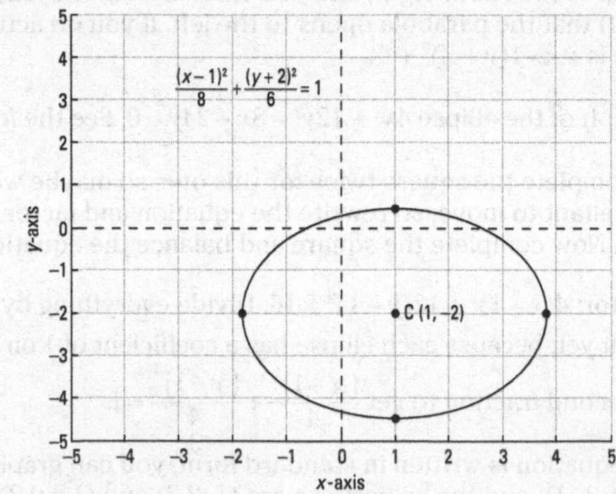

10 State the ordered pair for the vertices, the co-vertices, and the foci of the ellipse $(x+1)^2+\frac{(y-4)^2}{16}=1$. The vertices are (–1, 8) and (–1, 0); the co-vertices are (–2, 4) and (0, 4); and the foci are $\left(-1, 4 \pm \sqrt{15}\right)$.

The sneakiest part of this problem is that the x half of the equation isn't written as a fraction, but that's easy to remedy by writing the denominator of 1: $\frac{(x+1)^2}{1}+\frac{(y-4)^2}{16}=1$. This tells you that the center is (–1, 4), $a = 4$, and $b = 1$. The vertices are (–1, 4 ± 4) = (–1, 8) and (–1, 0). The co-vertices are (–1 ± 1, 4) = (–2, 4) and (0, 4). Lastly, $f^2 = a^2 - b^2$, so $f = \sqrt{15}$, which gives you the foci at $\left(-1, 4\pm\sqrt{15}\right)$.

11 Find the equation of the hyperbola that meets the following criteria: It has its center at (4, 1), one of its vertices is at (7, 1), and one of its asymptotes is $3y = 2x - 5$. The answer is $\frac{(x-4)^2}{9}-\frac{(y-1)^2}{4}=1$.

You're given the center (4, 1) and the equation of the asymptote, which you can rewrite in slope-intercept form by dividing by 3 to get $y=\frac{2}{3}x-\frac{5}{3}$. Because $a = 3$, you see that the vertex is 3 units to the right and this is a horizontal hyperbola. The slope of the asymptote, $\frac{2}{3}$, is the value of $\frac{b}{a}$. If we lost you there, you have to write the equation of the asymptote in point-slope form because you know the point is the center (4, 1) and the slope is $m = \frac{2}{3}$. This gives you the equation $y-1=\frac{2}{3}(x-4)$. Adding 1 to both sides makes the equation look like the equation of the asymptote for a horizontal hyperbola and helps you identify $\frac{b}{a}$ at the same time. Now that you know the center, a, and b, you can write the equation: $\frac{(x-4)^2}{9}-\frac{(y-1)^2}{4}=1$.

12 Sketch the graph of the equation $(2x - y)(x + 5y) - 9xy = 10$. See the following graph for the answer.

Does the term $-9xy$ make you a little nervous? Don't be. Just FOIL out the binomials and get $2x^2 + 10xy - xy - 5y^2 - 9xy = 10$. Notice that all the xy terms cancel to give you $2x^2 - 5y^2 = 10$. You can divide everything by 10 to write this equation in its form and get $\frac{x^2}{5}-\frac{y^2}{2}=1$. The center of this hyperbola is at (0, 0); $a=\sqrt{5}$, or ≈ 2.24; and $b=\sqrt{2}$, or ≈ 1.41.

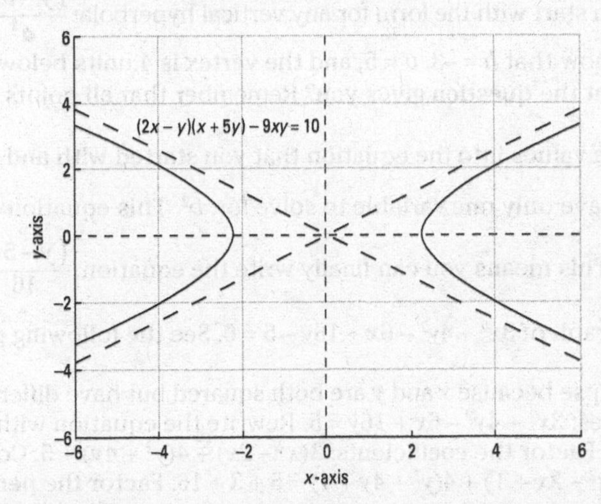

13 Sketch the graph of the equation $x^2 + 2x - 4y^2 + 32y = 59$. See the following graph for the answer.

You just knew we were gonna say, "Complete the square," didn't you? Get going by factoring out the coefficients, including the 1 in front of the x^2: $1(x^2 + 2x) - 4(y^2 - 8y) = 59$. Complete the square and balance away: $1(x^2 + 2x + 1) - 4(y^2 - 8y + 16) = 59 + 1 - 64$. Factor and simplify: $1(x + 1)^2 - 4(y - 4)^2 = -4$.

Divide everything by –4: $\frac{(x+1)^2}{-4}+(y-4)^2=1$. Did you notice how it suddenly became a vertical hyperbola because the y fraction is positive? You have to rewrite it to put it in its correct form: $(y-4)^2-\frac{(x+1)^2}{4}=1$. That gives you the following graph.

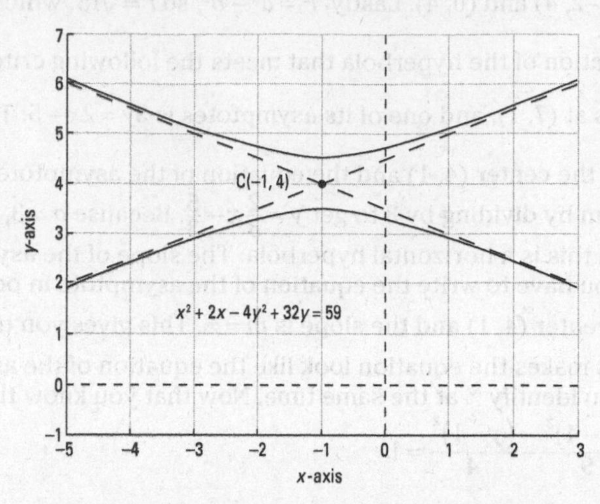

14 Write the equation of the hyperbola that has its center at (–3, 5), one vertex at (–3, 1), and passes through the point $\left(1,5-4\sqrt{2}\right)$. The answer is $\frac{(y-5)^2}{16}-\frac{(x+3)^2}{16}=1$.

This one requires a thinking cap, that's for sure. We suggest drawing it out on a sheet of graph paper and marking the center, the vertex, and the point, but that's it. Doing so shows you that it's a vertical hyperbola, so start with the form for any vertical hyperbola: $\frac{(y-v)^2}{a^2}-\frac{(x+h)^2}{b^2}=1$. Based on what you're given, you know that $h=-3$, $v=5$, and the vertex is 4 units below the center, so $a=4$. What do you do with the point the question gives you? Remember that all points are (x, y), so $x=1$ and $y=5-4\sqrt{2}$.

Plug all these values into the equation that you started with and get $\frac{\left(5-4\sqrt{2}-5\right)^2}{4^2}-\frac{(1+3)^2}{b^2}=1$, which means you have only one variable to solve for: b^2. This equation simplifies to $2-\frac{16}{b^2}=1$. Solving it gets you $b^2=16$. This means you can finally write the equation: $\frac{(y-5)^2}{16}-\frac{(x+3)^2}{16}=1$.

15 Sketch the graph of $3x^2+4y^2-6x+16y-5=0$. See the following graph for the answer.

This is an ellipse because x and y are both squared but have different coefficients of the same sign. Add 5 to both sides: $3x^2+4y^2-6x+16y=5$. Rewrite the equation with the x and the y together: $3x^2-6x+4y^2+16y=5$. Factor the coefficients: $3(x^2-2x)+4(y^2+4y)=5$. Complete the square and balance the equation: $3(x^2-2x+1)+4(y^2+4y+4)=5+3+16$. Factor the perfect square trinomials and simplify: $3(x-1)^2+4(y+2)^2=24$. Divide everything by 24: $\frac{(x-1)^2}{8}+\frac{(y+2)^2}{6}=1$, which turns out to be the same ellipse as in Question 9. This gives you the following graph.

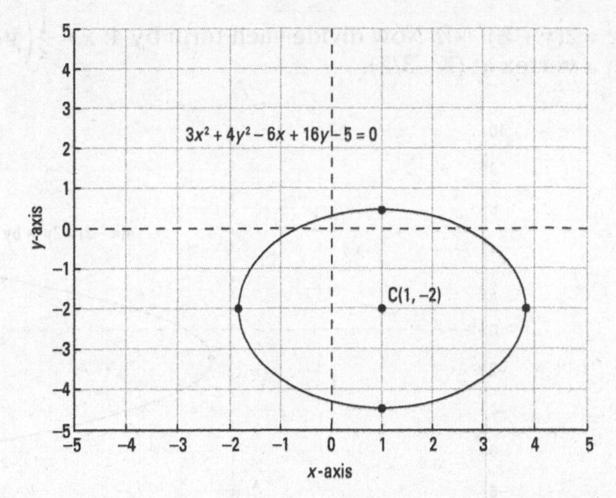

16 Sketch the graph of $4x^2 - 8x - 1 = 4y^2 - 4y$. See the following graph for the answer.

Rewrite the equation first: $4x^2 - 8x - 4y^2 + 4y = 1$. You should recognize that you have a hyperbola on your hands because you have an x^2 and a y^2 where exactly one has a negative coefficient. Factor out the coefficients: $4(x^2 - 2x) - 4(y^2 - y) = 1$. Complete the square and balance the equation: $4(x^2 - 2x + 1) - 4(y^2 - y + \frac{1}{4}) = 1 + 4 - 1$. Factor the perfect squares: $4(x - 1)^2 - 4(y - \frac{1}{2})^2 = 4$. Divide everything by 4: $(x-1)^2 - \left(y - \frac{1}{2}\right)^2 = 1$. In this particular hyperbola, the values of a and b are both 1. Knowing this and the center gives you the following graph.

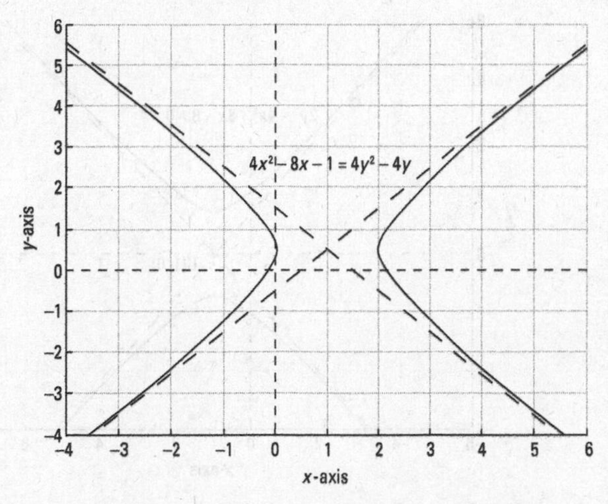

17 Sketch the graph of $4(x - 2) = 2y^2 + 6y$. See the following graph for the answer.

Notice right away that the equation doesn't have an x^2, so this is a horizontal parabola. Go ahead and distribute the 4 first: $4x - 8 = 2y^2 + 6y$. Now, factor the coefficient on the y^2 variable: $4x - 8 = 2(y^2 + 3y)$. Completing the square for this one gets you fractions; half of 3 is $\frac{3}{2}$ and that value squared is $\frac{9}{4}$. Add this inside the parentheses, and don't forget to add $2 \cdot \frac{9}{4}$ to the other side to keep the equation balanced: $4x - 8 + \frac{9}{2} = 2(y^2 + 3y + \frac{9}{4})$. Simplify and factor: $4x - \frac{7}{2} = 2(y + \frac{3}{2})^2$. Begin to solve for x by adding over the ½: $4x = 2(y + \frac{3}{2})^2 + \frac{1}{2}$. Now divide each term by 4: $x = \frac{1}{2}\left(y+\frac{3}{2}\right)^2 + \frac{7}{8}$. This gives you a horizontal parabola with a vertex at (7/8, –3/2).

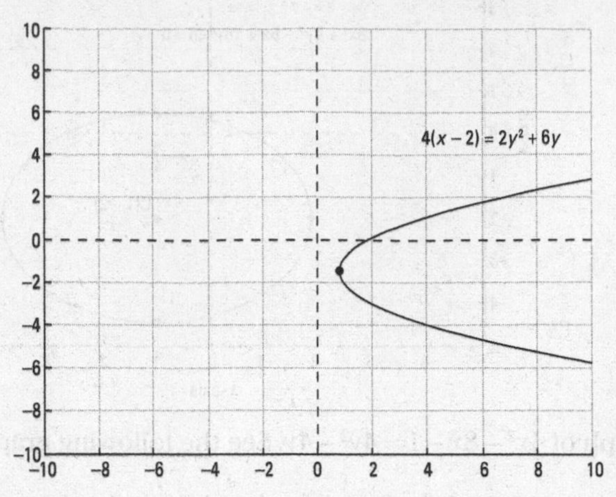

18 Sketch the graph of $2y^2 - 4x^2 + 8x - 8 = 0$. See the following graph for the answer.

You have a hyperbola to graph this time. We'll skip the narrative on how to complete the square and show the steps only (we're confident you're a pro at it by now): $2y^2 - 4x^2 + 8x = 8$; $2y^2 - 4(x^2 - 2x) = 8$; $2y^2 - 4(x^2 - 2x + 1) = 8 - 4$; $2y^2 - 4(x - 1)^2 = 4$; $\frac{y^2}{2} - (x-1)^2 = 1$. This gives you the hyperbola shown in the graph.

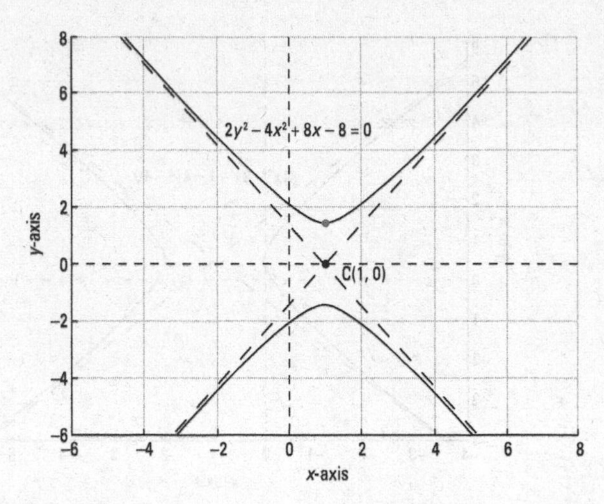

19 Sketch the graph of $4x^2 + 4y^2 - 8y + 16x - 4 = 0$. See the following graph for the answer.

You should recognize that this is a circle because of the x^2 and y^2 with equal coefficients on both. Here are the steps to complete the square: $4x^2 + 4y^2 - 8y + 16x = 4$; $4x^2 + 16x + 4y^2 - 8y = 4$; $4(x^2 + 4x) + 4(y^2 - 2y) = 4$; $4(x^2 + 4x + 4) + 4(y^2 - 2y + 1) = 4 + 16 + 4$; $4(x + 2)^2 + 4(y - 1)^2 = 24$. Because this is a circle, you need to get coefficients of 1 in front of both sets of parentheses by dividing by 4: $(x + 2)^2 + (y - 1)^2 = 6$. This circle has its center at (–2, 1) and its radius is $\sqrt{6}$.

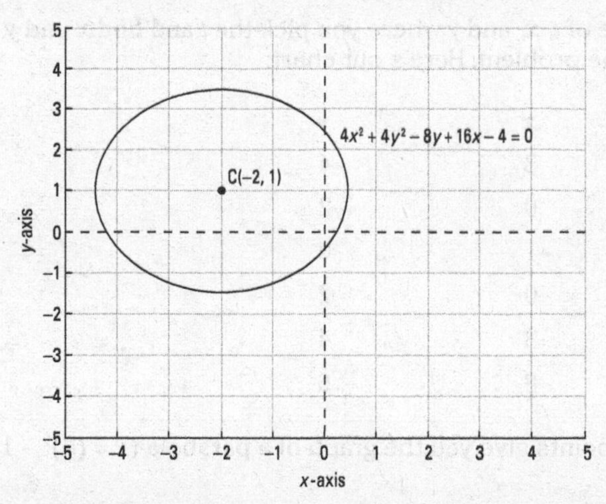

20 Sketch the graph of $3x^2 - 4y^2 + 3x - 2y - \frac{23}{2} = 0$. See the following graph for the answer.

This is another hyperbola because the coefficient on y^2 is negative while the coefficient on x^2 is positive. Here are the usual steps: $3x^2 + 3x - 4y^2 - 2y = \frac{23}{2}$; $3(x^2 + x) - 4(y^2 + (\frac{1}{2})y) = \frac{23}{2}$; $3(x^2 + x + \frac{1}{4}) - 4(y^2 + (\frac{1}{2})y + \frac{1}{16}) = \frac{23}{2} + \frac{3}{4} - \frac{1}{4}$; $3(x + \frac{1}{2})^2 - 4(y + \frac{1}{4})^2 = 12$; $\frac{\left(x+\frac{1}{2}\right)^2}{4} - \frac{\left(y+\frac{1}{4}\right)^2}{3} = 1$. This gives you the hyperbola shown in the graph.

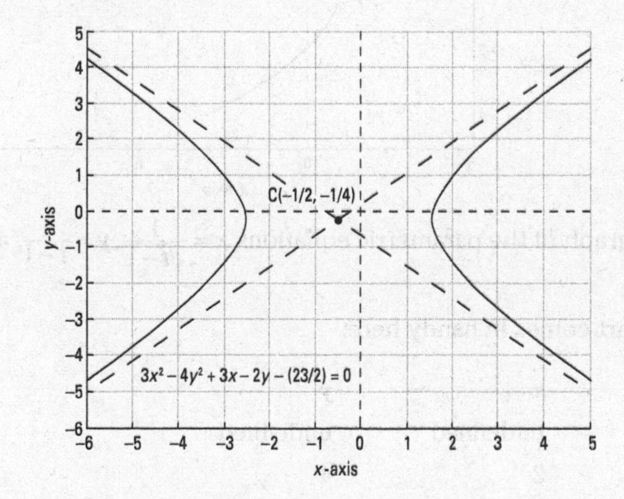

21 Sketch the graph of the parametric equations $x = t^2 - 1$, $y = 2t$, and $-2 \le t \le 3$. See the following graph for the answer.

Set up a table of t, x, and y where you pick the t and find x and y. Be sure to stay within the interval defined by the problem. Here's our chart:

t	x	y
–2	3	–4
–1	0	–2
0	–1	0
1	0	2
2	3	4
3	8	6

These (x, y) points give you the graph of a parabola ($x = (y/2)^2 - 1$).

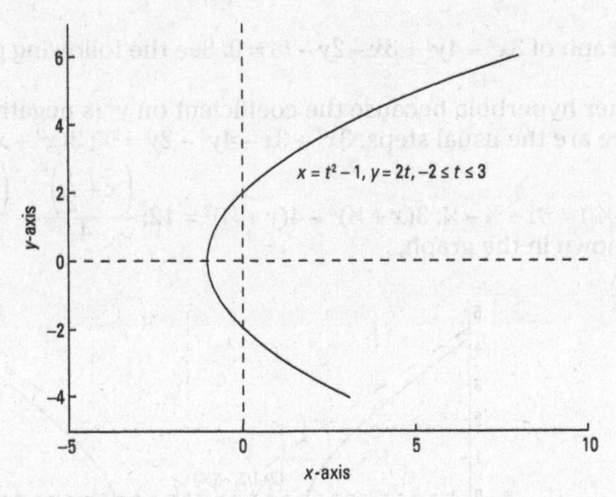

22 Sketch the graph of the parametric equations $x = \frac{1}{\sqrt{t-1}}, y = \frac{1}{t-1}$, and $t > 1$. See the following graph for the answer.

Another chart comes in handy here:

t	x	y
1	undefined	undefined
1.25	2	4
2	1	1
3	$\frac{1}{\sqrt{2}}$	½
4	$\frac{1}{\sqrt{3}}$	⅓
5	½	¼

These points give you the graph ($y = x^2$, $x > 0$), which is shown.

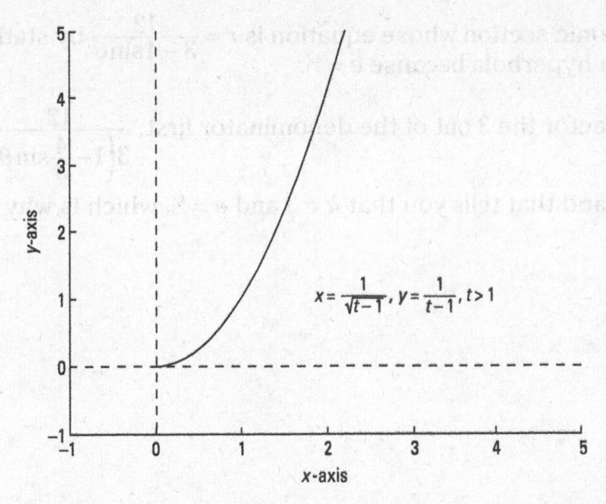

23 Eliminate the parameter and find an equation in x and y whose graph contains the curve of the parametric equations $x = t^2$, $y = 1 - t$, and $t \geq 0$. The answer is $y = 1 - \sqrt{x}$.

Solve for t in the first equation by taking the square root of both sides: $\sqrt{x} = t$. Substitute this value into the other equation: $y = 1 - \sqrt{x}$. Some teachers may not allow you to leave the square root in this equation and instead require you to write it in a form that looks more like a conic section. Subtract 1 from both sides: $y - 1 = -\sqrt{x}$. Then square both sides to get $(y - 1)^2 = x$. That looks more like a parabola now, doesn't it?

24 Eliminate the parameter of the parametric equations $x = t - 5$ and $y = \sqrt{t}$. The answer is $y = \sqrt{x+5}$.

The first equation is easy to solve for t by adding 5 to both sides: $x + 5 = t$. Substitute this value into the other equation and get $y = \sqrt{x+5}$. If you're required by your teacher to write this equation without the square root, square both sides to get $y^2 = x + 5$. This, too, looks like a portion of a parabola.

25 Graph the equation of $r = \frac{8}{1+\cos\theta}$ and label any vertices. See the following graph for the answer.

Notice that the given equation is the same thing as $r = \frac{8 \cdot 1}{1 + 1\cos\theta}$, which tells you that $k = 8$ and $e = 1$. This makes it a parabola.

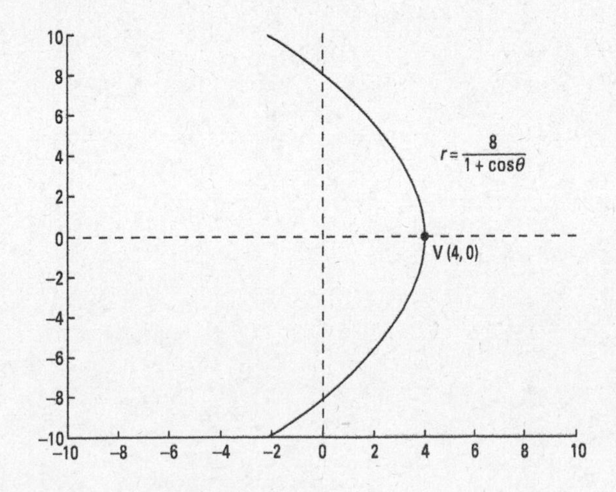

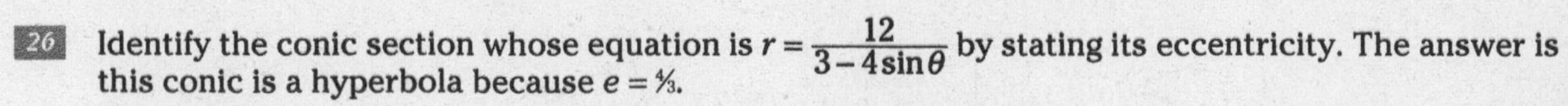

26 Identify the conic section whose equation is $r = \dfrac{12}{3-4\sin\theta}$ by stating its eccentricity. The answer is this conic is a hyperbola because $e = 4/3$.

You have to factor the 3 out of the denominator first: $\dfrac{12}{3\left(1-\frac{4}{3}\sin\theta\right)} = \dfrac{4}{1-\frac{4}{3}\sin\theta}$. This is the same thing as $\dfrac{3\cdot\frac{4}{3}}{1-\frac{4}{3}\sin\theta}$, and that tells you that $k = 3$ and $e = 4/3$, which is why this one is a hyperbola.

Chapter 13

Finding Solutions for Systems of Equations

In This Chapter

- Using the substitution and elimination methods to solve linear equations
- Solving larger systems of equations
- Graphing inequalities
- Discovering decomposing partial fractions
- Mastering matrices to solve systems of equations

No, a system of equations is *not* a way to organize, arrange, or classify equations. A *system of equations* is a collection of equations involving the same set of variables. The point is to find all solutions, if they exist, that work in all the equations. Solving one equation for one variable is almost always possible and usually pretty easy to do.

It goes without saying that the bigger the system of equations becomes, the longer and harder it may be to solve. Solving a system involves several techniques, and sometimes it may be easier to solve certain systems certain ways. That's why math textbooks show all the techniques — so you know when to use each technique.

Of course, you can choose to always solve all systems using one technique, but another technique may require fewer steps, which will save you time, not to mention money (money for aspirin, that is, for all the headaches you'd get if you solved all equations one specific way!).

A Quick-and-Dirty Technique Overview

Here's a handy guide to all the techniques we cover in this chapter and when it's best to use them:

- If a system has two or three variables, you can use *substitution* or *elimination* to solve.
- If a system has four or more variables, you should use *matrices,* in which case you have the following choices:
  - The Gaussian method
  - Inverse matrices
  - Cramer's Rule

We discuss all these techniques in detail in this chapter. And if that's not enough information for you, we also discuss *systems of inequalities* and how to solve them by graphing.

Solving Two Linear Equations with Two Variables

When you're presented with a system of linear equations with two variables, the best methods to solve them are known as *substitution* and *elimination.* As we mention earlier in this chapter, you can use either method to solve any system of this type, but textbooks usually show you both methods because each one has its unique advantages. In keeping with this spirit, we show you both methods, but we also explain *when* to use each one as well.

Just remember that with each system of equations with two variables, you have to find the value of *both* variables, usually x and y. Don't stop until you've solved for both, or you may only get half the credit on your next test for doing half the work.

Also remember that sometimes, systems of equations don't have a solution. Remember when you used to graph two straight lines on one graph to determine the point of intersection? That's actually the least accurate way to find the solution, so mathematicians came up with the other methods that we talk about in this chapter. However, you may recall that sometimes the two lines were parallel to each other — without an intersection! This meant that there was *no solution.* The fact that there may be no solution may pop up from time to time using these other methods as well.

So how do you recognize a system of equations with no solution without using a graph? That's easy — you end up with an equation that just doesn't make sense. It may say $2 = 7$ or $-1 = 10$, but you'll know right away that it has no solution. It's also possible to be given the same line (in disguise) twice. If you were to graph that system, you'd end up with one line on top of another.

These two lines share infinitely many points, so you say that the system has infinitely many solutions. These equations boil down at some point to an *identity* — the left and right sides of the equation are exactly the same (such as $2 = 2$, $10x = 10x$, or $4y - 3 = 4y - 3$), and these, too, are easy to recognize.

The substitution method

In the *substitution method,* you solve one equation for one variable and then substitute this expression for that variable in the other equation. If one of the two equations you're given has already been solved for one variable, huge bells and whistles should go off inside your head. You know you've got a winner for the substitution method. Of course, if one equation can be easily solved for one variable (one variable has a coefficient of 1), you also know that substitution is a good bet.

Q. Solve the system of equations:

$$\begin{cases} x = 4y - 1 \\ 2x + 5y = 11 \end{cases}$$

A. $x = 3, y = 1$. Notice how the first equation says "$x = \ldots$"? This tells you to use substitution. You can take this expression and substitute it for the x in the other equation. This gets you $2(4y - 1) + 5y = 11$. The substitution method makes your job easier because you end up with one equation in one variable — and this one is easy to solve! When you do, you get $y = 1$. Now that you know half of your answer (y in this case), you can substitute that value into the original equation to get the other half (x).

Save yourself some time and steps by substituting the first answer you get into the equation that has already been solved for a variable. For this example, because you know that $x = 4y - 1$ and you figure out that $y = 1$, it takes very few steps to figure out that $x = 3$.

1. Use substitution to solve the system

$$\begin{cases} r + s = 6 \\ s = 13 - 2r \end{cases}$$

Solve It

2. The sum of two numbers is 14 and their difference is 2. Find the numbers.

Solve It

The elimination method

Elimination is the method of choice when both of the linear equations given to you are written in *standard form:*

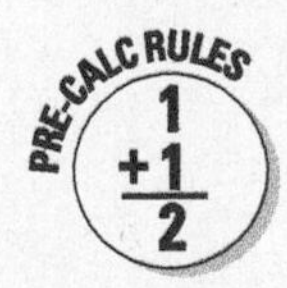

$$\begin{aligned} Ax + By = C \\ Dx + Ey = F \end{aligned}$$

where A, B, C, D, E, and F are all real numbers. It's called *standard form* because it's supposed to be *the* standard way that textbooks depict linear equations. But the truth is that textbook authors like to keep you on your toes so they write linear equations in all kinds of forms.

In the two equations, if the x and y terms are opposites of each other, you should choose elimination. In the *elimination method,* you add the two equations together so that one of the variables disappears (is eliminated). Sometimes, however, you must multiply one or both equations by a constant in order for the terms to have opposite signs. This way, when you add the two equations together, one of the variables will be eliminated.

Q. Solve the system $\begin{cases} 2x - y = 6 \\ 3x + y = 4 \end{cases}$

A. $x = 2$, $y = -2$. Notice that you *could* solve this system using substitution, because in the second equation, you can solve for y by subtracting $3x$ from both sides. But notice that the y terms are exact opposites of each other. If you add the two equations, you get $5x = 10$. This means that you can divide both sides by 5 to easily solve for x and get $x = 2$. Substituting this back into either equation, you find that $y = -2$.

Q. Solve the system $\begin{cases} 2x - 3y = 5 \\ 4x + 5y = -1 \end{cases}$

A. $x = 1$, $y = -1$. To solve this system by substitution is possible, but it would eventually mean dividing one of the coefficients and creating those ugly fractions we all hate. Instead, you can avoid the fractions by using the elimination method. The fact that both equations are written in standard form is another vote in favor of the elimination method. Notice that the y terms have opposite signs, so you can eliminate them (you can eliminate any variable you choose, but it's all about taking the fewest steps). It's a little like finding the least common multiple of both coefficients, in this case the 3 and the 5. The smallest number that both of those go into is 15, so you have to multiply the top equation by 5 and the bottom equation by 3. This gives you $\begin{cases} 10x - 15y = 25 \\ 12x + 15y = -3 \end{cases}$. Adding these two equations together gives you $22x = 22$, which gives you the solution $x = 1$. You then have to substitute this value back into one of the two original equations to solve for y. In this example, $y = -1$.

3. Solve the system $\begin{cases} \frac{x}{2} - \frac{y}{3} = -3 \\ \frac{2x}{3} + \frac{y}{2} = -2 \end{cases}$

Solve It

4. Solve the system $\begin{cases} 3x - 2y = 4 \\ 6x - 4y = 8 \end{cases}$

Solve It

Not-So-Straight: Solving Nonlinear Systems

The substitution and elimination methods are common tools for systems of equations that include *nonlinear equations.* Yes, now at least one of your two given equations is often a quadratic equation (it may also be a rational function or some other type). The method you choose to use for these types of systems depends on the types of equations that you're given. We break the following sections into those types and show you how to best solve each one.

One equation that's linear and one that isn't

When one equation is linear and the other equation isn't linear, it's best to use the substitution method. That's because the linear equation can be easily solved for one variable. You can then substitute this value into the other equation to solve. Most often, that means solving a quadratic equation at some point. If you need to brush up on those techniques, see Chapter 4.

Q. Solve the system of equations

$$\begin{cases} x^2 + y = 0 \\ 2x - y = 3 \end{cases}$$

A. $x = -3, y = -9; x = 1, y = -1$. As we mention earlier, it's usually easier to solve the linear equation first. The second given equation is the linear one, and it's easier to solve for y (no pesky coefficients to divide). Doing so gets you $y = 2x - 3$.

After you substitute this expression into the first equation for y, you get $x^2 + 2x - 3 = 0$. This quadratic polynomial factors to $(x + 3)(x - 1) = 0$. Then, using the zero product property (for more information, see Chapter 4), you get two solutions: $x = -3$ or $x = 1$. Two solutions for x means twice the substitution and twice the y answers. If $x = -3$, then $y = -9$, and if $x = 1$, $y = -1$.

5. Solve $\begin{cases} x^2 - y = 1 \\ x + y = 5 \end{cases}$

Solve It

6. Solve $\begin{cases} x + y = 9 \\ xy = 20 \end{cases}$

Solve It

Two nonlinear equations

In this section, we explore what happens when both of the given equations are nonlinear. In the context of this workbook, these steps will most likely create one final quadratic equation to solve.

Q. Solve the system $\begin{cases} x^2 + y^2 = 25 \\ x^2 - y = 5 \end{cases}$

A. $(x, y) = (0, -5), (\pm 3, 4)$. Notice right away that the x^2 terms in both equations have the same coefficient. If you multiply the second equation by –1 and then add the two equations together, you get $y^2 + y = 20$. Subtracting the 20 from both sides gets you the quadratic equation, $y^2 + y - 20 = 0$, that you have to factor: $(y + 5)(y - 4) = 0$. Solve and get $y = -5$ and $y = 4$. Substituting $y = -5$ into the second equation gets you $x^2 + 5 = 5$ or $x^2 = 0$, which means that $x = 0$. Substituting $y = 4$ into the same equation gets you $x^2 - 4 = 5$ or $x^2 = 9$, which gives you $x = \pm 3$. Both of these solutions work.

7. Solve $\begin{cases} x^2 + y^2 = 1 \\ x + y^2 = -5 \end{cases}$

Solve It

8. Solve $\begin{cases} 27x^2 - 16y^2 = -400 \\ -9x^2 + 4y^2 = 36 \end{cases}$

Solve It

Systems of equations disguised as rational equations

Sometimes you'll see two equations that look like they're rational equations. As we discuss in Chapter 3, sometimes rational functions have undefined values. Keep this in mind with the final solutions you find to the equation — they may not really work! *Always* check the solutions to these types of equations because you never know which ones are actually solutions and which ones aren't until you double-check. We recommend that you always start off by substituting a new variable for the rational expressions into the other equation so you can deal with a more normal-looking system (as if there were such a thing).

Q. Solve the system $\begin{cases} \dfrac{3}{x}+\dfrac{2}{y}=11 \\ \dfrac{1}{x}-\dfrac{1}{y}=-11 \end{cases}$

A. $x = -\frac{5}{11}$, $y = \frac{5}{44}$. Okay, so you see all those fractions right off the bat and you decide to throw in the towel and walk away, right? Wrong! Make the fractions go away by starting off with a substitution. Notice, first of all, that you can rewrite the given system as $\begin{cases} 3\dfrac{1}{x}+2\dfrac{1}{y}=11 \\ \dfrac{1}{x}-\dfrac{1}{y}=-11 \end{cases}$

By letting $u = \frac{1}{x}$ and $v = \frac{1}{y}$, you can conveniently rewrite the entire system as $\begin{cases} 3u+2v=11 \\ u-v=-11 \end{cases}$. You can then use any of the methods you're already familiar with to solve the system. For this example, if you multiply the second equation by 2, you get $2u - 2v = -22$. You can then add the two equations to eliminate v and get $5u = -11$. This means that $u = -\frac{11}{5}$. Then you can work your way backwards to get $x = -\frac{5}{11}$. We then suggest substituting your u value to find your v, which gets you on your way to finding that $y = \frac{5}{44}$.

9. Solve $\begin{cases} \dfrac{14}{x+3}+\dfrac{7}{4-y}=9 \\ \dfrac{21}{x+3}-\dfrac{3}{4-y}=0 \end{cases}$

Solve It

10. Solve $\begin{cases} \dfrac{12}{x+1}-\dfrac{12}{y-1}=8 \\ \dfrac{6}{x+1}+\dfrac{6}{y-1}=-2 \end{cases}$

Solve It

Solving More than Two (Equations and/or Variables)

Now that you have a grip on solving a system of equations in two variables, we'll change gears to systems with three variables. Why change to three variables, though? Well, for one reason, we live in a three-dimensional world, so we need three variables to represent it.

Most of the time when you're given systems larger than 2×2, you want to use elimination. You have to take two equations at a time and eliminate one variable. Then, you have to take another two equations and eliminate the same variable. If you start with a 3×3 system, this knocks you down to a 2×2 system, which you then must solve. If you start with a 4×4 system, you work down to a 3×3 system, which you must then work down to 2×2 and solve. Sounds fun, doesn't it?

Q. Solve the system of equations
$$\begin{cases} x-y+z=6 \\ x-z=-2 \\ x+y=-1 \end{cases}$$

A. $x=1$, $y=-2$, $z=3$. To make things generally easier in this section, we always label the given equations from top to bottom with capital letters. For this example, the top equation, $x-y+z=6$, is equation A, the middle equation is equation B, and the bottom equation is equation C. Notice that equation C is already missing the z variable. Also notice that equations A and B have z variables that are exact opposites of each other.

If you add those two together (A + B), you get a brand new equation (we call it D) which is $2x-y=4$. If you then take equation D and add C to it (D + C), you get one more equation, E, which is $3x=3$. This equation is solved easily to get $x=1$. You can then substitute this value into equation C and get $1+y=-1$, which means that $y=-2$. You can also use the fact that $x=1$ in equation B to get $1-z=-2$, or $-z=-3$, or — finally — $z=3$.

Q. Solve the system
$$\begin{cases} x+y+z+w=-1 \\ 2x+y+z=0 \\ 2y+z-w=-6 \\ x-z+2w=7 \end{cases}$$

A. $x=2$, $y=-1$, $z=-3$, and $w=1$. Most students panic with systems that are bigger than 3×3, so we decided to show one example of a 4×4 so you can see how it works. A: $x+y+z+w=-1$; B: $2x+y+z=0$; C: $2y+z-w=-6$; D: $x-z+2w=7$. Notice that B is missing its w variable, so if you can use the others to eliminate w as well, that'll be the first success.

A + C gives you a new equation, E: $x+3y+2z=-7$. −2A + D gives a new equation, F: $-x-2y-3z=9$. So now you have three equations with three variables: B, E, and F. Next, you have to pick another variable to eliminate — it doesn't matter which one; we eliminate x. E + F gives you equation G: $y-z=2$.

Meanwhile, B + 2F also eliminates x with equation H: $-3y-5z=18$. Finally, 3G + H eliminates y with one last equation, J: $-8z=24$; $z=-3$.

Using this solution in equation H gives you $-3y-5(-3)=18$; $-3y+15=18$; $-3y=3$; $y=-1$. Now that you know z and y, use them in equation E: $x+3(-1)+2(-3)=-7$; $x-3-6=-7$; $x=2$. Then use all those answers in equation A: $2-1-3+w=-1$; $w=1$.

11. Solve the system
$$\begin{cases} 3x-2y=17 \\ x-2z=1 \\ 3y+2z=1 \end{cases}$$

Solve It

12. Solve
$$\begin{cases} 2x-y+z=1 \\ x+y-z=2 \\ -x-y+z=2 \end{cases}$$

Solve It

13. Solve $\begin{cases} 2x+3y+4z=37 \\ 4x-3y+2z=17 \\ x+2y-3z=-5 \\ 3x-2y+z=11 \end{cases}$

Solve It

14. Solve $\begin{cases} 3a+b+c+d=0 \\ 4a+5b+2c=15 \\ 4a+2b+5d=-10 \\ -5a+3b-d=8 \end{cases}$

Solve It

Graphing Systems of Inequalities

A *system of inequalities* is more than one inequality in more than one variable. Up until now in your math career, you've probably seen systems of inequalities that have all been linear. In pre-calc, you continue with those types of problems but then move up to nonlinear systems of inequalities. That's right, you'll be seeing quadratics, too. For a review of how to graph one inequality, see Chapter 1. If you're quacky on quadratics, see Chapter 3. And if conics sound crazy, see Chapter 12. If, on the other hand, you're raring to go, read on.

One way to solve a system of inequalities is to graph it. You end up with (hopefully) two overlapping shaded regions — the overlap is the solution. Every single point in the overlap is a solution to the system. What happens if there's no overlap? Well, there's no solution!

When you multiply or divide an inequality by a negative number, the inequality sign changes: $<$ becomes $>$, $\leq$ becomes $\geq$, and vice versa. This is a pretty important fact to remember because it affects your shading in the end.

Q. Sketch the graph of the system of inequalities $\begin{cases} 3x+y\leq 5 \\ x+2y\leq 4 \end{cases}$

A. See the following figure. Because both of these inequalities are linear, you have to put them in slope-intercept form to graph them. The top inequality in slope-intercept form is $y\leq -3x+5$, and the bottom inequality becomes $y\leq -\frac{1}{2}x+2$. Graphing both on the same coordinate plane gives you the figure.

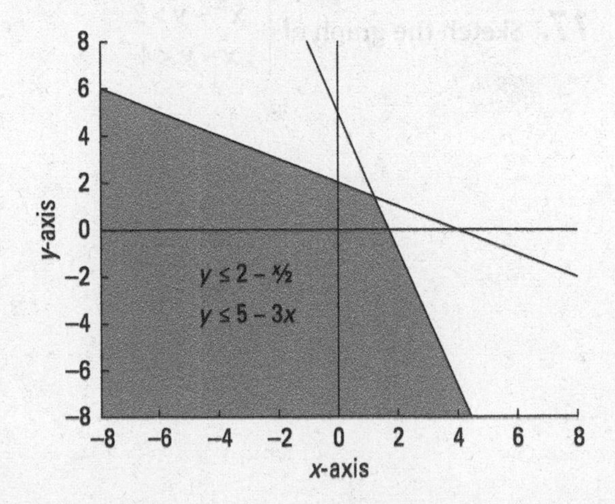

Q. Sketch the graph of the system of inequalities $\begin{cases} x^2+y^2<16 \\ x-2y>-4 \end{cases}$

A. See the following figure. This time, the top inequality represents the interior of a circle, and the bottom inequality represents a half-plane (bounded by a line). The circle is in the proper form to graph, so you don't have to do any work there (other than to graph, that is), while the bottom inequality in slope-intercept form is $y<\frac{1}{2}x+2$. For these types of problems, we recommend that you pick test points to see where to shade. For example, the origin (0, 0) is a great point to try in the original inequalities to see whether it works. Is $0^2+0^2<16$? Yes, so you shade inside the circle. In the second inequality, is $0-2(0)>-4$? Yes, so you also shade below the line.

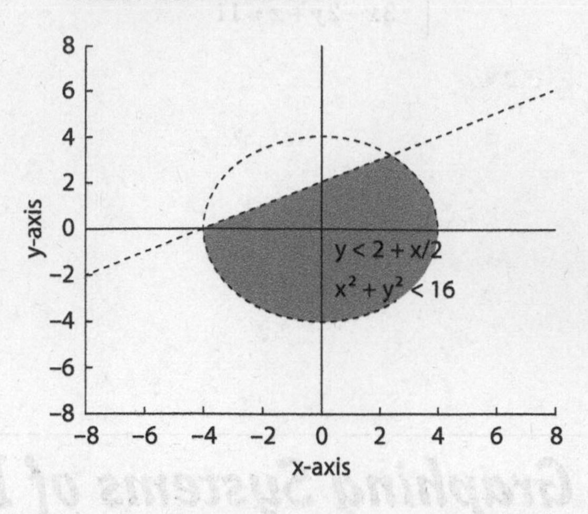

15. Sketch the graph of $\begin{cases} 2x+y\geq 9 \\ 2x-y\geq 1 \\ x\leq 7 \end{cases}$

Solve It

16. Sketch the graph of $\begin{cases} x^2+y^2\geq 9 \\ x^2+(y-3)^2\geq 9 \end{cases}$

Solve It

17. Sketch the graph of $\begin{cases} x^2-y>2 \\ x-y<4 \end{cases}$

Solve It

18. Sketch the graph of $\begin{cases} y\geq 0 \\ x+y<4 \\ y\leq\sqrt{x-1} \end{cases}$

Solve It

Breaking Down Decomposing Partial Fractions

Call us crazy, but we're partial to partial fractions! (We know, but we really couldn't resist.) The process known as decomposing into partial fractions takes one fraction and breaks it down as the sum or difference of two (or more) other fractions. The method is often called *partial fractions.* This requires being an expert at factoring, so if you need a review, turn to Chapter 4 and read up on how to do it. If you're a pro by now, you know to always follow these general steps if the degree of the numerator is less than that of the denominator (otherwise, do long division first):

1. **Factor the denominator.**
2. **Write separate fractions, one for each factor of the denominator based on these rules:**
   - If the factor is linear, it has some constant in the numerator.
   - If the factor is quadratic and irreducible (doesn't have real roots), it has a linear expression in the numerator.

Note: If any factor has a power on it, you have to create one fraction for each power, from 1 on up to the highest degree. This is probably best shown with an example. Suppose that you're able to factor the polynomial in a particularly long denominator into $(x-4)(x+1)^3(2x-1)(3x^2-4x+2)^2$. You would have to create the sum of seven different fractions: $\frac{A}{x-4}+\frac{B}{x+1}+\frac{C}{(x+1)^2}+\frac{D}{(x+1)^3}+\frac{E}{2x-1}+\frac{Fx+G}{3x^2-4x+2}+\frac{Hx+J}{(3x^2-4x+2)^2}$

The first denominator factor is linear, so its numerator is a constant. The second denominator factor is linear with multiplicity 3, so you need to create three different constant numerators: one for the first degree, the second for the second degree, and the third for the third degree. The third factor is also linear, so it gets one fraction with a constant on the top. Lastly, the final term is quadratic with multiplicity 2, so it gets two linear terms in the numerator: one for the first degree and the second for the second degree.

Q. Find the partial fraction decomposition of $\frac{7x+5}{x^2+x-2}$; then solve.

A. $\frac{4}{x-1}+\frac{3}{x+2}$. Begin by factoring the denominator of the given fraction into $\frac{7x+5}{x^2+x-2}=\frac{7x+5}{(x+2)(x-1)}$. Because both factors in the denominator are linear, you break them into two different fractions with constant numerators: $\frac{7x+5}{(x+2)(x-1)}=\frac{A}{x-1}+\frac{B}{x+2}$. Now, multiply every fraction by the factored denominator, which results in some big-time canceling: $\frac{7x+5}{\cancel{(x+2)(x-1)}}\cancel{(x+2)(x-1)}=\frac{A}{\cancel{x-1}}(x+2)\cancel{(x-1)}+\frac{B}{\cancel{x+2}}\cancel{(x+2)}(x-1)$. You now have the simplified equation $7x+5=A(x+2)+B(x-1)$. When you multiply this out, you get the equation $7x+5=Ax+2A+Bx-B$. Now, gather like terms: $7x+5=Ax+Bx+2A-B$. Factor the x out on the right side: $7x+5=(A+B)x+2A-B$. Notice how both sides match up, which means that $7x=(A+B)x$, or $7=A+B$ *and* $5=2A-B$. This gives you a system of equations to solve: $\begin{cases}7=A+B\\5=2A-B\end{cases}$. (Now you know why the textbooks usually include this material in the chapter with systems of equations (as well as why we include it here — aren't we clever?). Add these two equations to eliminate B and get $12=3A$, or $4=A$. Substituting this into the top equation gets you $7=4+B$, or $3=B$. You can now use these values to write the sum of two fractions: $\frac{4}{x-1}+\frac{3}{x+2}$

19. Find the constants A and B such that $\frac{x-38}{x^2+x-12}=\frac{A}{x+4}+\frac{B}{x-3}$

Solve It

20. Find the form of the partial fraction decomposition for $\frac{5x-4}{(x-1)^2}$, but don't find the constants.

Solve It

21. Find the partial fraction decomposition for $\frac{2x^2-21x+18}{(x-1)(x^2-4x+4)}$

Solve It

22. Find the partial fraction decomposition for $\frac{11x^2-7x+14}{2x^3-4x^2+3x-6}$

Solve It

Working with a Matrix

A *matrix* is a collection of numbers arranged in rows and columns. Each number inside the matrix is called an *entry* or *element.* A matrix comes in handy when you have a bunch of data that you need to keep track of. Usually, a matrix is named by its *dimensions,* or how big it is. This is also sometimes known as the *order* of the matrix and is always the number of rows by the number of columns. For example, if matrix M is 4×3, it has four rows and three columns. After the data is organized in this fashion, you can add, subtract, and even multiply matrices. There's also an operation known as *scalar multiplication,* which means you multiply the entire matrix by a constant.

To add or subtract matrices, you have to operate on their corresponding elements. In other words, you add or subtract the first row/first column in one matrix to or from the exact same element in another matrix. The two matrices must have the same dimensions; otherwise, an element in one matrix won't have a corresponding element in the other.

The following example shows two matrices and what their sum and differences are:

$$A=\begin{bmatrix}-5 & 1 & -3\\ 6 & 0 & 2\\ 2 & 6 & 1\end{bmatrix} \qquad A+B=\begin{bmatrix}-3 & 5 & 2\\ -2 & 10 & 5\\ 0 & 3 & -8\end{bmatrix}$$

$$B=\begin{bmatrix}2 & 4 & 5\\ -8 & 10 & 3\\ -2 & -3 & -9\end{bmatrix} \qquad A-B=\begin{bmatrix}-7 & -3 & -8\\ 14 & -10 & -1\\ 4 & 9 & 10\end{bmatrix}$$

Next, we show the scalar multiplication 3A:

$$3A=3\begin{bmatrix}-5 & 1 & -3\\ 6 & 0 & 2\\ 2 & 6 & 1\end{bmatrix}=\begin{bmatrix}-15 & 3 & -9\\ 18 & 0 & 6\\ 6 & 18 & 3\end{bmatrix}$$

Multiplying matrices is another can of worms. First of all, to multiply two matrices AB (the matrices are written right next to each other, with no symbol in between), the number of columns in matrix A *must* match the number of rows in matrix B. If matrix A is $m \times n$ and matrix B is $n \times p$, the product AB has dimensions $m \times p$. And remember, when it comes to matrix multiplication, AB doesn't equal BA; in fact, just because AB exists doesn't even mean that BA does as well.

For all problems in this section,

$$M=\begin{bmatrix}-5 & -1 & 3 & 6\\ 0 & 2 & -2 & 6\end{bmatrix} \qquad N=\begin{bmatrix}2 & 4 & 5 & -8\\ 10 & 3 & -2 & -3\end{bmatrix} \qquad P=\begin{bmatrix}-1 & 2 & -1\\ 4 & 4 & 0\\ 2 & 3 & 1\\ -5 & 2 & -1\end{bmatrix}$$

Q. Find 3M – 2N.

A. $\begin{bmatrix}-19 & -11 & -1 & 34\\ -20 & 0 & -2 & 24\end{bmatrix}$

First, substitute M and N in the expression of 3M – 2N:

$3\begin{bmatrix}-5 & -1 & 3 & 6\\ 0 & 2 & -2 & 6\end{bmatrix}-2\begin{bmatrix}2 & 4 & 5 & -8\\ 10 & 3 & -2 & -3\end{bmatrix}$. Follow the order of operations and multiply in the scalars: $\begin{bmatrix}-15 & -3 & 9 & 18\\ 0 & 6 & -6 & 18\end{bmatrix}-\begin{bmatrix}4 & 8 & 10 & -16\\ 20 & 6 & -4 & -6\end{bmatrix}$. Then subtract the two matrices, watching the negative signs: $\begin{bmatrix}-19 & -11 & -1 & 34\\ -20 & 0 & -2 & 24\end{bmatrix}$.

Q. Find MP.

A. $\begin{bmatrix} -23 & 7 & 2 \\ -26 & 14 & -8 \end{bmatrix}$

You need to multiply each element of each row of the left matrix by each element of each column of the right matrix.

The sum of the first row times the first column: $-5(-1) - 1(4) + 3(2) + 6(-5) = -23$. This is the first row, first column answer.

The sum of the first row times the second column: $-5(2) - 1(4) + 3(3) + 6(2) = 7$. This is the first row, second column answer.

The sum of the first row times the third column: $-5(-1) - 1(0) + 3(1) + 6(-1) = 2$. This is the first row, third column answer.

The sum of the second row times the first column: $0(-1) + 2(4) - 2(2) + 6(-5) = -26$. This is the second row, first column answer.

The sum of the second row times the second column: $0(2) + 2(4) - 2(3) + 6(2) = 14$. This is the second row, second column answer.

The sum of the second row times the third column: $0(-1) + 2(0) - 2(1) + 6(-1) = -8$. This is the second row, third column answer.

Putting these all into a matrix gives you the answer $\begin{bmatrix} -23 & 7 & 2 \\ -26 & 14 & -8 \end{bmatrix}$

23. Find 4N.

Solve It

24. Find 4N + 5M.

Solve It

25. Find 3M – P.

Solve It

26. Find NP.

Solve It

Getting It in the Right Form: Simplifying Matrices

You can write any linear system of equations in *matrix form.* To do so, follow these steps:

1. **Write all the coefficients in one matrix, called the *coefficient matrix.*** Each equation gets its own row in the matrix, and each variable gets its own column, written in the same order as the equations.
2. **Multiply this times another column matrix with all the variables in it, called the *variable matrix,* in order from top to bottom.**
3. **Set this product equal to a column matrix with the answers in it, sometimes called the *answer matrix.***

Row echelon form is similar to matrix form; however, you only work with the coefficient matrix. Across any row, the first number element (besides 0) that you run into is called the *leading coefficient.* For a coefficient matrix to be in row echelon form,

- Any row with all 0s in it must be the bottom row.
- The leading coefficient in any row must be to the right of the leading coefficient in the row above it.

Reduced row echelon form takes row echelon form and makes all the leading coefficients the number 1. Also, each element above or below a leading coefficient must be 0. The following matrix A is in reduced row echelon form, while the following matrix B is in row echelon but not in *reduced* row echelon form.

$$A=\begin{bmatrix} 0 & 1 & 0 & 0 \\ 0 & 0 & 1 & 0 \\ 0 & 0 & 0 & 1 \end{bmatrix} \qquad B=\begin{bmatrix} 0 & 1 & 2 & 7 \\ 0 & 0 & 1 & 0 \\ 0 & 0 & 0 & 1 \end{bmatrix}$$

Finally, *augmented form* takes the coefficient matrix and tacks on an extra column — a column with the answers in it so that you can look at the entire system in one convenient package.

These ways of writing systems of equations in matrices come in handy when dealing with systems that are 4×4 or larger. Your goal is to get the matrix into row echelon form using *elementary row operations.* These operations are different from the operations we discuss in the preceding section because they're done on only one row at a time. Here are three row operations you can perform:

- Multiply each element of a row by a constant.
- Interchange any two rows.
- Add two rows together.

We stay consistent with *Pre-Calculus For Dummies* and use the same notation that we use there to represent these elementary row operations. This means that $4r_2 \rightarrow r_2$ multiplies the second row by 4 to change the second row. $r_1 \leftrightarrow r_3$ swaps row one with row three. $r_3 + r_1 \rightarrow r_1$ adds row three to row one and changes row one. $4r_2 + r_1 \rightarrow r_1$ first multiplies row two by 4 and then adds that to row one to change row one.

You can use any combination of these row operations to get the given matrix into row echelon form. Use reduced row echelon form only if you're specifically told to do so by your teacher or textbook, because it takes more steps.

We only focus on the forms and the row operations in this section. To really dig in deep and discover how to get a matrix in row echelon form, read on to the next section.

Q. Write the system of equations $\begin{cases} 3x - y = 6 \\ 2x + 3y = 3 \end{cases}$ as a matrix system.

A. $\begin{bmatrix} 3 & -1 \\ 2 & 3 \end{bmatrix}\begin{bmatrix} x \\ y \end{bmatrix} = \begin{bmatrix} 6 \\ 3 \end{bmatrix}$. The matrix on the left is the coefficient matrix, containing all the coefficients from the system. The second matrix is the variable matrix, and the third one, on the right, is the answer matrix. This completes the job for this question.

Q. Write the system from the preceding question as an augmented matrix.

A. $\left[\begin{array}{cc|c} 3 & -1 & 6 \\ 2 & 3 & 3 \end{array}\right]$. Just take the coefficient matrix and add on the answer matrix. *Voilà!* You have an augmented matrix. Notice that the vertical line separates the two and lets you know that this is an augmented matrix and not a normal 2×3 matrix.

27. Using the augmented matrix from the last example, use elementary row operations to find $-3r_2 \rightarrow r_2$.

Solve It

28. Now, using your answer from Problem 27, find $r_1 \leftrightarrow r_2$.

Solve It

29. Now, keep going and find $r_1 + r_2 \rightarrow r_2$.

Solve It

30. Lastly, find $3r_2 + r_1 \rightarrow r_1$.

Solve It

Solving Systems of Equations Using Matrices

You can solve a system of equations using matrices in three ways. Putting a matrix in row echelon form using the techniques described in the last section is called *Gaussian elimination.* The second way uses a method called *inverse matrices,* and the third method is called *Cramer's Rule.* Your book may cover all these techniques, it may cover only one, or it may not even cover this material at all. But we're here for you if you need us! We dedicate one section to each of these ways, to keep it nice and simple.

Gaussian elimination

The process of putting a matrix in row echelon form is called *Gaussian elimination.* We focus in this section on matrices in augmented form because that's most commonly what you'll be asked to do, but know that the rules don't change if you're asked to do this with some other form of matrix. The goals of using the elementary row operations are simple: Get a 1 in the upper-left corner of the matrix, get 0s in all positions underneath this 1, get 1s for all leading coefficients diagonally from the upper-left to the lower-right corners, and then get 0s below each of them. When you get to that point, you use a process called *back substitution* to solve for all the variables in the system.

Q. Put the system of equations $\begin{cases} 3x - y = 6 \\ 2x + 3y = 3 \end{cases}$ in augmented form; then write the matrix in row echelon form and solve the system.

A. $\left[\begin{array}{cc|c} 1 & -1/3 & 2 \\ 0 & 11/3 & -1 \end{array}\right]$, $x = {}^{21}\!/\!_{11}$, $y = -\frac{3}{11}$.

In the last section, you wrote this system as an augmented matrix: $\left[\begin{array}{cc|c} 3 & -1 & 6 \\ 2 & 3 & 3 \end{array}\right]$. Now you need to get it into row echelon form. First, you need to get a 0 below the element in the upper-left corner. This is easiest if you get a 1 in the upper-left corner first. In fact, some textbooks may say that row echelon form has leading coefficients of 1 for this reason. Follow the elementary row operation $\frac{1}{3}r_1 \leftrightarrow r_1$ to get $\left[\begin{array}{cc|c} 1 & -1/3 & 2 \\ 2 & 3 & 3 \end{array}\right]$. Now, to make the first element of row two 0, you need to add –2. So take $-2r_1 + r_2 \rightarrow r_2$ to get $\left[\begin{array}{cc|c} 1 & -1/3 & 2 \\ 0 & 11/3 & -1 \end{array}\right]$. This gives you an equation in the second row that's easy to solve for y: $\frac{11}{3}y = -1$, or $y = -\frac{3}{11}$. Now you can work backwards using back substitution in the top equation: $x - \frac{1}{3}y = 2$. You know the value of y, so substitute that in and get $x - \frac{1}{3}\left(-\frac{3}{11}\right) = 2$, or $x = {}^{21}\!/\!_{11}$.

31. Solve the system of equations $\begin{cases} 2x + 5y = 7 \\ 3x - 5y = 2 \end{cases}$ by writing it in augmented form and then putting the matrix in row echelon form.

Solve It

32. Use Gaussian elimination to solve

$$\begin{cases} 3x - 2y + 6z = 7 \\ x - 2y - z = -2 \\ -3x + 10y + 11z = 18 \end{cases}$$

Solve It

Inverse matrices

Another way to solve a system is by using an *inverse matrix.* This process is based on the idea that if you write a system in matrix form, you'll have the coefficient matrix multiplying the variable matrix on the left side — if only you could divide a matrix, you'd have it made in the shade! Well, if you look at the simple equation $3x = 12$, you can solve it by dividing both sides by 3, $\frac{3x}{3} = \frac{12}{3}$, to get $x = 4$. This is the same thing as multiplying by its multiplicative inverse on both sides, which turns out to be the same in matrices! You have to use an inverse matrix. Remember from Chapter 3 that if $f(x)$ is a function, its inverse is denoted by $f^{-1}(x)$. This is true for matrices as well: If A is the matrix, A^{-1} is its inverse. If you have three matrices (A, B, and C) and you know that AB = C, then you can solve for B by multiplying the inverse matrix A^{-1} on both sides:

$A^{-1}[AB] = A^{-1}C$, which simplifies to $B = A^{-1}C$

Finding a matrix's inverse

But how do you find a matrix's inverse? Realize first that only square matrices have inverses. The number of rows must be equal to the number of columns. Even then, not every square matrix has an inverse. If the determinant (which we talk more about in the next section) of a matrix is 0, it doesn't have an inverse. The definition of a determinant involves a lot of math mumbo jumbo which, in our humble opinion, won't help you much with finding the answer. We'd rather cut to the chase and simply show you how to find the determinant of a 2 × 2 matrix:

$$\begin{vmatrix} a & b \\ c & d \end{vmatrix} = ad - bc$$

where $\begin{vmatrix} a & b \\ c & d \end{vmatrix}$ denotes the determinant of the matrix $\begin{bmatrix} a & b \\ c & d \end{bmatrix}$. When a matrix does have an inverse, you can use several ways to find it depending on how big the matrix is. If it's a 2 × 2, you can find it by hand using a simple formula. If it's 3 × 3 or bigger, you *can* find it by hand, but it will be much more involved.

In the meantime, if matrix A is the 2 × 2 matrix $\begin{bmatrix} a & b \\ c & d \end{bmatrix}$, its inverse, $\begin{bmatrix} a & b \\ c & d \end{bmatrix}^{-1}$, is found using $\begin{bmatrix} a & b \\ c & d \end{bmatrix}^{-1} = \frac{1}{ad - bc}\begin{bmatrix} d & -b \\ -c & a \end{bmatrix}$, provided that $ad - bc \neq 0$.

Using an inverse matrix to solve a system

Now that you can find the inverse matrix, all you have to do to solve is follow these steps:

1. **Write the system as a matrix equation.**
2. **Create the inverse matrix.**
3. **Multiply this inverse in front of both sides of the equation.**
4. **Cancel on the left side; multiply the matrices on the right.**
5. **Multiply the scalar.**

Q. Set up the matrix equation for the system $\begin{cases} 3x - 2y = -1 \\ x + y = 3 \end{cases}$ by using inverse matrices.

A. $x = 1$, $y = 2$. First, set up the matrix equation: $\begin{bmatrix} 3 & -2 \\ 1 & 1 \end{bmatrix}\begin{bmatrix} x \\ y \end{bmatrix} = \begin{bmatrix} -1 \\ 3 \end{bmatrix}$. Now, find the inverse matrix using the formula from earlier in this chapter: $\frac{1}{(3)(1)-(-2)(1)}\begin{bmatrix} 1 & 2 \\ -1 & 3 \end{bmatrix} = \frac{1}{5}\begin{bmatrix} 1 & 2 \\ -1 & 3 \end{bmatrix}$.

Now, multiply this inverse on the left of both sides of the equation:

$\frac{1}{5}\begin{bmatrix} 1 & 2 \\ -1 & 3 \end{bmatrix}\begin{bmatrix} 3 & -2 \\ 1 & 1 \end{bmatrix}\begin{bmatrix} x \\ y \end{bmatrix} = \frac{1}{5}\begin{bmatrix} 1 & 2 \\ -1 & 3 \end{bmatrix}\begin{bmatrix} -1 \\ 3 \end{bmatrix}$. **To multiply a matrix by its inverse cancels everything on the left except for the variable matrix:**

$\frac{1}{5}\begin{bmatrix} 1 & 2 \\ -1 & 3 \end{bmatrix}\begin{bmatrix} 3 & -2 \\ 1 & 1 \end{bmatrix}\begin{bmatrix} x \\ y \end{bmatrix} = \frac{1}{5}\begin{bmatrix} 1 & 2 \\ -1 & 3 \end{bmatrix}\begin{bmatrix} -1 \\ 3 \end{bmatrix}$. **That means all you have to do is multiply the matrices on the right and then multiply the scalar:**

$\begin{bmatrix} x \\ y \end{bmatrix} = \frac{1}{5}\begin{bmatrix} 5 \\ 10 \end{bmatrix} = \begin{bmatrix} 1 \\ 2 \end{bmatrix}$. **This gives the solutions from top to bottom as $x = 1$ and $y = 2$.**

33. Solve the system $\begin{cases} 4x - y = -10 \\ 2x + 3y = 16 \end{cases}$ using the inverse matrix.

Solve It

34. Solve the system $\begin{cases} 4x + 3y = 17 \\ 2x - y = 11 \end{cases}$ using the inverse matrix.

Solve It

Cramer's Rule

***Cramer's Rule* is a method based on determinants of matrices that's used to solve systems of equations. The determinant of a 2×2 matrix $\begin{bmatrix} a & b \\ c & d \end{bmatrix}$ is $ad - bc$. The determinant of a 3×3 matrix is found using a process called *diagonals* in some textbooks. If $A = \begin{bmatrix} a_1 & b_1 & c_1 \\ a_2 & b_2 & c_2 \\ a_3 & b_3 & c_3 \end{bmatrix}$, then first rewrite the first two columns immediately following the third. Draw three diagonal lines from the upper left to the lower right and three diagonal lines from the lower left to the upper right, as shown in Figure 13-1.**

$$|A| = \begin{vmatrix} a_1 & b_1 & c_1 \\ a_2 & b_2 & c_2 \\ a_3 & b_3 & c_3 \end{vmatrix} \begin{matrix} a_1 & b_1 \\ a_2 & b_2 \\ a_3 & b_3 \end{matrix}$$

Figure 13-1: How to find a matrix's determinant.

Then multiply down the three diagonals from left to right and up the other three. Find the sum of the products on the top and the sum of the products on the bottom. Finally, find the difference of the top and bottom. This is the same thing as $(a_1b_2c_3 + b_1c_2a_3 + c_1a_2b_3) - (a_3b_2c_1 + b_3c_2a_1 + c_3a_2b_1)$.

For a 2×2 system, $\begin{cases} ax + by = c \\ dx + ey = f \end{cases}$:

$$x = \frac{\begin{vmatrix} c & b \\ f & e \end{vmatrix}}{\begin{vmatrix} a & b \\ d & e \end{vmatrix}} \qquad y = \frac{\begin{vmatrix} a & c \\ d & f \end{vmatrix}}{\begin{vmatrix} a & b \\ d & e \end{vmatrix}}$$

Observe that the numerator for x is simply the determinant of the matrix resulting from replacing the x coefficient column with the answer column in the coefficient matrix. The numerator for y is simply the determinant of the matrix resulting from replacing the y coefficient column with the answer column in the coefficient matrix.

Q. Solve the system of equations $\begin{cases} -2x + 3y = 17 \\ -5x - y = 17 \end{cases}$ using Cramer's Rule.

A. $x = -4$, $y = 3$. Using Cramer's Rule by substituting the coefficients and the constants, $x = \frac{\begin{vmatrix} 17 & 3 \\ 17 & -1 \end{vmatrix}}{\begin{vmatrix} -2 & 3 \\ -5 & -1 \end{vmatrix}}$. Find the determinants in the numerator and the denominator: $x = \frac{-17 - 51}{2 + 15} = \frac{-68}{17} = -4$. Do the same thing for $y = \frac{\begin{vmatrix} -2 & 17 \\ -5 & 17 \end{vmatrix}}{\begin{vmatrix} -2 & 3 \\ -5 & -1 \end{vmatrix}} = \frac{-34 + 85}{2 + 15} = \frac{51}{17} = 3$. This is why, if you're looking for a pain-free way of solving a system, we recommend Cramer's Rule. See how easy that was?

35. Find the determinant of $\begin{pmatrix} 2 & -1 & 4 \\ -3 & 4 & 6 \\ -2 & -1 & 5 \end{pmatrix}$

Solve It

36. Use Cramer's Rule to solve $\begin{cases} 2x - y + 4z = 7 \\ -3x + 4y + 6z = -1 \\ -2x - y + 5z = 4 \end{cases}$

Solve It

Answers to Problems on Systems of Equations

Following are particular answers to problems dealing with systems of equations. Note that there may be better, alternative ways to solve these problems. In addition to providing these answers, we also provide guidance on getting the answers if you need to review where you went wrong.

1 Use substitution to solve the system $\begin{cases} r+s=6 \\ s=13-2r \end{cases}$. The answer is $r=7$ and $s=-1$.

Substituting the fact that $s = 13 - 2r$, change the first equation to, say, $r + 13 - 2r = 6$. This simplifies to $-r + 13 = 6$, or $-r = -7$, which means that $r = 7$. Now that you know this value, you can substitute it into the second equation: $s = 13 - 2(7) = 13 - 14 = -1$. The final answer: $r = 7$, $s = -1$.

2 The sum of two numbers is 14 and their difference is 2. Find the numbers. The answer is $x = 8$ and $y = 6$.

First, you need to change the given words into a system of equations using variables. The sum of two numbers being 14 becomes $x + y = 14$, and their difference of 2 becomes $x - y = 2$. The first equation has an x variable with a coefficient of 1, so you can solve for it easily by subtracting y from both sides: $x = 14 - y$. Now, substitute this expression for x in the other equation: $14 - y - y = 2$. Combine like terms: $14 - 2y = 2$. Solve for y: $-2y = -12$; $y = 6$. Now that you've got that on lockdown, substitute it into the other equation to solve for x: $x = 14 - 6$; $x = 8$.

3 Solve the system $\begin{cases} \frac{x}{2}-\frac{y}{3}=-3 \\ \frac{2x}{3}+\frac{y}{2}=-2 \end{cases}$. The answer is $(x,y)=\left(\frac{-78}{17},\frac{36}{17}\right)$.

We didn't do it on purpose, but sometimes the answers to these questions just aren't pretty, so don't expect them to be. Even though you know that the answer is going to fractionville, we still recommend that the first thing you do in any equation of this type is to get rid of the fractions by multiplying every term by the LCD. The LCD for both equations in this problem turns out to be 6, so get multiplying: $\begin{cases} 6\cdot\frac{x}{2}-6\cdot\frac{y}{3}=6\cdot(-3) \\ 6\cdot\frac{2x}{3}+6\cdot\frac{y}{2}=6\cdot(-2) \end{cases} \rightarrow \begin{cases} 3x-2y=-18 \\ 4x+3y=-12 \end{cases}$. Now that this looks more like all the other systems you've been dealing with, which variable would you like to eliminate? y? Excellent choice. Multiply the top equation by 3 and the bottom by 2: $\begin{cases} 9x-6y=-54 \\ 8x+6y=-24 \end{cases}$. Adding these two equations eliminates y: $17x = -78$, which means that $x=\frac{-78}{17}$. Substitute this value to solve for y: $3\cdot\frac{-78}{17}-2y=-18$ or $\frac{-234}{17}-2y=-18$; $2y=\frac{72}{17}$; $y=\frac{36}{17}$.

4 Solve the system $\begin{cases} 3x-2y=4 \\ 6x-4y=8 \end{cases}$. The answer is $x=k$, $y=\frac{3k-4}{2}$.

We haven't shown you this type of answer before, so stick with us and we'll explain what happened. Notice first of all that all you have to do is multiply the top equation by -2 to get $-6x + 4y = -8$, which is the exact opposite of the bottom equation, $6x - 4y = 8$. If you add these two, you get $0 = 0$, which is always true. Therefore, this system has infinitely many solutions.

Lots of answers will work in this system (actually, an infinite number of them). If you graph this system on a coordinate plane, you get two lines that lie on top of each other. How many points do those two lines share in common? All of them. Some books ask you to write this out using variables to represent constants. For example, if you arbitrarily pick that $x = k$, you can plug that into the top equation to get $3k - 2y = 4$, which means that $-2y = 4 - 3k$, or $y = \frac{3k-4}{2}$.

5 Solve $\begin{cases} x^2 - y = 1 \\ x + y = 5 \end{cases}$. The answer is $x = 2$ and $y = 3$ or $x = -3$ and $y = 8$.

As we explain in the section this problem was presented in, sometimes you end up with quadratics that, when solved, have two solutions. First, solve the linear equation for a variable, like x in the second equation: $x = 5 - y$. Now substitute this into the first equation: $(5 - y)^2 - y = 1$. FOIL out the binomial to get $25 - 10y + y^2 - y = 1$. Combine like terms: $25 - 11y + y^2 = 1$. Now get 0 on one side of the equation: $24 - 11y + y^2 = 0$. This factors to $(3 - y)(8 - y) = 0$, which, when you use the zero product property, gets you two solutions for y: $y = 3$ and $y = 8$. Accept that both of these are true and substitute them, one at a time, into the original quadratic equation to get the most *possible* solutions for x. First: If $y = 3$, then $x^2 - 3 = 1$; $x^2 = 4$; $x = \pm 2$. $2 + 3 = 5$ works in the second equation, but notice that $-2 + 3 = 5$ doesn't. That means that when y is 3, x only equals 2. Now do the same thing for $y = 8$: $x^2 - 8 = 1$; $x^2 = 9$; $x = \pm 3$. In the second equation, $3 + 8 = 5$ is false, but $-3 + 8 = 5$ is true, so the other solution is $x = -3$, $y = 8$.

6 Solve the system of equations $\begin{cases} x + y = 9 \\ xy = 20 \end{cases}$. The answer is $x = 4$ and $y = 5$ or $x = 5$ and $y = 4$.

First, solve the linear equation for x: $x = 9 - y$. Plug this into the second given equation: $(9 - y)y = 20$. Distribute to get $9y - y^2 = 20$ and get a quadratic to solve. Get 0 on one side: $0 = y^2 - 9y + 20$, which factors to $0 = (y - 5)(y - 4)$. This means that y is 5 or 4. Plug them both into either original equation: $y = 5$: $x + 5 = 9$, $x = 4$. $y = 4$: $x + 4 = 9$, $x = 5$.

7 Solve $\begin{cases} x^2 + y^2 = 1 \\ x + y^2 = -5 \end{cases}$. The answer is that it has complex solutions: $\left(-2,\ \pm i\sqrt{3}\right)$; $\left(3,\ \pm i2\sqrt{2}\right)$.

This is also a first for you, but don't be surprised if your textbook or teacher throws these monkey wrenches at you, too. First, notice that both given equations have y^2 in them, with the same signs.

If you multiply the second equation by -1, you get $\begin{cases} x^2 + y^2 = 1 \\ -x - y^2 = 5 \end{cases}$. Now add both equations together to get $x^2 - x = 6$. Next, get the equation to equal 0: $x^2 - x - 6 = 0$. This factors to $(x - 3)(x + 2) = 0$, which does give two solutions, $x = 3$ or $x = -2$. When you plug $x = 3$ into $x + y^2 = -5$, you obtain $y = \pm i2\sqrt{2}$. When you plug $x = -2$ into $x + y^2 = -5$, you obtain $y = \pm i\sqrt{3}$, which satisfies both of the original equations. Thus, the solutions are $\left(-2,\ \pm i\sqrt{3}\right)$; $\left(3,\ \pm i2\sqrt{2}\right)$.

8 Solve $\begin{cases} 27x^2 - 16y^2 = -400 \\ -9x^2 + 4y^2 = 36 \end{cases}$. The answer is $\left(\pm\frac{16}{3},\ \pm\sqrt{73}\right)$ or $\left(\pm\frac{16}{3},\ \mp\sqrt{73}\right)$.

The y terms have opposite signs, so it's easier to eliminate them after you multiply the second equation by 4: $\begin{cases} 27x^2 - 16y^2 = -400 \\ -36x^2 + 16y^2 = 144 \end{cases}$. Adding these two equations gets you $-9x^2 = -256$, or $x^2 = 256/9$, which finally means that $x = \pm\frac{16}{3}$. Now notice that both of the original equations have x^2 in them, but no x term. If you square $16/3$ or $-16/3$, you get the same result: $256/9$. This means the positive and negative signs don't really matter when it comes to solving for y: $27\left(\frac{256}{9}\right) - 16y^2 = -400$.

Next, simplify: $768 - 16y^2 = -400$. Subtract 768 from both sides: $-16y^2 = -1{,}168$. Now, by dividing -16, you get $y^2 = 73$, or $y = \pm\sqrt{73}$. Wow, that made even our heads hurt!

9 Solve $\begin{cases} \dfrac{14}{x+3} + \dfrac{7}{4-y} = 9 \\ \dfrac{21}{x+3} - \dfrac{3}{4-y} = 0 \end{cases}$. The solution is $x = 4$ and $y = 3$.

First, rewrite the system by letting $u = \dfrac{1}{x+3}$ and $v = \dfrac{1}{4-y}$ and getting $\begin{cases} 14u + 7v = 9 \\ 21u - 3v = 0 \end{cases}$. Now, multiply the first equation by 3 and the second equation by 7: $\begin{cases} 42u + 21v = 27 \\ 147u - 21v = 0 \end{cases}$. Adding these two equations gets you $189u = 27$, which means that $u = 1/7$. Here's where you have to pay attention though! If this were a multiple choice test, most people would pick $1/7$ for the solution to this system, and they'd be wrong. That's why you have us, to remind you that you're not done yet! That's because you found u but not x or y. Work your way backwards: $\dfrac{1}{7} = \dfrac{1}{x+3}$. If the numerators are equal, the denominators have to be as well: $7 = x + 3$ means that $4 = x$. Now you can use that to get $y = 3$ from the following steps:

$$42 \cdot \frac{1}{7} + 21v = 27$$

$$6 + 21v = 27$$

$$v = 1$$

$$1 = \frac{1}{4-y}$$

$$1 = 4 - y$$

$$y = 3$$

Because this is a rational expression, also be sure to always check your solution to see whether it's extraneous. In other words, if $x = 4$ or $y = 3$, do you get 0 in the denominator of either given equation? In this case, the answer is no — so these answers are legit!

10 Solve $\begin{cases} \dfrac{12}{x+1} - \dfrac{12}{y-1} = 8 \\ \dfrac{6}{x+1} + \dfrac{6}{y-1} = -2 \end{cases}$. The answer is $x = 5$ and $y = -1$.

If you let $u = \dfrac{1}{x+1}$ and $v = \dfrac{1}{y-1}$, you can rewrite the system as $\begin{cases} 12u - 12v = 8 \\ 6u + 6v = -2 \end{cases}$. Now, multiply the second equation by 2 and get $\begin{cases} 12u - 12v = 8 \\ 12u + 12v = -4 \end{cases}$. Add them to get $24u = 4$, or $u = 1/6$. Work your way backwards from there: $\dfrac{1}{6} = \dfrac{1}{x+1}$; $6 = x + 1$; $x = 5$. The following steps get you to the solution for y:

$$12 \cdot \frac{1}{6} - 12v = 8$$

$$2 - 12v = 8$$

$$-12v = 6$$

$$v = -\frac{1}{2}$$

$$-\frac{1}{2} = \frac{1}{y-1}$$

$$-2 = y - 1$$

$$y = -1$$

11 Solve the system $\begin{cases} 3x-2y=17 \\ x-2z=1 \\ 3y+2z=1 \end{cases}$. The answer is $(x, y, z) = (5, -1, 2)$.

A is the top equation: $3x - 2y = 17$. B is the middle equation: $x - 2z = 1$. C is $3y + 2z = 1$. C has no x variable, so you use the other two to eliminate x. $A - 3\times B$ will do just that. $\begin{cases} 3x-2y=17 \\ -3(x-2z=1) \end{cases}$ becomes $\begin{cases} 3x-2y=17 \\ -3x+6z=-3 \end{cases}$. Add these two equations to get $-2y + 6z = 14$, which you can call equation D. $2C + 3D$ eliminates y: $\begin{cases} 2(3y+2z=1) \\ 3(-2y+6z=14) \end{cases}$ becomes $\begin{cases} 6y+4z=2 \\ -6y+18z=42 \end{cases}$. Add these two to get $22z = 44$, or $z = 2$. Plug this into equation C: $3y + 2(2) = 1$; $3y + 4 = 1$; $3y = -3$; $y = -1$. Lastly, plug $z = 2$ into equation B: $x - 2(2) = 1$; $x - 4 = 1$; $x = 5$.

12 Solve $\begin{cases} 2x-y+z=1 \\ x+y-z=2 \\ -x-y+z=2 \end{cases}$. The answer is no solution.

Right away, notice that all the coefficients on the middle and bottom equations are exact opposites of each other. When you add these two equations you get $0 = 4$. Because this equation is false, there's no solution.

13 Solve $\begin{cases} 2x+3y+4z=37 \\ 4x-3y+2z=17 \\ x+2y-3z=-5 \\ 3x-2y+z=11 \end{cases}$. The answer is $x = 4$, $y = 3$, and $z = 5$.

This system has three variables, so you only need three independent equations. Because you have four, you just have an additional one to play with: A: $2x + 3y + 4z = 37$; B: $4x - 3y + 2z = 17$; C: $x + 2y - 3z = -5$; and D: $3x - 2y + z = 11$.

A + B eliminates y: $6x + 6z = 54$. Divide by 6 to get equation E: $x + z = 9$.

C + D also eliminates y: $4x - 2z = 6$. Divide by 2 to get equation F: $2x - z = 3$.

E + F eliminates z: $3x = 12$; $x = 4$.

Work backwards by plugging $x = 4$ into equation F: $2(4) - z = 3$; $8 - z = 3$; $-z = -5$; $z = 5$.

Now $x = 4$ and $z = 5$ go into equation A: $2(4) + 3y + 4(5) = 37$; $8 + 3y + 20 = 37$; $3y + 28 = 37$; $3y = 9$; $y = 3$.

Note: It's not "typical" to have a solution in cases when a system has more equations than variables. It happens only because this system happens to be redundant rather than inconsistent.

14 Solve $\begin{cases} 3a+b+c+d=0 \\ 4a+5b+2c=15 \\ 4a+2b+5d=-10 \\ -5a+3b-d=8 \end{cases}$. The answer is $a = 1$, $b = 3$, $c = -2$, and $d = -4$.

To avoid confusion between A and a, we're naming the equations in this system a little differently: T: $3a + b + c + d = 0$; S: $4a + 5b + 2c = 15$; N: $4a + 2b + 5d = -10$; L: $-5a + 3b - d = 8$. Now start eliminating!

-2T + S: $\left\{\begin{array}{c}-2(3a+b+c+d=0)\\4a+5b+2c=15\end{array}\right\}\rightarrow\left\{\begin{array}{c}-6a-2b-2c-2d=0\\4a+5b+2c=15\end{array}\right\}$ gives you equation M: $-2a + 3b - 2d = 15$. That way, you now have three equations with *a*, *b*, and *d* in them: N, L, and the new one, M. Start over and eliminate another variable.

5M + 2N: $\left\{\begin{array}{c}5(-2a+3b-2d=15)\\2(4a+2b+5d=-10)\end{array}\right\}\rightarrow\left\{\begin{array}{c}-10a+15b-10d=75\\8a+4b+10d=-20\end{array}\right\}$ gives you equation R: $-2a + 19b = 55$.

M – 2L: $\left\{\begin{array}{c}-2a+3b-2d=15\\-2(-5a+3b-d=8)\end{array}\right\}\rightarrow\left\{\begin{array}{c}-2a+3b-2d=15\\10a-6b+2d=-16\end{array}\right\}$ gives you equation P: $8a - 3b = -1$.

You're down to two equations and two variables.

4R + P: $\left\{\begin{array}{c}4(-2a+19b=55)\\8a-3b=-1\end{array}\right\}\rightarrow\left\{\begin{array}{c}-8a+76b=220\\8a-3b=-1\end{array}\right\}$. Add these two equations to get $73b = 219$. This gives you your first solution, $b = 3$. Work your way backwards to get the other solutions.

P: $8a - 3(3) = -1$; $8a - 9 = -1$; $8a = 8$; $a = 1$.

M: $-2(1) + 3(3) - 2d = 15$; $-2 + 9 - 2d = 15$; $7 - 2d = 15$; $-2d = 8$; $d = -4$.

T: $3(1) + 3 + c - 4 = 0$; $3 + 3 + c - 4 = 0$; $2 + c = 0$; $c = -2$.

15 Sketch the graph of $\left\{\begin{array}{l}2x+y\geq 9\\2x-y\geq 1\\x\leq 7\end{array}\right.$. See the graph for the answer.

Put the first two inequalities in slope-intercept form first. The top inequality is $y \geq -2x + 9$; the second inequality is $y \leq 2x - 1$. Put these inequalities and $x \leq 7$ on the same graph.

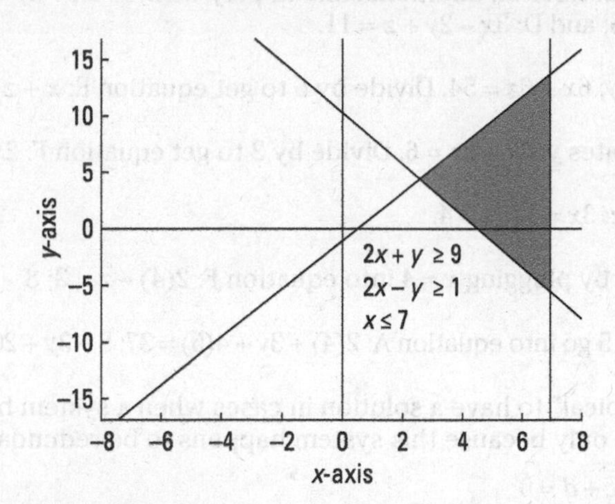

16 Sketch the graph of $\left\{\begin{array}{l}x^2+y^2\geq 9\\x^2+(y-3)^2\geq 9\end{array}\right.$. See the graph for the answer.

Both of these inequalities describe regions bounded by circles. The answer is the area outside of both these two circles. If you don't recognize them as such, turn to Chapter 12 to read up on conic sections. Graph them both on the same graph.

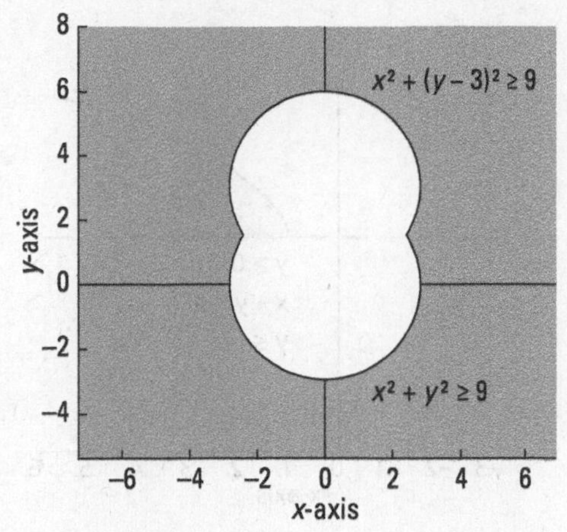

17 Sketch the graph of $\begin{cases} x^2 - y > 2 \\ x - y < 4 \end{cases}$. See the graph for the answer.

The first inequality represents the region below the parabola, $y = x^2 - 2$. The second inequality represents the region above the line, $y = x - 4$. Graph them both on the same graph.

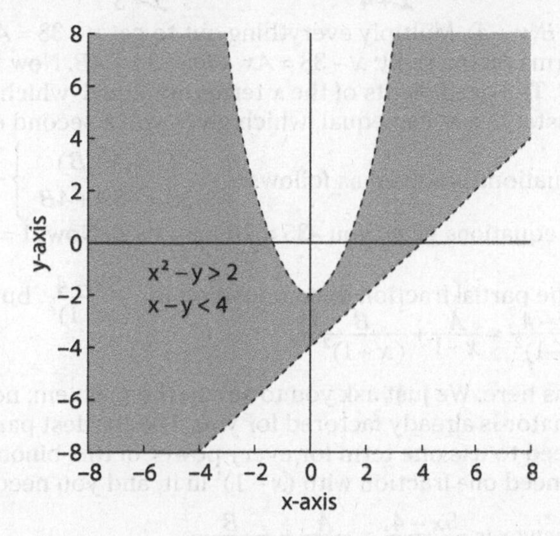

18 Sketch the graph of $\begin{cases} y \geq 0 \\ x + y < 4 \\ y \leq \sqrt{x-1} \end{cases}$. See the graph for the answer.

The expression on the right-hand side of the third inequality is a square root function. If you don't remember how to graph it, turn to Chapter 3 for a refresher.

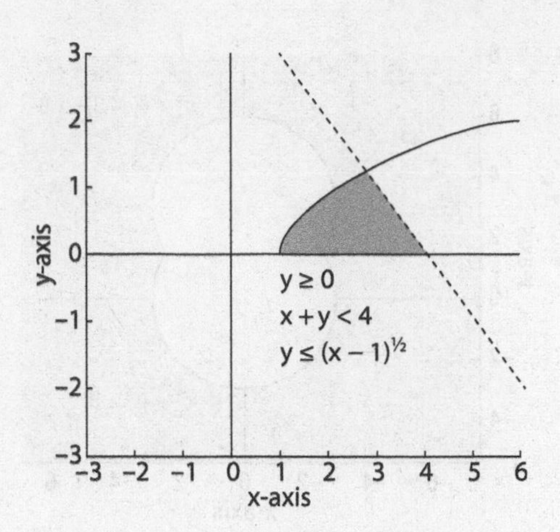

19 Find the constants A and B: $\frac{x-38}{x^2+x-12}=\frac{A}{x+4}+\frac{B}{x-3}$. The answer is $A = 6$ and $B = -5$.

We've just started the decomposition process for you. The first thing you should do is multiply everything by the factored denominator $\frac{x-38}{\cancel{(x+4)(x-3)}}\cancel{(x+4)(x-3)}=\frac{A}{\cancel{x+4}}\cancel{(x+4)}(x-3)+\frac{B}{\cancel{x-3}}(x+4)\cancel{(x-3)}$; in other words, $x - 38 = A(x - 3) + B(x + 4)$. Multiply everything out to get $x - 38 = Ax - 3A + Bx + 4B$. Collect the like terms on the right: $x - 38 = Ax + Bx - 3A + 4B$. Now factor out the x: $x - 38 = (A + B)x - 3A + 4B$. The coefficients of the x terms are equal, which gives you one equation: $1 = A + B$. The constants are also equal, which gives you a second equation: $-38 = -3A + 4B$.

This system of equations is solved as follows: $\left\{\begin{array}{c} 3(1=A+B) \\ -38=-3A+4B \end{array}\right\}\rightarrow\left\{\begin{array}{c} 3=3A+3B \\ -38=-3A+4B \end{array}\right\}$.

Adding these two equations gives you $-35 = 7B$, or $-5 = B$. Now, $1 = A - 5$ tells you that $6 = A$.

20 Find the form of the partial fraction decomposition for $\frac{5x-4}{(x-1)^2}$, but don't find the constants. The answer is $\frac{5x-4}{(x-1)^2}=\frac{A}{x-1}+\frac{B}{(x-1)^2}$.

Note the directions here. We just ask you to set up the problem, not to perform the decomposition. The denominator is already factored for you. The hardest part of this problem is remembering that you need to use one term for every power of this binomial, up to its degree of 2. In other words, you need one fraction with $(x - 1)^1$ in it, and you need another with $(x - 1)^2$ in it.

This is why the answer is $\frac{5x-4}{(x-1)^2}=\frac{A}{x-1}+\frac{B}{(x-1)^2}$.

21 Find the partial fraction decomposition for $\frac{2x^2-21x+18}{(x-1)(x^2-4x+4)}$. The answer is $\frac{-1}{x-1}+\frac{3}{x-2}-\frac{16}{(x-2)^2}$.

First, factor the given denominator to become $\frac{2x^2-21x+18}{(x-1)(x-2)^2}$. Because each factor is linear, set up three different fractions with constants on the top: one for the $(x - 1)$ factor, one for the $(x - 2)$ factor, and one for the $(x - 2)^2$ factor: $\frac{2x^2-21x+18}{(x-1)(x-2)^2}=\frac{A}{x-1}+\frac{B}{x-2}+\frac{C}{(x-2)^2}$.

Multiply every term by the factored denominator, cancel, and get $2x^2 - 21x + 18 = A(x - 2)^2 + B(x - 1)(x - 2) + C(x - 1)$. Multiply this all out to get $2x^2 - 21x + 18 = Ax^2 - 4Ax + 4A + Bx^2 - 3Bx + 2B + Cx - C$. Collect the like terms: $2x^2 - 21x + 18 = Ax^2 + Bx^2 - 4Ax - 3Bx + Cx + 4A + 2B - C$.

Factor out the x^2 and the x on the right side: $2x^2 - 21x + 18 = (A + B)x^2 + (-4A - 3B + C)x + 4A + 2B - C$. Set the coefficients of x^2 equal to each other: $2 = A + B$. Set the coefficients of x equal to each other: $-21 = -4A - 3B + C$. Lastly, set the constants equal to each other: $18 = 4A + 2B - C$. Solve this system of equations using elimination to get $A = -1$, $B = 3$, and $C = -16$. This is where the answer $\frac{-1}{x-1} + \frac{3}{x-2} - \frac{16}{(x-2)^2}$ comes from.

22 Find the partial fraction decomposition for $\frac{11x^2-7x+14}{2x^3-4x^2+3x-6}$. The answer is $\frac{3x-1}{2x^2+3} + \frac{4}{x-2}$.

Factor the denominator by using grouping: $2x^2(x - 2) + 3(x - 2)$ becomes $(x - 2)(2x^2 + 3)$. Use this to make two different fractions with a constant on top of the linear factor and a linear expression on top of the quadratic factor: $\frac{11x^2-7x+14}{2x^3-4x^2+3x-6} = \frac{Ax+B}{2x^2+3} + \frac{C}{x-2}$. Multiply every term by the factored denominator, cancel, and get $11x^2 - 7x + 14 = (Ax + B)(x - 2) + C(2x^2 + 3)$. Multiply it all out to get $11x^2 - 7x + 14 = Ax^2 - 2Ax + Bx - 2B + 2Cx^2 + 3C$. Collect the like terms and factor to get $11x^2 - 7x + 14 = (A + 2C)x^2 + (-2A + B)x - 2B + 3C$. This gives you a system with three equations: $\begin{cases} 11 = A+2C \\ -7 = -2A+B \\ 14 = -2B+3C \end{cases}$. Solving this system tells you that $A = 3$, $B = -1$, and $C = 4$, which gives you the answer $\frac{3x-1}{2x^2+3} + \frac{4}{x-2}$.

23 Find 4N. The answer is $\begin{bmatrix} 8 & 16 & 20 & -32 \\ 40 & 12 & -8 & -12 \end{bmatrix}$.

First, write out the problem by substituting the given matrix N in $4\begin{bmatrix} 2 & 4 & 5 & -8 \\ 10 & 3 & -2 & -3 \end{bmatrix}$.

Distribute the 4 to every element inside the matrix to get the answer: $\begin{bmatrix} 8 & 16 & 20 & -32 \\ 40 & 12 & -8 & -12 \end{bmatrix}$

24 Find 4N + 5M. The answer is $\begin{bmatrix} -17 & 11 & 35 & -2 \\ 40 & 22 & -18 & 18 \end{bmatrix}$

First, substitute the given matrices into the expression: $4\begin{bmatrix} 2 & 4 & 5 & -8 \\ 10 & 3 & -2 & -3 \end{bmatrix} + 5\begin{bmatrix} -5 & -1 & 3 & 6 \\ 0 & 2 & -2 & 6 \end{bmatrix}$

Distribute both scalars to every element of their matrix:

$\begin{bmatrix} 8 & 16 & 20 & -32 \\ 40 & 12 & -8 & -12 \end{bmatrix} + \begin{bmatrix} -25 & -5 & 15 & 30 \\ 0 & 10 & -10 & 30 \end{bmatrix}$. Add these two matrices by adding their corresponding elements to get $\begin{bmatrix} -17 & 11 & 35 & -2 \\ 40 & 22 & -18 & 18 \end{bmatrix}$.

25 Find 3M – P. The answer is no solution.

These matrices aren't the same dimensions, so you can't add them. There's no solution.

26 Find NP. The answer is $\begin{bmatrix} 64 & 19 & 11 \\ 13 & 20 & -9 \end{bmatrix}$

Substitute the given matrices into the expression $\begin{bmatrix} 2 & 4 & 5 & -8 \\ 10 & 3 & -2 & -3 \end{bmatrix}\begin{bmatrix} -1 & 2 & -1 \\ 4 & 4 & 0 \\ 2 & 3 & 1 \\ -5 & 2 & -1 \end{bmatrix}$. Check to see whether you can even multiply them. The matrix on the left is 2×4 and the one on the right is 4×3, so you can multiply them. Multiply every row from the left matrix by every column from the right matrix.

The sum of the first row times the first column: $2(-1) + 4(4) + 5(2) - 8(-5) = 64$.

The sum of the first row times the second column: 2(2) + 4(4) + 5(3) – 8(2) = 19.

The sum of the first row times the third column: 2(–1) + 4(0) + 5(1) – 8(–1) = 11.

The sum of the second row times the first column: 10(–1) + 3(4) – 2(2) – 3(–5) = 13.

The sum of the second row times the second column: 10(2) + 3(4) – 2(3) – 3(2) = 20.

The sum of the second row times the third column: 10(–1) + 3(0) – 2(1) – 3(–1) = –9.

Putting these all into a matrix gives you the answer: $\begin{bmatrix} 64 & 19 & 11 \\ 13 & 20 & -9 \end{bmatrix}$

27 Using the augmented matrix from the last example, use elementary row operations to find $-3r_2 \to r_2$. The answer is $\left[\begin{array}{cc|c} 3 & -1 & 6 \\ -6 & -9 & -9 \end{array}\right]$

Just multiply the second row of the given equation by –3 to get $\left[\begin{array}{cc|c} 3 & -1 & 6 \\ -6 & -9 & -9 \end{array}\right]$

28 Now, using your answer from Problem 27, find $r_1 \leftrightarrow r_2$. The answer is $\left[\begin{array}{cc|c} -6 & -9 & -9 \\ 3 & -1 & 6 \end{array}\right]$

Swap the first row with the second row and you get $\left[\begin{array}{cc|c} -6 & -9 & -9 \\ 3 & -1 & 6 \end{array}\right]$

29 Now, keep going and find $r_1 + r_2 \to r_2$. The answer is $\left[\begin{array}{cc|c} -6 & -9 & -9 \\ -3 & -10 & -3 \end{array}\right]$

Add each element from row one to its corresponding element in row two to change row two and get $\left[\begin{array}{cc|c} -6 & -9 & -9 \\ -3 & -10 & -3 \end{array}\right]$

30 Lastly, find $3r_2 + r_1 \to r_1$. The answer is $\left[\begin{array}{cc|c} -15 & -39 & -18 \\ -3 & -10 & -3 \end{array}\right]$

Temporarily multiply the second row by 3 to get [–9 –30 | –9]. Add these to the corresponding elements in row one to change row one and you get $\left[\begin{array}{cc|c} -15 & -39 & -18 \\ -3 & -10 & -3 \end{array}\right]$

31 Solve the system of equations $\begin{cases} 2x + 5y = 7 \\ 3x - 5y = 2 \end{cases}$ by writing it in augmented form and then putting the matrix in row echelon form. The answer is $x = 9/5$ and $y = 17/25$.

The matrix in augmented form is $\left[\begin{array}{cc|c} 2 & 5 & 7 \\ 3 & -5 & 2 \end{array}\right]$. Multiply the top row by ½ to get a 1 in the upper left corner: $\left[\begin{array}{cc|c} 1 & 5/2 & 7/2 \\ 3 & -5 & 2 \end{array}\right]$. You need to add a –3 to the second row to get a 0 under the one: $-3r_1 + r_2 \to r_2$ gives you $\left[\begin{array}{cc|c} 1 & 5/2 & 7/2 \\ 0 & -25/2 & -17/2 \end{array}\right]$; this is in row echelon form. Set up an equation from the second row: $\frac{-25y}{2} = \frac{-17}{2}$. Solve this equation to get $y = 17/25$. Use that answer and back substitute — $x + \frac{5}{2}\left(\frac{17}{25}\right) = \frac{7}{2}$ — to get that $x = 9/5$.

32 Use Gaussian elimination to solve $\begin{cases} 3x-2y+6z=7 \\ x-2y-z=-2 \\ -3x+10y+11z=18 \end{cases}$. The answer is $x = 1$, $y = 1$, and $z = 1$.

Set up the system as an augmented matrix: $\left[\begin{array}{ccc|c} 3 & -2 & 6 & 7 \\ 1 & -2 & -1 & -2 \\ -3 & 10 & 11 & 18 \end{array}\right]$

$r_1 \leftrightarrow r_2$ gets a 1 in the upper-left corner: $\left[\begin{array}{ccc|c} 1 & -2 & -1 & -2 \\ 3 & -2 & 6 & 7 \\ -3 & 10 & 11 & 18 \end{array}\right]$

$-3r_1 + r_2 \rightarrow r_2$ gets a 0 under the 1 in the second row: $\left[\begin{array}{ccc|c} 1 & -2 & -1 & -2 \\ 0 & 4 & 9 & 13 \\ -3 & 10 & 11 & 18 \end{array}\right]$

$3r_1 + r_3 \rightarrow r_3$ gets a 0 under the 1 in the third row: $\left[\begin{array}{ccc|c} 1 & -2 & -1 & -2 \\ 0 & 4 & 9 & 13 \\ 0 & 4 & 8 & 12 \end{array}\right]$

$\frac{1}{4}r_2 \rightarrow r_2$ gets a 1 in the next position along the diagonal: $\left[\begin{array}{ccc|c} 1 & -2 & -1 & -2 \\ 0 & 1 & 9/4 & 13/4 \\ 0 & 4 & 8 & 12 \end{array}\right]$

$-4r_2 + r_3 \rightarrow r_3$ gets a 0 under the 1 you just created: $\left[\begin{array}{ccc|c} 1 & -2 & -1 & -2 \\ 0 & 1 & 9/4 & 13/4 \\ 0 & 0 & -1 & -1 \end{array}\right]$. This means that $-z = -1$, or $z = 1$. Back substitute — $y + \frac{9}{4}(1) = \frac{13}{4}$ — to get that $y = 1$. Back substitute again: $x - 2(1) - 1(1) = -2$, or $x = 1$.

33 Solve the system $\begin{cases} 4x-y=-10 \\ 2x+3y=16 \end{cases}$ using inverse matrices. The answer is $x = -1$ and $y = 6$.

First, write the system as a matrix equation: $\begin{bmatrix} 4 & -1 \\ 2 & 3 \end{bmatrix}\begin{bmatrix} x \\ y \end{bmatrix} = \begin{bmatrix} -10 \\ 16 \end{bmatrix}$. Now, find the inverse matrix using the handy formula we show you in the section in which this question appears:

$\frac{1}{4(3)-(-1)(2)}\begin{bmatrix} 3 & 1 \\ -2 & 4 \end{bmatrix} = \frac{1}{14}\begin{bmatrix} 3 & 1 \\ -2 & 4 \end{bmatrix}$. Multiply this inverse on both sides of the equation:

$\frac{1}{14}\begin{bmatrix} 3 & 1 \\ -2 & 4 \end{bmatrix}\begin{bmatrix} 4 & -1 \\ 2 & 3 \end{bmatrix}\begin{bmatrix} x \\ y \end{bmatrix} = \frac{1}{14}\begin{bmatrix} 3 & 1 \\ -2 & 4 \end{bmatrix}\begin{bmatrix} -10 \\ 16 \end{bmatrix}$. Multiply the two matrices:

$\begin{bmatrix} x \\ y \end{bmatrix} = \frac{1}{14}\begin{bmatrix} -14 \\ 84 \end{bmatrix}$. Multiply the scalar: $\begin{bmatrix} x \\ y \end{bmatrix} = \begin{bmatrix} -1 \\ 6 \end{bmatrix}$. Your solutions from top to bottom are $x = -1$ and $y = 6$.

34 Solve the system $\begin{cases} 4x+3y=17 \\ 2x-y=11 \end{cases}$ using inverse matrices. The answer is $x = 5$ and $y = -1$.

Write the system as a matrix equation: $\begin{bmatrix} 4 & 3 \\ 2 & -1 \end{bmatrix}\begin{bmatrix} x \\ y \end{bmatrix} = \begin{bmatrix} 17 \\ 11 \end{bmatrix}$. Find the inverse:

$\frac{1}{4(-1)-(2)(3)}\begin{bmatrix} -1 & -3 \\ -2 & 4 \end{bmatrix} = \frac{-1}{10}\begin{bmatrix} -1 & -3 \\ -2 & 4 \end{bmatrix}$. Multiply this on both sides:

$\frac{-1}{10}\begin{bmatrix} -1 & -3 \\ -2 & 4 \end{bmatrix}\begin{bmatrix} 4 & 3 \\ 2 & -1 \end{bmatrix}\begin{bmatrix} x \\ y \end{bmatrix} = \frac{-1}{10}\begin{bmatrix} -1 & -3 \\ -2 & 4 \end{bmatrix}\begin{bmatrix} 17 \\ 11 \end{bmatrix}$. Multiply the matrices:

$\begin{bmatrix} x \\ y \end{bmatrix} = \frac{-1}{10}\begin{bmatrix} -50 \\ 10 \end{bmatrix}$. Multiply the scalar: $\begin{bmatrix} x \\ y \end{bmatrix} = \begin{bmatrix} 5 \\ -1 \end{bmatrix}$.

35 Find the determinant of $\begin{pmatrix} 2 & -1 & 4 \\ -3 & 4 & 6 \\ -2 & -1 & 5 \end{pmatrix}$. The answer is 93.

Because this is a 3×3 matrix, you have to use diagonals. First, rewrite the first two columns after the third one: $\left|\begin{array}{ccc|cc} 2 & -1 & 4 & 2 & -1 \\ -3 & 4 & 6 & -3 & 4 \\ -2 & -1 & 5 & -2 & -1 \end{array}\right|$. The sum of the diagonals from bottom-left to top-right is: $(-2)(4)(4) + (-1)(6)(2) + (5)(-3)(-1) = -32 - 12 + 15 = -29$. The sum of the diagonals from top-left to bottom-right is: $(2)(4)(5) + (-1)(6)(-2) + 4(-3)(-1) = 40 + 12 + 12 = 64$. The difference of the bottom sum minus the top sum is $64 - (-29) = 93$.

36 Use Cramer's Rule to solve $\begin{cases} 2x - y + 4z = 7 \\ -3x + 4y + 6z = -1 \\ -2x - y + 5z = 4 \end{cases}$. The answer is $x = 1$, $y = -1$, and $z = 1$.

Set up the quotient to find x first: $x = \dfrac{\begin{vmatrix} 7 & -1 & 4 \\ -1 & 4 & 6 \\ 4 & -1 & 5 \end{vmatrix}}{\begin{vmatrix} 2 & -1 & 4 \\ -3 & 4 & 6 \\ -2 & -1 & 5 \end{vmatrix}} = \dfrac{93}{93} = 1$. The determinants are computed by using the process of diagonals (see the section "Cramer's Rule" earlier in this chapter).

Solve for y: $y = \dfrac{\begin{vmatrix} 2 & 7 & 4 \\ -3 & -1 & 6 \\ -2 & 4 & 5 \end{vmatrix}}{\begin{vmatrix} 2 & -1 & 4 \\ -3 & 4 & 6 \\ -2 & -1 & 5 \end{vmatrix}} = \dfrac{-93}{93} = -1$.

Solve for z: $z = \dfrac{\begin{vmatrix} 2 & -1 & 7 \\ -3 & 4 & -1 \\ -2 & -1 & 4 \end{vmatrix}}{\begin{vmatrix} 2 & -1 & 4 \\ -3 & 4 & 6 \\ -2 & -1 & 5 \end{vmatrix}} = \dfrac{93}{93} = 1$.

Chapter 14

Spotting Patterns in Sequences, Series, and Binomials

In This Chapter

- Finding general formulas for sequences and series
- Dealing with arithmetic sequences and series
- Solving geometric sequences and series
- Using the binomial theorem

This chapter is all about patterns. No, we're not making quilts, although we could . . . nah! We'll stick with patterns of numbers, not cloth. Namely, we explore sequences, series, and the binomial theorem.

A *sequence* is an ordered list of numbers. A *series* is the sum of the terms in a sequence. Sequences and series often follow a pattern. The *binomial theorem* is the result of discovering the pattern of an expanded binomial.

Hmm . . . we think we're sensing a pattern of patterns here.

One mathematical term that comes up in this chapter is *factorial,* which you may remember from your previous math classes. A factorial, n!, read "n factorial," is defined as $1 \cdot 2 \cdot 3 \cdot \ldots \cdot (n-1) \cdot n$.

Major General Sequences and Series: Calculating Terms

Mathematically, a sequence is usually written in the following form: $\{a_i\}_{i=1,\ldots,n} = a_1, a_2, a_3, \ldots, a_n$. Here, n is the number of terms, a_i is the i-th term of the sequence, a_1 is the first term, a_2 is the second term, and so on. Similarly, a series can be written as the sum of the terms: $a_1 + a_2 + a_3 + \ldots + a_n$. The pattern of sequences and series can usually be described by a general expression or rule. Because sequences and series can have any finite number of terms, this expression allows you to find any term in the list without having to find all the terms. If you're not given the rule for the general term, you can sometimes find it if you're given the first few terms of a sequence or series.

Sometimes a term in a sequence depends on the term(s) before it. These are called *recursive sequences.* A famous example of a recursive sequence is the Fibonacci Sequence: 1, 1, 2, 3, 5, 8, 13, . . . where each term is the sum of the two before it, after the first two terms are given.

Q. Write the first five terms of the sequence whose rule is $a_n = n^2 + 3$.

A. 4, 7, 12, 19, 28. To find each term, you just plug the number of the term (n) into the formula: $a_1 = (1)^2 + 3 = 4$; $a_2 = (2)^2 + 3 = 7$; $a_3 = (3)^2 + 3 = 12$; $a_4 = (4)^2 + 3 = 19$; $a_5 = (5)^2 + 3 = 28$.

Q. Write a general expression for the sequence to find the nth term: $-\frac{1}{2}, \frac{2}{3}, -\frac{3}{4}, \frac{4}{5}$.

A. $a_n = (-1)^n \frac{n}{n+1}$. First, notice that the sign alternates between negative and positive. To deal with this, multiply by powers of -1: $(-1)^n$. Next, notice that the sequence's numerator is the same as the term number (n), so n becomes your numerator. Finally, you can see that the denominator is simply one number larger than the numerator (and term number), so it can be written as $n + 1$. Putting these pieces together, you get $a_n = (-1)^n \frac{n}{n+1}$.

1. Write the first five terms of the sequence whose rule is $a_n = \frac{n-3}{2n}$.

Solve It

2. Write a general expression for the sequence to find the nth term: 2, 4, 10, 28, 82, . . .

Solve It

3. Write a general expression for the sequence to find the nth term: 2, 2, $\frac{8}{3}$, 4, $\frac{32}{5}$, . . .

Solve It

4. Write the next two terms of the sequence 1, 1, 2, 3, 7, 16, . . .

Solve It

5. Find the sum of the first five terms of the series whose n-th term is described by $2^{n-1} + 1$.

Solve It

6. Find the sum of the first five terms of the series whose n-th term is described by $3^n + 2n$.

Solve It

Working Out the Common Difference: Arithmetic Sequences and Series

One special type of sequence is called an *arithmetic sequence.* In these sequences, each term differs from the one before it by a *common difference, d.* As a result, you have a formula for finding the *n*th term of an arithmetic sequence:

$$a_n = a_1 + (n-1)d$$

where a_1 is the first term, n is the number of terms, and d is the common difference.

To find the sum of some terms of an arithmetic sequence, also called an *arithmetic series,* you have to add a given number of terms together. This you can write in summation notation:

$$S_k = \sum_{n=1}^{k} a_n = \frac{k}{2}(a_1 + a_k)$$

This is read as "the kth partial sum of a_n" where $n = 1$ is the lower limit, k is the sum's upper limit, a_1 is the first term, and a_k is the last term to be added. To find a partial sum of an arithmetic series, find a_1 and a_k, and then use the formula, S_k, given earlier.

Q. Find the 60th term of the arithmetic sequence: 4, 7, 10, 13, . . .

A. $a_{60} = 181$. The easiest way to begin this problem is to find the formula for the nth term. To do so, you need a_1, which is 4. You also need the common difference, d, which can be found by subtracting two sequential terms, for example, $a_2 - a_1 = 7 - 4 = 3$. Plugging these into the general formula and simplifying, you get: $a_n = a_1 + (n-1)d = 4 + (n-1)3 = 4 + 3n - 3 = 1 + 3n$. Now you can find the 60th term by plugging in 60 for n: $a_{60} = 1 + 3(60) = 1 + 180 = 181$.

Q. Find $\sum_{n=1}^{8}(-3n+5)$

A. −68. Notice that the differences of any two consecutive terms are the same number, −3. Hence, it's an arithmetic series. To find the sum, you just have to use the arithmetic series formula. For this, you need k (which is 8), a_1, and a_k. Start by finding a_1: $-3(1)+5=2$. Then find a_8: $-3(8)+5=-19$. Finally, plug these into the formula: $S_8=(8/2)(2+-19)=(4)(-17)=-68$.

7. Find the 50th term of the arithmetic sequence: −6, −1, 4, 9, . . .

Solve It

8. Find the formula for the n-th term of an arithmetic sequence where $a_1=-3$ and $a_{15}=53$.

Solve It

9. Find the formula for the nth term of an arithmetic sequence where $a_5=-5$ and $a_{20}=-35$.

Solve It

10. Find $\sum_{n=1}^{5}\left(\frac{1}{2}n+2\right)$

Solve It

11. Find $\sum_{n=4}^{10}(2n-3)$

Solve It

12. Write the arithmetic series $2+7/3+8/3+3+10/3$ in summation notation and find the result.

Solve It

Be Fruitful and Multiply: Simplifying Geometric Sequences and Series

When consecutive terms in a sequence have a common ratio, the sequence is called a *geometric sequence.* To find that ratio, r, you divide each term by the term before it, and the quotient should be the same. Just like the other sequences, a_1 denotes the first term. To find the next term, multiply by the common ratio, r. Another pattern! The formula for the nth term of a geometric sequence is $a_n = a_1 \cdot r^{n-1}$.

As with other sequences, you can find the sum of geometric sequences, called *geometric series.* To find a partial sum of a geometric sequence, you can use the following formula: $S_k = \sum_{n=1}^{k} a_n = a_1\left(\frac{1-r^k}{1-r}\right)$, where $r \neq 1$.

Here, $n = 1$ is the lower limit, k is the sum's upper limit, r is the common ratio, and a_1 is the first term.

You can actually find the value of an infinite sum of many geometric series. As long as r lies within the range $-1 < r < 1$, you can find the infinite sum. If $r > 1$ or $r < -1$, a_n will become arbitrarily large in absolute value, so the sum won't converge. To find the infinite sum of a geometric series where r is within the range $-1 < r < 1$, use the following formula:

$$S = \sum_{n=1}^{\infty} a_n = \frac{a_1}{1-r}$$

Because we just plug and chug, geometric sequences and series are pretty easy to deal with. You just need to remember your rules for simplifying fractions. You can do it!

Q. Find the 10th term of the geometric sequence: 3, –6, 12, –24, . . .

A. $a_{10} = -1{,}536$. For the formula for the nth term of a geometric sequence, you need a_1 and r. a_1 is given in the problem: 3. To find r, all you need to do is divide a_2 by a_1: $-6/3 = -2$. Now you can simply plug these values into the formula:
$a_{10} = 3\cdot(-2)^{10-1} = 3\cdot(-2)^9 = 3\cdot(-512) =$ -1536

Q. Find the sum: $\sum_{n=1}^{5} 6\left(\frac{1}{3}\right)^{n-1}$.

A. $242/27$. To use the partial sum formula, you need to know a_1, r, and k. From the problem, you can identify r as $1/3$ and k as 5. To find a_1, simply plug in 1 for n: $a_1 = 6(1/3)^{1-1} = 6 \cdot (1/3)^0 = 6 \cdot 1 = 6$. Now all you have to do is plug and chug:

$$S_5 = 6\left(\frac{1-\left(\frac{1}{3}\right)^5}{1-\frac{1}{3}}\right) = 6\frac{\frac{242}{243}}{\frac{2}{3}} = 6\cdot\frac{242}{243}\cdot\frac{3}{2} = \frac{242}{27}$$

13. Find the 16th term of a geometric sequence given $a_1 = 5$ and $a_2 = -15$.

Solve It

14. Find the 8th term of a geometric sequence given $a_2 = 6$ and $a_6 = 486$.

Solve It

15. Find the sum: $\sum_{n=1}^{6} 4\left(-\frac{1}{2}\right)^{n-1}$

Solve It

16. Find the partial sum of the geometric series: $\frac{1}{6} + \frac{1}{3} + \frac{2}{3} + \ldots + \frac{32}{3}$.

Solve It

17. Find the sum of the infinite geometric series: $\frac{2}{3} + \frac{1}{3} + \frac{1}{6} + \ldots$

Solve It

18. Find the sum: $\sum_{n=1}^{\infty} 3\left(-\frac{2}{3}\right)^{n-1}$

Solve It

Expanding Polynomials Using the Binomial Theorem

Binomials are polynomials with exactly two terms. Often, binomials are raised to powers to complete computations, and when you multiply out a binomial so that it doesn't have any parentheses, it's called a *binomial expansion.* One way to complete binomial expansions is to distribute terms, but if the power is high, this method can be tedious.

An easier way to expand binomials is to use the *binomial theorem:*

$$(a+b)^n =$$

$$\binom{n}{0}a^n b^0 + \binom{n}{1}a^{n-1}b^1 + \binom{n}{2}a^{n-2}b^2 + \ldots + \binom{n}{n-2}a^2 b^{n-2} + \binom{n}{n-1}a^1 b^{n-1} + \binom{n}{n}a^0 b^n$$

Here, a is the first term, b is the second term, and the coefficient for the $r+1$th term $\binom{n}{r} = \frac{n!}{r!(n-r)!}$ is the *combinations* formula. For example, to find the binomial coefficient given by $\binom{6}{2}$, plug the values into the formula and simplify: $\frac{6!}{2!(6-2)!} = \frac{6!}{2!4!} = \frac{720}{2 \cdot 24} = 15$.

This seems like a lot of work, but trust us — if you just take it one step at a time, this method will save you an immense amount of time!

Q. Write the expansion of $(3x-2)^4$.

A. $81x^4 - 216x^3 + 216x^2 - 96x + 16$. To expand, simply replace a with $3x$, b with -2, and n with 4 to get

$$\binom{4}{0}(3x)^4(-2)^0 + \binom{4}{1}(3x)^3(-2)^1 + \binom{4}{2}(3x)^2(-2)^2 + \binom{4}{3}(3x)^1(-2)^3 + \binom{4}{4}(3x)^0(-2)^4.$$

Now, to simplify this mess, start with the combinations formula for each term:
$(1)(3x)^4(-2)^0 + (4)(3x)^3(-2)^1 + (6)(3x)^2(-2)^2 + (4)(3x)^1(-2)^3 + (1)(3x)^0(-2)^4$.

Then, raise the monomials to the specified powers:
$(1)81x^4 + (4)(27x^3)(-2) + (6)(9x^2)(4) + (4)(3x)(-8) + (1)(1)(16)$.

Finally, combine like terms and simplify: $81x^4 - 216x^3 + 216x^2 - 96x + 16$.

19. Find the coefficient of $x^8 y^4$ in $(x + y)^{12}$.

Solve It

20. Find the coefficient of $x^3 y^7$ in $(2x - 3y)^{10}$.

Solve It

21. Expand $(k - 4)^5$.

Solve It

22. Expand $(y + 4z)^6$.

Solve It

Answers to Problems on Sequences, Series, and Binomials

This section contains the answers for the practice problems presented in this chapter. We suggest you read the following explanations if your answers don't match up with ours (or if you just want a refresher on solving a particular type of problem).

1 Write the first five terms of the sequence: $a_n = \frac{n-3}{2n}$. The answer is –1, –¼, 0, ⅛, ⅕.

To find each term, you plug the number of the term (n) into the formula:
$a_1 = \frac{1-3}{2\cdot1} = \frac{-2}{2} = 1$, $a_2 = \frac{2-3}{2\cdot2} = \frac{-1}{4}$, $a_3 = \frac{3-3}{2\cdot3} = 0$, $a_4 = \frac{4-3}{2\cdot4} = \frac{1}{8}$, $a_5 = \frac{5-3}{2\cdot5} = \frac{1}{5}$

2 Write a general expression for the sequence to find the nth term: 2, 4, 10, 28, 82, . . . The answer is $a_n = 3^{n-1} + 1$.

At first, this one seems tricky. But if you subtract 1 from each term, the pattern becomes apparent: 1, 3, 9, 27, 81 . . . these are powers of 3! In fact, they can be found by taking 3 to sequential powers: $3^0, 3^1, 3^2, 3^3, 3^4$, where the power is 1 less than the term, or $n - 1$. So, putting that all together, we get $a_n = 3^{n-1} + 1$.

3 Write a general expression for the sequence to find the nth term: 2, 2, ⁸⁄₃, 4, ³²⁄₅, . . . The answer is $a_n = \frac{2^n}{n}$

Here's the hint for this one: The third term has a 3 in the denominator, and the fifth term has a 5 in the denominator. If you write each as an unreduced fraction over the term, n, the pattern for the denominator reveals itself: ²⁄₁, ⁴⁄₂, ⁸⁄₃, ¹⁶⁄₄, ³²⁄₅. The denominator is n. Now you just have to figure out the pattern for the numerator. Easy! They're all powers of 2. The first term is 2^1, the second is 2^2, and so on. So, your general expression is simply $\frac{2^n}{n}$.

4 Write the next two terms of the sequence: 1, 1, 2, 3, 7, 16, . . . The answer is 65, 321.

Ah! This is one of those recursive sequences we were mentioning. This pattern is found by adding the first term squared to the second term, and so on: $1^2 + 1 = 2$, $1^2 + 2 = 3$, $2^2 + 3 = 7$. . . . So, to find the next two terms, you just need to continue the pattern: $7^2 + 16 = 65$ and $16^2 + 65 = 321$.

5 Find the sum of the first five terms of the series whose nth term is described by $2^{n-1} + 1$. The answer is 36.

This one is an easy plug and chug. Simply plug in values 1 through 5 for n and simplify: $2^{1-1} + 1 = 2$, $2^{2-1} + 1 = 3$, $2^{3-1} + 1 = 5$, $2^{4-1} + 1 = 9$, $2^{5-1} + 1 = 17$. Now you just need to add the terms: $2 + 3 + 5 + 9 + 17 = 36$.

6 Find the sum of the first five terms of the series whose nth term is described by $3^n + 2n$. The answer is 393.

Just like the last one, plug and chug away: $3^1 + 2(1) = 5$, $3^2 + 2(2) = 13$, $3^3 + 2(3) = 33$, $3^4 + 2(4) = 89$, $3^5 + 2(5) = 253$. Now just add 'em up: $5 + 13 + 33 + 89 + 253 = 393$.

7 Find the 50th term of the arithmetic sequence: –6, –1, 4, 9, . . . The answer is $a_{50} = 239$.

To start, find the formula for the nth term. For this, you need a_1, which is –6. Next, you need the common difference, d, found by subtracting two sequential terms: $a_2 - a_1 = -1 - (-6) = 5$. From here, simply plug these into the general formula and simplify: $a_n = -6 + (n - 1)5 = -6 + 5n - 5 = 5n - 11$. Now that you have the general formula, you can find the 50th term by plugging in 50 for n: $a_{50} = 5(50) - 11 = 250 - 11 = 239$.

8 Find the general formula of an arithmetic sequence where $a_1 = -3$ and $a_{15} = 53$. The answer is $a_n = 4n - 7$.

Here, you have the first term, a_1, so you just need to find d. You can use a_{15} to find it. Simply plug in a_1, a_{15}, and $n = 15$ into the general formula and solve using algebra: $53 = -3 + (15 - 1)d$; $53 = -3 + 14d$; $56 = 14d$; $4 = d$. Now you can plug it in to find the general formula: $a_n = -3 + (n - 1)4 = -3 + 4n - 4 = 4n - 7$.

9 Find the general formula of an arithmetic sequence where $a_5 = -5$ and $a_{20} = -35$. The answer is $a_n = -2n + 5$.

This one is a bit more complicated but completely doable! To start, recognize that you have two terms, which means you can create two equations. Then you have a system of equations (check out Chapter 13 for a refresher). Solve these to find the missing variables and you can write your general formula. Start by writing your two equations with your given values and simplify: $-5 = a_1 + (5 - 1)d = -5 = a_1 + 4d$ and $-35 = a_1 + (20 - 1)d = -35 = a_1 + 19d$. Next, use elimination to solve the system by multiplying the first equation by -1 and adding the two together to get $-30 = 15d$. Therefore, $d = -2$. Now, simply substitute this back into either equation to find a_1: $-5 = a_1 + 4(-2)$; $-5 = a_1 - 8$; $3 = a_1$. Finally, plug in a_1 and d to find the general formula: $a_n = 3 + (n - 1)(-2)$; $a_n = 3 - 2n + 2$; $a_n = -2n + 5$.

10 Find $\sum_{n=1}^{5}\left(\frac{1}{2}n+2\right)$. The answer is $\frac{35}{2}$.

To find the sum, you just have to use the arithmetic series formula. For this, you need k (which is 5), a_1, and a_k. Start by finding a_1: $\frac{1}{2} \cdot (1) + 2 = \frac{5}{2}$. Then find a_5: $\frac{1}{2} \cdot (5) + 2 = \frac{9}{2}$. Finally, plug these into the formula: $S_5 = (\frac{5}{2})(\frac{5}{2} + \frac{9}{2}) = (\frac{5}{2})(\frac{14}{2}) = (\frac{5}{2})(7) = \frac{35}{2}$.

11 Find $\sum_{n=4}^{10}(2n-3)$. The answer is 77.

Follow the same steps as in Question 10. However, notice that the lower limit is 4, so you need to start by finding a_4: $2(4) - 3 = 5$, and this is like your a_1. The number of terms from 4 to 10 is 7, so $k = 7$. You also need $a_{10} = 2(10) - 3 = 17$. Finally, just plug in the values: $S_7 = (\frac{7}{2})(5 + 17) = (\frac{7}{2})(22) = 7(11) = 77$.

12 Write the arithmetic series $2 + \frac{7}{3} + \frac{8}{3} + 3 + \frac{10}{3}$ in summation notation and find the result. The answer is $\sum_{n=1}^{5}\left(\frac{5}{3}+\frac{1}{3}n\right)=\frac{40}{3}$

For summation notation, you need to find the general formula and know how many terms you're dealing with. In this case, you have five terms. Therefore, you know your upper limit, k, is 5. For the general formula, you need the first term, $a_1 = 2$, and the common difference, d, which is found by subtracting two sequential terms: $a_2 - a_1 = \frac{7}{3} - 2 = \frac{1}{3}$. Plug these in and simplify to find your general formula: $a_n = 2 + (n - 1)\frac{1}{3} = 2 + \frac{1}{3} \cdot n - \frac{1}{3} = \frac{5}{3} + \frac{1}{3} \cdot n$. Then, plug the general formula into the summation notation and add the values given in the original problem to find the result:

$$\sum_{n=1}^{5}\left(\frac{5}{3}+\frac{1}{3}n\right)=\frac{40}{3}$$

13 Find the 16th term of a geometric sequence given $a_1 = 5$ and $a_2 = -15$. The answer is $a_{16} = -71744535$.

To find the 16th term, you need to find the general formula. For that, you need a_1 and r. a_1 is given in the problem: 5. To find r, all you need to do is divide a_2 by a_1: $-\frac{15}{5} = -3$. Now you can simply plug these values into the formula: $a_{16} = 5 \cdot (-3)^{16-1} = 5 \cdot (-3)^{15} = -5 \cdot 14348907 = -71744535$.

14 Find the 8th term of a geometric sequence given $a_2 = 6$ and $a_6 = 486$. The answer is 4374.

This time you don't have the first term, so you have to set up a system of equations and simplify: $6 = a_1 \cdot r^{2-1}$; $6 = a_1r^1$ and $486 = a_1 \cdot r^{6-1}$; $486 = a_1r^5$. Isolate a_1 and use substitution to solve for r. $a_1 = 6/r$; $486 = (6/r)r^5$; $486 = 6r^4$; $81 = r^4$; $3 = r$. Then substitute r back into either equation to find a_1: $6 = a_1(3)$; $2 = a_1$. Now you can set up the general formula: $a_n = 2 \cdot 3^{n-1}$. Finally, to find the 8th term, plug in $n = 8$: $a_8 = 2 \cdot 3^{8-1} = 2 \cdot 3^7 = 2 \cdot 2187 = 4374$.

15 Find the sum: $\sum_{n=1}^{6} 4\left(-\frac{1}{2}\right)^{n-1}$. The answer is $^{21}/_8$.

Start by finding $a_1 = 4(-1/2)^{1-1} = 4(-1/2)^0 = 4 \cdot 1 = 4$. Because $r = -1/2$, you have everything you need to plug into the partial sum formula: $\sum_{n=1}^{6} 4\left(-\frac{1}{2}\right)^{n-1} = 4\left(\frac{1-\left(-\frac{1}{2}\right)^6}{1-\left(-\frac{1}{2}\right)}\right) = 4\left(\frac{1-\frac{1}{64}}{1+\frac{1}{2}}\right) = 4\frac{63}{64}\cdot\frac{2}{3} = \frac{21}{8}$

16 Find the partial sum of the geometric series: $1/6 + 1/3 + 2/3 + \ldots\ 32/3$. The answer is $^{127}/_6$.

Here you have a_1 and you can find r by dividing a_2 by a_1: $1/3 \div 1/6 = 2$. The trick here is that you need to find n. To do so, plug the last term into the general formula and use properties of exponents to solve for n: $32/3 = 1/6 \cdot 2^{n-1}$; $64 = 2^{n-1}$; $2^6 = 2^{n-1}$; $6 = n - 1$; $7 = n$. Now that you have all the variables, plug them in to find the partial sum: $S_7 = \frac{1}{6}\left(\frac{1-2^7}{1-2}\right) = \frac{1}{6}\cdot\frac{-127}{-1} = \frac{127}{6}$

17 Find the sum of the infinite geometric series: $2/3 + 1/3 + 1/6 + \ldots$ The answer is $4/3$.

You need to start by finding r, which is a_2 divided by a_1: $1/3 \div 2/3 = 1/2$. a_1 is $2/3$. So all you need to do is plug these values into the appropriate formula and use what you know about fractions to simplify: $S = \frac{\frac{2}{3}}{1-\frac{1}{2}} = \frac{\frac{2}{3}}{\frac{1}{2}} = 2\cdot\frac{2}{3} = \frac{4}{3}$

18 Find the sum: $\sum_{n=1}^{\infty} 3\left(-\frac{2}{3}\right)^{n-1}$. The answer is $9/5$.

To start, find a_1 by plugging 1 into the general formula: $3\left(-\frac{2}{3}\right)^{1-1} = 3\left(-\frac{2}{3}\right)^0 = 3$. Next, notice that you have r, $-2/3$. From here, plug in these values to find the infinite sum: $S = \frac{3}{1-\left(-\frac{2}{3}\right)} = \frac{3}{\frac{5}{3}} = 3\cdot\frac{3}{5} = \frac{9}{5}$

19 Find the coefficient of x^8y^4 in $(x + y)^{12}$. The answer is $495x^8y^4$.

All you have to do to find the 5th term is use the handy binomial theorem. In this case, $r = 5 - 1 = 4$, $a = x$, and $b = y$: $\binom{12}{4}x^{12-4}y^4$. Because the original binomial doesn't have any coefficients, the coefficient just comes from the combinations formula: $\binom{12}{4} = \frac{12!}{4!(12-4)!} = \frac{479001600}{24\cdot 40320} = 495$. Multiply the other terms by this and you get $495x^8y^4$.

20 Find the coefficient of x^3y^7 in $(2x - 3y)^{10}$. The answer is $-2099520x^3y^7$.

In this case, $r = 8 - 1 = 7$, $a = 2x$, $b = -3y$. Plug these into the binomial theorem and simplify:

$$\binom{10}{7}(2x)^{10-7}(-3y)^7 = \frac{10!}{7!(10-7)!}(2x)^3(-3y)^7 = 120\cdot 8x^3\cdot\left(-2187y^7\right) = -2099520x^3y^7$$

21 Expand $(k-4)^5$. The answer is $k^5 - 20k^4 + 160k^3 - 640k^2 + 1280k - 1024$.

To expand, simply replace a with k, b with -4, and n with 5 to get

$$\binom{5}{0}(k)^5(-4)^0+\binom{5}{1}(k)^4(-4)^1+\binom{5}{2}(k)^3(-4)^2+\binom{5}{3}(k)^2(-4)^3+\binom{5}{4}(k)^1(-4)^4+\binom{5}{5}(k)^0(-4)^5$$

To simplify, start with the combinations formula for each term:

$$(1)(k)^5(-4)^0+(5)(k)^4(-4)^1+(10)(k)^3(-4)^2+(10)(k)^2(-4)^3+(5)(k)^1(-4)^4+(1)(k)^0(-4)^5.$$

Next, raise the monomials to the specified powers:

$$(1)(k)^5(1)+(5)(k)^4(-4)+(10)(k)^3(16)+(10)(k)^2(-64)+(5)(k)^1(256)+(1)(k)^0(-1024).$$

Last, combine like terms and simplify: $k^5 - 20k^4 + 160k^3 - 640k^2 + 1280k - 1024$.

22 Expand $(y+4z)^6$. The answer is $y^6 + 24y^5z + 240y^4z^2 + 1280y^3z^3 + 3840y^2z^4 + 6144yz^5 + 4096z^6$.

Here, replace a with y, b with $4z$, and n with 6 to get:

$$\binom{6}{0}(y)^6(4z)^0+\binom{6}{1}(y)^5(4z)^1+\binom{6}{2}(y)^4(4z)^2+\binom{6}{3}(y)^3(4z)^3+\binom{6}{4}(y)^2(4z)^4+\binom{6}{5}(y)^1(4z)^5+\binom{6}{6}(y)^0(4z)^6$$

Then, to simplify this mess, start with the combinations formula for each term:

$$(1)(y)^6(4z)^0+(6)(y)^5(4z)^1+(15)(y)^4(4z)^2+(20)(y)^3(4z)^3+(15)(y)^2(4z)^4+(6)(y)^1(4z)^5+(1)(y)^0(4z)^6$$

Raise the monomials to the specified powers:

$$(1)(y)^6(1)^0+(6)(y)^5(4z)+(15)(y)^4(16z^2)+(20)(y)^3(64z^3)+(15)(y)^2(256z^4)+(6)(y)^1(1024z^5)+(1)(y)^0(4096z^6)$$

Finally, combine like terms and simplify: $y^6 + 24y^5z + 240y^4z^2 + 1280y^3z^3 + 3840y^2z^4 + 6144yz^5 + 4096z^6$.

Chapter 15

Previewing Calculus

In This Chapter

- Using different techniques to find limits of functions
- Applying limit laws to find limits of combined functions
- Discovering continuity and discontinuity in functions

Your high school math career probably began with Algebra I. Some of the information you learned there was likely repeated in Algebra II and presented again in pre-calculus (albeit a bit morphed and more complicated than it was in your previous courses). The end of pre-calc is the beginning of calculus. Calculus teachers assume you've been paying attention during your math years (or they assume that you bought this book and are now a math genius) and that most of the material you've learned has stuck with you. Because of this, calc teachers move pretty quickly into new material.

Calculus is the study of change. Until now, all the information you've used to solve problems has been constant, so your answers have always been constant. For example, up until calc, in a distance problem, the rate at which a car is moving remains constant. The slope of a straight line is always a constant. The volume of a shape is always a constant. But in calculus, all these can move and grow and change. For example, the car can accelerate, decelerate, and accelerate again, all within the same problem, which changes the whole outcome. The line can now be a curve so that its slope changes over time. The shape that you're trying to find the volume of can get bigger or smaller, so that the volume changes over time.

All this change may send you screaming in terror to the nearest antianxiety salt lick, but chances are if you've made it this far in your math career, you can handle it. So just in case your pre-calc course ends with a preview of calc, we include the first couple of topics here.

Finding Limits: Graphically, Analytically, and Algebraically

If you haven't noticed by now, graphing functions has slowly become more complex and intricate. The more complicated the function is, the more complicated the graph tends to be. By now you've seen functions that are undefined at certain values; the graph has either a hole or a vertical asymptote, which affects your domain. The end of pre-calc (and the beginning of calc) looks at the *limit* of a function at a point — what the function would do if it could.

In symbols, a limit is written as $\lim_{x \to n} f(x) = L$, which is read as "the limit of $f(x)$ as x approaches n is L." L is the limit that you're looking for. The limit of $f(x)$ as x approaches n is L if the values of $f(x)$ can be made as close as we like to L by taking x sufficiently close to (but not equal to) n. For the limit of a function to exist, the left limit and the right limit must be equal.

- A left limit of a given function $f(x)$ as x approaches n is a value that $f(x)$ tends to no matter how x approaches n from the left side of n. This is written as $\lim_{x \to n^-} f(x) = L$.
- A right limit of a given function $f(x)$ as x approaches n is a value that $f(x)$ tends to no matter how x approaches n from the right side of n. This is written as $\lim_{x \to n^+} f(x) = L$.

Only when the left limit and right limit are the same does the function have a limit. When $\lim_{x \to n^-} f(x) = \lim_{x \to n^+} f(x) = L$, then $\lim_{x \to n} f(x) = L$.

You can find a limit in three different ways: graphically, analytically, and algebraically. In the following sections, we take a look at each one so you know how to handle them.

Before you try any of the techniques for finding a limit, always try plugging in the value that x is approaching into the function. We also recommend using only the graphing method when you've been given the graph and asked to find the limit. The analytical method works for most functions, and sometimes it's the only method you have in your pocket that will work. However, if the algebraic method works, then go with it — the analytical method is just too long and tedious.

Graphically

When you're given the graph of a function and asked to find the limit, just read the graph as x approaches the given value and see what the y value would have been (or was if the function is defined).

Q. In the given graph for $f(x)$, find $\lim_{x \to -3} f(x)$, $\lim_{x \to 4} f(x)$, and $\lim_{x \to 6} f(x)$.

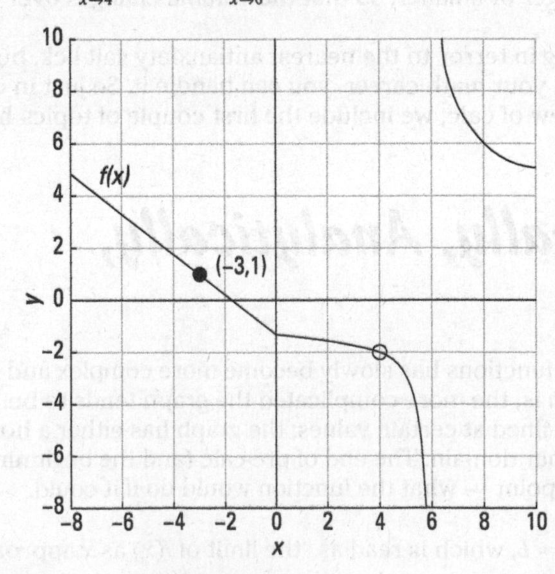

A. $\lim_{x \to -3} f(x) = 1$, $\lim_{x \to 4} f(x) = -2$, and $\lim_{x \to 6} f(x)$ doesn't exist. The function is defined at $x = -3$ because you can see a dot there. The limit of the function as x tends to -3 exists and is 1. The y-values are near 1 when x is close to -3, so $\lim_{x \to -3} f(x) = 1$.

The function isn't defined at $x = 4$ because the graph has a hole, but if you move along the graph from the left as x approaches 4, y also approaches a value of -2. Because this is the same from the right side, $\lim_{x \to 4} f(x) = -2$. The graph has an asymptote at $x = 6$, and you can see that the function values are increasingly positive as x approaches 6 from the right and increasingly negative as x approaches 6 from the left. Thus, the limit as x approaches 6 doesn't exist. Some teachers and books write this as *DNE* (does not exist) because we see it frequently (and we're lazy as mathematicians).

1. In the given graph for $g(x)$, find $\lim_{x \to -5} g(x)$, $\lim_{x \to -2} g(x)$, and $\lim_{x \to 1} g(x)$.

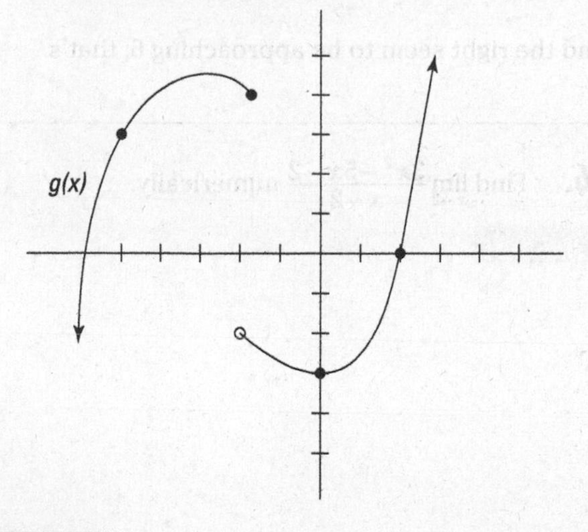

Solve It

2. In the given graph for $h(x)$, find $\lim_{x \to -3} h(x)$, $\lim_{x \to 5} h(x)$, and $\lim_{x \to 0} h(x)$.

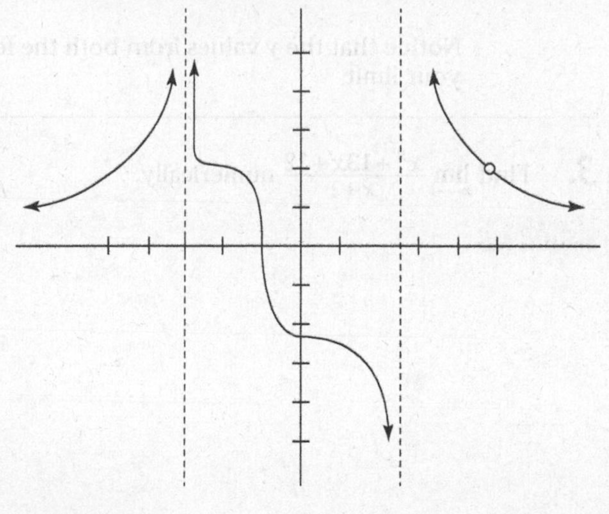

Solve It

Analytically

Analytically means *systematically,* and that's exactly how you find the limit using this method. This method is also called *numerically* estimating a limit. It should be pointed out that this method is not fool-proof. You set up a chart, and the value that x is approaching goes in the middle of the top row. On the left you put values that get closer to the value that x is approaching, and you do the same thing on the right. The second row should be the y values when you plug the top row into the function. Hopefully, the bottom row values approach the same number from the left and right, and *voilà* — you have your limit!

Q. Find $\lim_{x \to 2} \frac{x^2+2x-8}{x-2}$

A. $\lim_{x \to 2} \frac{x^2+2x-8}{x-2} = 6$. Notice that if you plug 2 into the function for x, you get 0 in the denominator. That means the function is undefined there. If the directions ask you to find this limit analytically, set up a chart. Here's the table that we set up. ***Note:*** Your chart doesn't have to look like this; no one way works all the time in finding the limit analytically, but most of the tables we've seen are set up similarly. Try to get those x values really close to the value you're approaching — that's usually the best way to find the limit.

x	1	1.9	1.99	1.999	2	2.001	2.01	2.1	3
y	5	5.9	5.99	5.999	???	6.001	6.01	6.1	7

Notice that the y values from both the left and the right seem to be approaching 6; that's your limit.

3. Find $\lim_{x\to -1} \frac{x^2+13x+12}{x+1}$ numerically.

Solve It

4. Find $\lim_{x\to 2} \frac{3x^2-5x-2}{x-2}$ numerically.

Solve It

Algebraically

To find a limit algebraically, you can use four different techniques: plugging in, factoring, rationalizing the numerator or denominator, and finding the lowest common denominator. Always start by plugging the given number into the function just to see whether it works. If the answer is undefined, then move on to one of the other three techniques — each one depending on the given function.

Plugging in

This first technique asks you to substitute the given value into the function. If you get an undefined value, like 0 in the denominator of the fraction, try something else, unless the numerator approaches something nonzero, in which case, you automatically know that the limit doesn't exist. But when substitution works, this technique is the quickest way to find a limit. We like short methods, and we hope you do, too!

Factoring

When the function is a rational function with polynomials in the numerator and the denominator, try factoring them. If you've forgotten how to factor a polynomial and need a refresher, see Chapter 4. Chances are some factors will cancel from both the numerator and the denominator. You can then substitute the given value into the cancelled version and usually get an answer that's also your limit.

If you still get an undefined function when you follow the steps of factoring the rational function on the top and bottom, canceling, and plugging in the given value, then the limit does not exist (DNE).

Rationalizing the numerator

When you see square roots in the rational function in the numerator and plugging in doesn't work, always try to rationalize the numerator. That's right — you multiply by the conjugate of the numerator on both the top and bottom of the fraction. When you do, you usually see a bunch of terms cancel, and the function simplifies down to a point where you can plug in the given value and find the limit.

Finding the lowest common denominator

When the rational function is a *complex rational function,* find the common denominators and add or subtract terms, then cancel and simplify. You can then plug in the given value to find the limit.

Q. Find $\lim\limits_{x\to 0}\dfrac{\frac{1}{x+5}-\frac{1}{5}}{x}$

A. $\lim\limits_{x\to 0}\dfrac{\frac{1}{x+5}-\frac{1}{5}}{x}=-\dfrac{1}{25}$. Find the common denominator of the fractions on the top first:

$\lim\limits_{x\to 0}\dfrac{\frac{5\cdot 1}{5(x+5)}-\frac{1(x+5)}{5(x+5)}}{x}=\lim\limits_{x\to 0}\dfrac{\frac{5}{5(x+5)}-\frac{x+5}{5(x+5)}}{x}$. Subtract and then simplify the top now that you have a common denominator: $\dfrac{\frac{5-x-5}{5(x+5)}}{x}=\dfrac{\frac{-x}{5(x+5)}}{x}=\dfrac{-1}{5(x+5)}$. Notice that you can now plug 0 into the last expression and get the limit: $-\dfrac{1}{25}$.

5. Find $\lim\limits_{x\to 2}\dfrac{3x^2-5x-2}{x-2}$ algebraically.

Solve It

6. Find $\lim\limits_{x\to 5}\dfrac{\sqrt{x-1}-2}{x-5}$

Solve It

Knowing Your Limits

Calculus also provides you with a few limit laws that help you find the limits of combined functions: added, subtracted, multiplied, divided, and even raised to powers. If you can find the limit of each individual function, you can find the limit of the combined function as well.

If $\lim_{x \to n} f(x) = L$ and $\lim_{x \to n} g(x) = M$, the limit laws are:

- **Addition law:** $\lim_{x \to n}\left(f(x)+g(x)\right) = L + M$
- **Subtraction law:** $\lim_{x \to n}\left(f(x)-g(x)\right) = L - M$
- **Multiplication law:** $\lim_{x \to n}\left(f(x)g(x)\right) = L \cdot M$
- **Division law:** $\lim_{x \to n}\frac{f(x)}{g(x)} = \frac{L}{M}$, provided $M \neq 0$
- **Power law:** $\lim_{x \to n}\left(f(x)\right)^p = L^p$

For this whole section, use the following to answer the questions:

$\lim_{x \to 1} f(x) = -5$ $\lim_{x \to 1} g(x) = 2$ $\lim_{x \to 1} h(x) = 0$

Q. Find $\lim_{x \to 1}\left(g(x)-f(x)\right)$.

A. $\lim_{x \to 1}\left(g(x)-f(x)\right) = 7$. The limit of $g(x)$ is 2 and the limit of $f(x)$ is –5. To find $\lim_{x \to 1}\left(g(x)-f(x)\right)$, use the subtraction law: 2 – (–5) = 7. It really is that easy!

7. Find $\lim_{x \to 1}\frac{f(x)}{g(x)}$

Solve It

8. Find $\lim_{x \to 1}\left(f(x)^2 - 2h(x) + \frac{1}{g(x)}\right)$

Solve It

9. Find $\lim_{x \to 1} \frac{\sqrt{g(x)} - f(x)}{5g(x)}$

Solve It

10. Find $\lim_{x \to 1} \frac{f(x)}{h(x)}$

Solve It

Determining Continuity

The word "continuity" means the same thing in math as it does in your everyday life. Something that's continuous has a stability or a permanence to it . . . it never stops. In pre-calc, you've seen functions that have holes in their graph, jumps in their graph, or asymptotes — just to name a few. A graph that doesn't have holes, jumps, or vertical asymptotes keeps going forever, and we call that function *continuous.*

Polynomial functions, exponential functions, logarithmic functions, rational functions, and trigonometric functions are continuous at every point of their domain. If you're ever asked to determine the continuity of one of these types of functions, don't bother — the answer is that it's always continuous.

We usually look at specific values in the domain to determine continuity instead of looking at the entire function. Even discontinuous functions are discontinuous only at certain places. The discontinuity at a certain x value in any function is always either *removable* (a hole in the graph) or *nonremovable.* In the case of rational functions that have discontinuities due to a factor in the denominator that goes to zero at the value c, it all depends on the factored versions of the polynomials in the numerator and denominator. If all instances of the factor causing the discontinuity cancel out of the denominator, the discontinuity at c is removable. If not, the discontinuity is nonremovable.

Three things must be true for a function to be continuous at $x = c$:

- **$f(c)$ must be defined.** When you plug c into the function, you must get a value out either by definition or through a specific rule. For example, getting 0 in the denominator is unacceptable and therefore a discontinuity.
- **The limit of the function as x approaches c must exist.** The left and right limits must be the same. If they aren't, the function is discontinuous there.
- **The function's value and the limit must be the same.** If the value of the function is one thing and the limit is something different, that's not good; the function is discontinuous there.

Here's the graph of a function where each one of the preceding situations fails:

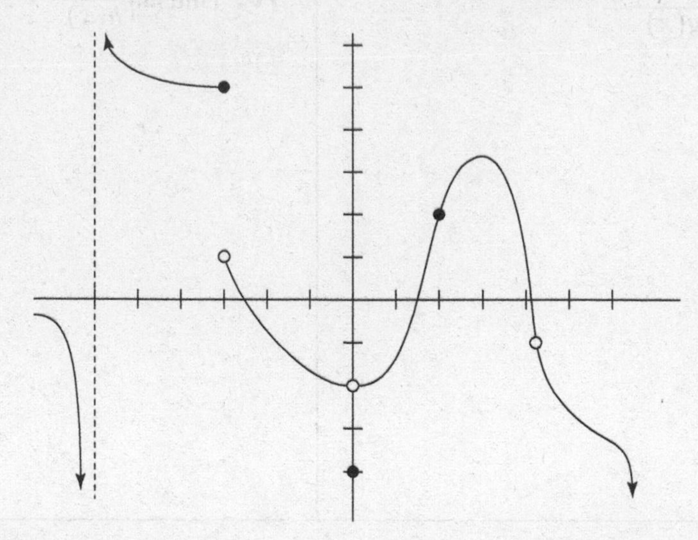

At $x = 4.3$, the graph has a hole. The function is undefined there, and therefore discontinuous at $x = 4.3$.

At $x = -3$ the function jumps. The limit as x approaches -3 from the left is 5, and from the right the limit is 1, so the limit doesn't exist, and the function is discontinuous.

At $x = 0$ the function is defined at one point: $f(0) = -4$, but the limit as x approaches 0 from the left and from the right is -2. These two values must be the same for the function to be continuous.

One point where the function is continuous is at $x = 2$: $\lim_{x \to 2} f(x) = 2 = f(2)$.

Q. Is $f(x) = \frac{3}{x+2}$ continuous at $x = 3$?

A. $f(x)$ is continuous at $x = 3$. This is a rational function and $x = 3$ is in its domain.

Q. Explain why $f(x) = \frac{3}{x+2}$ is not continuous at $x = -2$. Is this discontinuity removable or nonremovable?

A. The function is not defined at $x = -2$. The absolute value of this rational function tends to infinity as x tends to -2. Therefore the discontinuity that exists is nonremovable.

11. Determine whether $g(x)=\dfrac{x^2-4x-5}{x^2-2x-15}$ is continuous at $x=5$.

Solve It

12. Is $g(x)=\dfrac{x^2-4x-5}{x^2-2x-15}$ continuous at $x=0$?

Solve It

13. Is $g(x)=\dfrac{x^2-4x-5}{x^2-2x-15}$ continuous at $x=-3$?

Solve It

14. Determine whether

$$h(x)=\begin{cases} 3x-1 & \text{if } x\le -2 \\ x^2-11 & \text{if } x>-2 \end{cases}$$

is continuous at $x=-2$.

Solve It

15. Determine whether

$$p(x)=\begin{cases} \frac{1}{2}x-3 & \text{if } x\le 1 \\ 4x+3 & \text{if } x>1 \end{cases}$$

is continuous at $x=1$.

Solve It

16. Determine all numbers at which

$$q(x)=\begin{cases} x^2-3 & \text{if } x<-2 \\ x+3 & \text{if } -2\le x<1 \\ x^2-3x & \text{if } x\ge 1 \end{cases}$$

is continuous.

Solve It

Answers to Problems on Calculus

Following are the answers to problems dealing with calculus. We also provide guidance on getting the answers if you need to review where you went wrong.

1 In the given graph for $g(x)$, find $\lim_{x\to-5} g(x)$, $\lim_{x\to-2} g(x)$, and $\lim_{x\to1} g(x)$. The answers are $\lim_{x\to-5} g(x)=3$, $\lim_{x\to-2} g(x)$ DNE, and $\lim_{x\to1} g(x)=-2.6$.

Looking at the graph for $g(x)$, when x tends to –5, $g(x)$ tends to 3. When x approaches –2 from the left, the limit is 4, but when x approaches –2 from the right, the limit is –2. Because these two values aren't the same, $\lim_{x\to-2} g(x)$ does not exist (DNE). And when x approaches 1, the graph has a y value somewhere between –2 and –3. Because this isn't an exact science, just do your best to approximate the limit. It looks to us like it's about –2.6.

2 In the given graph for $h(x)$, find $\lim_{x\to-3} h(x)$, $\lim_{x\to5} h(x)$, and $\lim_{x\to0} h(x)$. The answers are $\lim_{x\to-3} h(x)=\infty$, $\lim_{x\to5} h(x)=2$, and $\lim_{x\to0} h(x)=-2.5$.

This answer also comes directly from the given graph. Trace your fingers along the graph as x gets closer and closer to –3 from the left and notice that the line keeps going up. Also notice that as x approaches this same value from the right, the graph is also going up — forever and ever up. That's why $\lim_{x\to-3} h(x)=\infty$. Meanwhile, as x approaches 5, $h(x)$ approaches 2, and as x approaches 0, it looks like $h(x)$ is very close to –2.5.

3 Find $\lim_{x\to-1} \frac{x^2+13x+12}{x+1}$ numerically. The answer is 11.

See the following chart for the analytical evaluation of this limit.

x	–2	–1.1	–1.01	–1.001	–1	–0.999	–0.99	–0.9	0
y	10	10.9	10.99	10.999	???	11.001	11.01	11.1	12

By looking at the y values in the second row, it looks like from both the left and the right y is approaching 11.

4 Find $\lim_{x\to2} \frac{3x^2-5x-2}{x-2}$ numerically. The answer is 7.

Here's the chart we used for this limit.

x	1	1.9	1.99	1.999	2	2.001	2.01	2.1	3
y	4	6.7	6.97	6.997	???	7.003	7.03	7.3	10

Looking at this one, we call it 7. What do you think?

5 Find $\lim_{x\to2} \frac{3x^2-5x-2}{x-2}$ algebraically. The answer is 7.

Plugging 2 into the function gives you a 0 in the denominator and also in the numerator, so you must try another technique. This rational function has a numerator that factors. (You found the limit numerically in Question 4.) When you factor it, you should get $\frac{(3x+1)(x-2)}{x-2}$. This reduces to $3x+1$, which gives you a function that you *can* plug 2 into: $3(2)+1=6+1=7$.

6 Find $\lim_{x \to 5} \frac{\sqrt{x-1}-2}{x-5}$. The answer is ¼.

Substitute 5 into this equation and you also get 0 in the denominator and in the numerator. Noticing that the numerator has a square root, you should find yourself thinking something along the lines of, "Perhaps I should multiply by the conjugate to rationalize the numerator." If you did think that or something close to it, give yourself a huge pat on the back.

Here's how to multiply by the conjugate: $\lim_{x \to 5} \frac{\sqrt{x-1}-2}{x-5} \cdot \frac{\sqrt{x-1}+2}{\sqrt{x-1}+2}$. FOIL out the numerators and watch the square roots disappear. However, don't multiply out the denominators — the final expression will cancel easier if you don't:

$$\frac{\left(\sqrt{x-1}-2\right)\left(\sqrt{x-1}+2\right)}{(x-5)\left(\sqrt{x-1}+2\right)} = \frac{\left(\sqrt{x-1}\right)^2-2^2}{(x-5)\left(\sqrt{x-1}+2\right)} = \frac{x-5}{(x-5)\left(\sqrt{x-1}+2\right)}.$$

The numerator and the factor on the left on the bottom both cancel and give you $\lim_{x \to 5} \frac{\sqrt{x-1}-2}{x-5} = \lim_{x \to 5} \frac{1}{\sqrt{x-1}+2}$. Now when you plug in 5, you find that the limit is ¼.

7 Find $\lim_{x \to 1} \frac{f(x)}{g(x)}$. The answer is –5⁄2.

Because you know both limits, to find the limit of their quotient, divide their limits as well: $\lim_{x \to 1} \frac{f(x)}{g(x)} = -\frac{5}{2}$.

8 Find $\lim_{x \to 1} \left(f(x)^2 - 2h(x) + \frac{1}{g(x)} \right)$. The answer is 51⁄2.

Plug in the information that you know based on the given limits:

$\lim_{x \to 1} f(x)^2 - 2h(x) + \frac{1}{g(x)} = (-5)^2 + \frac{1}{2} = \frac{51}{2}$

9 Find $\lim_{x \to 1} \frac{\sqrt{g(x)} - f(x)}{5g(x)}$. The answer is $\frac{5+\sqrt{2}}{10}$.

Plug and chug away: $\lim_{x \to 1} \frac{\sqrt{g(x)} - f(x)}{5g(x)} = \frac{\sqrt{2}-(-5)}{5(2)} = \frac{5+\sqrt{2}}{10}$

10 Find $\lim_{x \to 1} \frac{f(x)}{h(x)}$. The answer is DNE.

This time, putting the limit of $h(x)$ in the denominator also puts 0 in the denominator. The limit may not exist because the denominator approaches 0 and the numerator doesn't as x tends to 1. However, the limit may tend to +/– infinity if $h(x)$ stays negative or positive while tending to but not equaling to 1.

11 Determine whether $g(x) = \frac{x^2-4x-5}{x^2-2x-15}$ is continuous at $x = 5$. The answer is no, the function is not continuous at $x = 5$.

Factor the given equation first: $g(x) = \frac{x^2-4x-5}{x^2-2x-15} = \frac{(x-5)(x+1)}{(x-5)(x+3)}$. Cancel to get $\frac{x+1}{x+3}$. Notice that when you plug 5 into this simplified expression, you do get an answer of 6⁄8, or ¾. But this isn't the original, given equation. The graph is going to look and act like $\frac{x+1}{x+3}$, but because the original function is $\frac{x^2-4x-5}{x^2-2x-15}$, there's still going to be a hole in the graph (try plugging 5 into either of them and see what happens). This is why g has a removable discontinuity at $x = 5$.

12 Is $g(x)=\frac{x^2-4x-5}{x^2-2x-15}$ continuous at $x=0$? The answer is yes, the function $g(x)$ is continuous at $x=0$.

Observe that this is a rational function with 0 in its domain. You can simply plug 0 into this function and get out a value: $\frac{0^2-4\cdot 0-5}{0^2-2\cdot 0-15}=\frac{-5}{-15}=\frac{1}{3}$

13 Is $g(x)=\frac{x^2-4x-5}{x^2-2x-15}$ continuous at $x=-3$? The answer is no, the function is not continuous at $x=-3$. In fact, $x=-3$ is a nonremovable discontinuity.

Plugging –3 into the original function $\frac{x^2-4x-5}{x^2-2x-15}$ gives you 0 in the denominator, so you know right away that it's discontinuous. When you factor and simplify to $\frac{x+1}{x+3}$, you *still* get 0 in the denominator, so the discontinuity is nonremovable.

14 Determine whether $h(x)=\begin{cases}3x-1 & \text{if} \quad x\le -2\\ x^2-11 & \text{if} \quad x>-2\end{cases}$ is continuous at $x=-2$. The function is continuous at $x=-2$.

If you don't know how to deal with piece-wise functions like this, get a refresher from Chapter 3. First, look at $h(-2)=3(-2)-1=-6-1=-7$. The function exists at $x=-2$.

Now, look at $\lim_{x\to -2^-} h(x)=3(-2)-1=-6-1=-7$. Next, look at $\lim_{x\to -2^+} h(x)=(-2)^2-11=4-11=-7$. Because the left limit matches the right limit, the function has a limit as x approaches –2. Lastly, because the function value matches the limit value, the function is continuous at $x=-2$.

15 Determine whether $p(x)=\begin{cases}\frac{1}{2}x-3 & \text{if} \quad x\le 1\\ 4x+3 & \text{if} \quad x>1\end{cases}$ is continuous at $x=1$. The function isn't continuous at $x=1$.

$p(1)=\frac{1}{2}\cdot 1-3=-\frac{5}{2}$. The function exists at $x=1$.

$\lim_{x\to 1^-} p(x)=\frac{1}{2}-3=-\frac{5}{2}$, but $\lim_{x\to 1^+} p(x)=4(1)+3=7$. These two values aren't equal, so there is no limit and the function is discontinuous at $x=1$.

16 Determine all numbers at which $q(x)=\begin{cases}x^2-3 & \text{if} \quad x<-2\\ x+3 & \text{if} \quad -2\le x<1\\ x^2-3x & \text{if} \quad x\ge 1\end{cases}$ is continuous. The function is continuous everywhere except $x=1$. In interval notation, this is written as $(-\infty,1)\cup(1,\infty)$.

The only places this piece-wise function has potential discontinuities are where the function may break into pieces — where the interval begins or ends.

$\lim_{x\to -2^-} q(x)=(-2)^2-3=1$, and $\lim_{x\to -2^+} q(x)=-2+3=1$. $\lim_{x\to -2^+} q(x)=\lim_{x\to -2^-} q(x)=q(-2)=1$, so the graph is continuous there.

$\lim_{x\to 1^-} q(x)=1+3=4$, but $\lim_{x\to 1^+} q(x)=1^2-3(1)=-2$. $\lim_{x\to 1^+} q(x)\neq \lim_{x\to 1^-} q(x)$, so the graph is discontinuous there.

Part V
The Part of Tens

"Remember, basil is a polynomial and we'll be using two tablespoons of epsilon. But first, let's test to see if the vinegar is greater than or equal to the oil."

In this part . . .

This part has a summary of the parent graphs we cover in Chapter 3, including how to transform them. Think of it as a quick guide to all the topics we cover regarding graphing and transforming parent functions. This part also includes a chapter on the mistakes we commonly see in pre-calc and how to avoid them (please avoid them!).

Chapter 16

Ten Uses for Parent Graphs

In This Chapter

- Graphing polynomial functions
- Picturing absolute value and rational functions
- Visualizing exponential and logarithmic functions
- Sketching trig functions

A picture is worth a thousand words, and graphing is just math in pictures! These pictures can give you important information about the characteristics of a function. The most common graphs are called *parent graphs.* These graphs are in their original, unshifted form. Any parent graph can be stretched, shrunk, shifted, or flipped. They're extremely useful because you can use them to graph a more complicated version of the same function using transformations (see Chapter 3). That way, if you're given a complex function (say a crazy-looking quadratic), you automatically have a basic idea of what the graph will look like without having to plug in a whole bunch of numbers first. Essentially, by knowing what the parent looks like, you get a good idea about the kids — the apple doesn't fall far from the tree, right? In this chapter, we check out those family pictures!

Squaring Up with Quadratics

The basic quadratic is simplicity itself: $y = x^2$. Its graph is a parabola with a vertex at the origin, reflected over the *y*-axis (see Figure 16-1). You can find out more about graphing quadratics in Chapters 3 and 12.

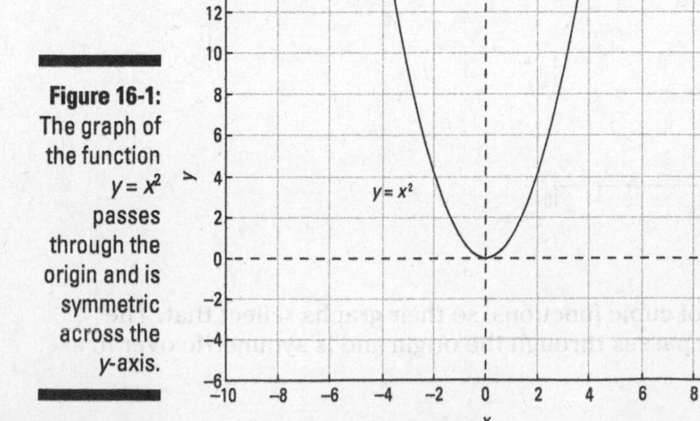

Figure 16-1: The graph of the function $y = x^2$ passes through the origin and is symmetric across the *y*-axis.

Cueing Up for Cubics

The parent graph of the cubic function, $y = x^3$, also passes through the origin. This graph is symmetric over the origin (see Figure 16-2). We cover cubics in Chapter 3.

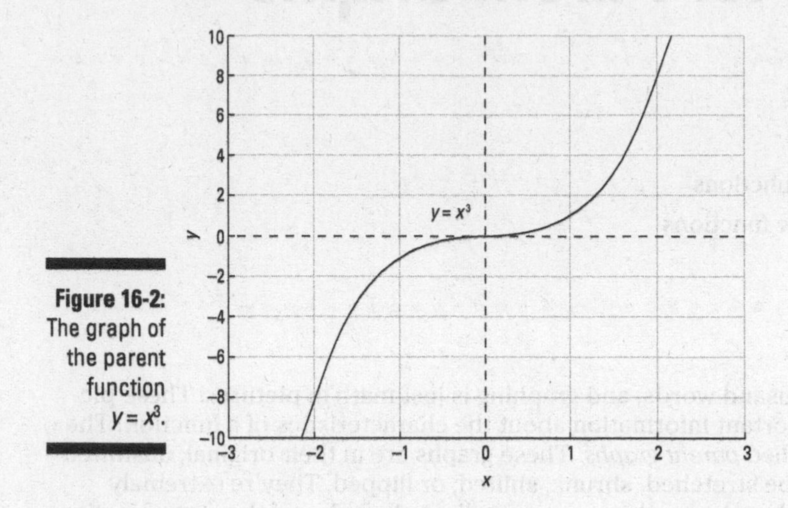

Figure 16-2: The graph of the parent function $y = x^3$.

Rooting for Square Roots and Cube Roots

A square root graph looks like a parabola that has been rotated clockwise 90 degrees and cut in half. It's cut in half (only positive) because you can't take the square root of a negative number. The parent graph is pictured in Figure 16-3.

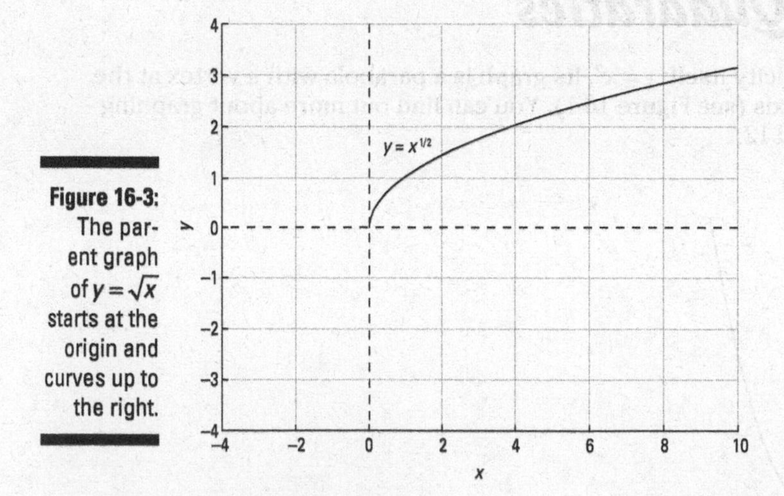

Figure 16-3: The parent graph of $y = \sqrt{x}$ starts at the origin and curves up to the right.

Cube root functions are the inverse of cubic functions, so their graphs reflect that. The parent graph of a cube root function passes through the origin and is symmetric over it, as shown in Figure 16-4.

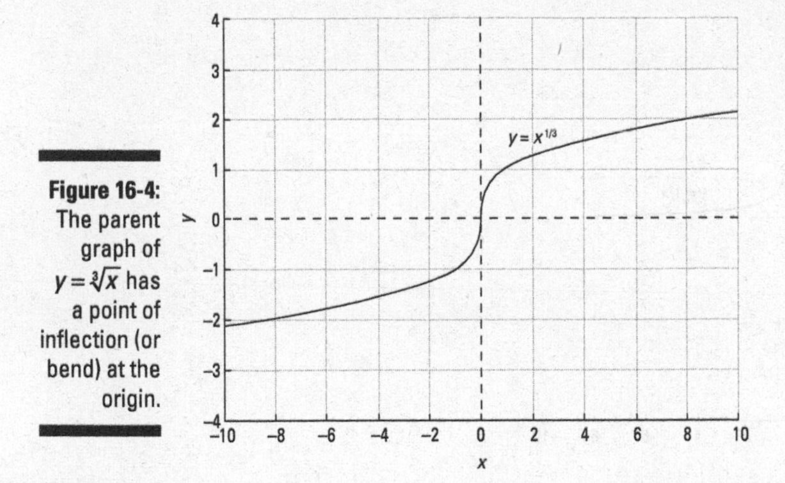

Figure 16-4: The parent graph of $y = \sqrt[3]{x}$ has a point of inflection (or bend) at the origin.

Graphing Absolutely Fabulous Absolute Value Functions

Because the absolute value function turns all input into non-negative values (0 or positive), the parent graph is only above the *x*-axis. Figure 16-5 shows the parent graph in its characteristic V shape.

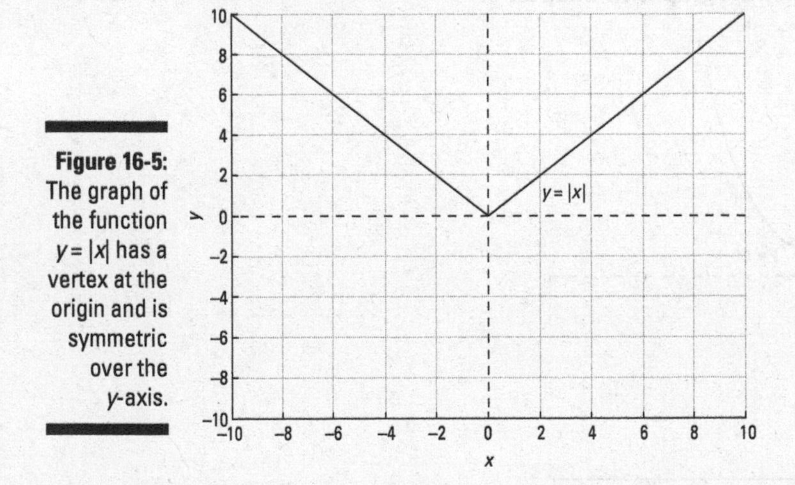

Figure 16-5: The graph of the function $y = |x|$ has a vertex at the origin and is symmetric over the *y*-axis.

Flipping over Rational Functions

In Chapter 3, we take you through the steps for graphing rational functions. These involve finding asymptotes, intercepts, and key points. Because these functions don't really have parent graphs per se, we thought we'd show you an example of the most basic rational function: $y = \frac{1}{x}$ (see Figure 16-6). To see more, flip back to Chapter 3.

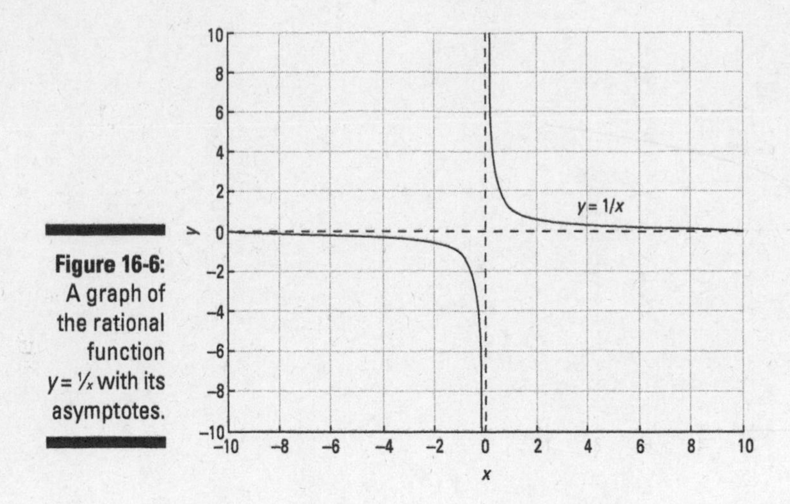

Figure 16-6: A graph of the rational function $y = \frac{1}{x}$ with its asymptotes.

Exploring Exponential Graphs and Logarithmic Graphs

The parent graph of an exponential function is $y = b^x$ where b is the base. Because b has to be some number to graph, we thought we'd show you the graph of $y = e^x$. This graph passes through the point (0, 1) and has a horizontal asymptote of the x-axis, as shown in Figure 16-7. We cover exponential graphs in Chapter 5.

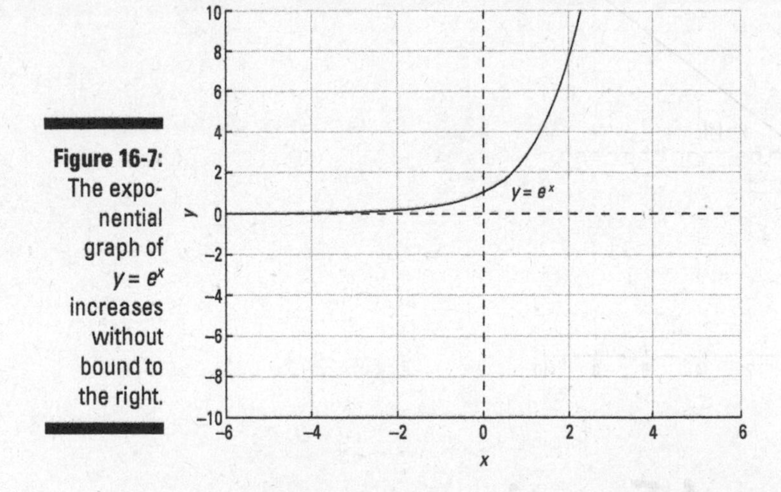

Figure 16-7: The exponential graph of $y = e^x$ increases without bound to the right.

The inverse (see Chapter 3) of an exponential function is a logarithmic function. So we show you the inverse of the graph of $y = e^x$, which is the graph of $y = \log_e x$, also known as the natural log, or $y = \ln x$ (see Chapter 5). This graph passes through the point (1, 0) and has a vertical asymptote of the y-axis, as shown in Figure 16-8. We also cover logarithmic graphs in Chapter 5. Observe that exponential functions with $b > 1$ increase very rapidly, while $y = \ln x$ increases very slowly. However, they all increase without bound.

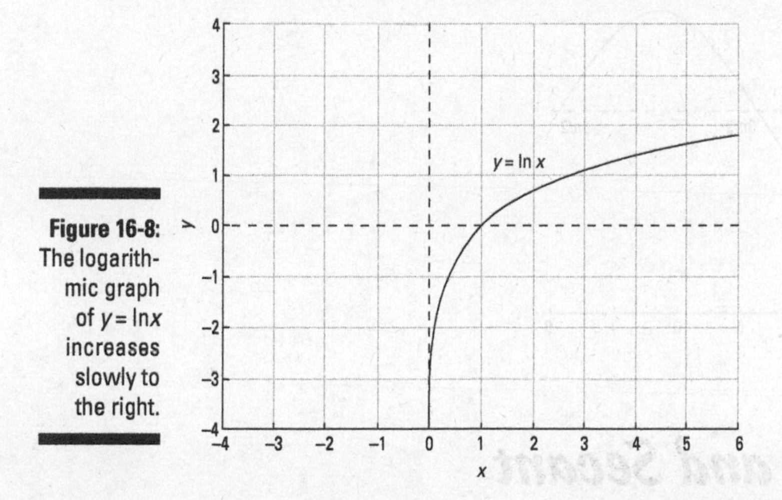

Figure 16-8: The logarithmic graph of $y = \ln x$ increases slowly to the right.

Seeing the Sine and Cosine

A sine graph looks like a wave. The parent graph passes through the origin and has an amplitude of 1. The period is 2π, which means that the wave repeats itself every 2π. Figure 16-9 shows one full period of the parent sine graph.

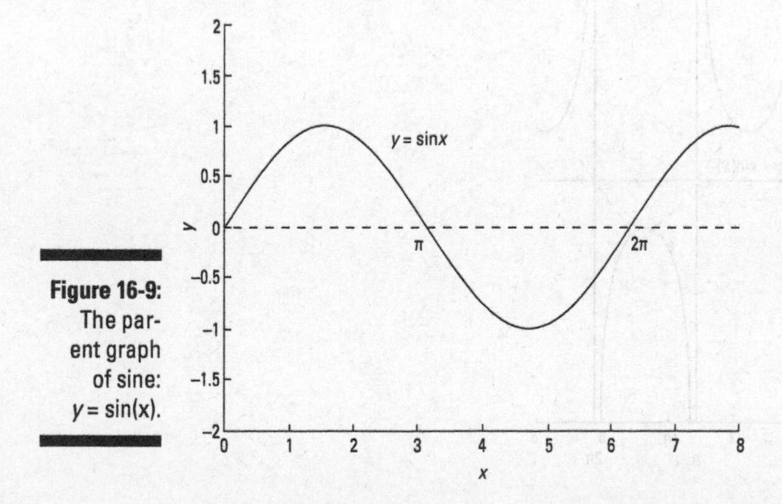

Figure 16-9: The parent graph of sine: $y = \sin(x)$.

Like sine, the graph of cosine is a wave. This parent graph passes through the point (0, 1) and also has an amplitude of 1 and a period of 2π. You can see the parent graph in Figure 16-10. For more information about graphing sine and cosine, turn to Chapter 7.

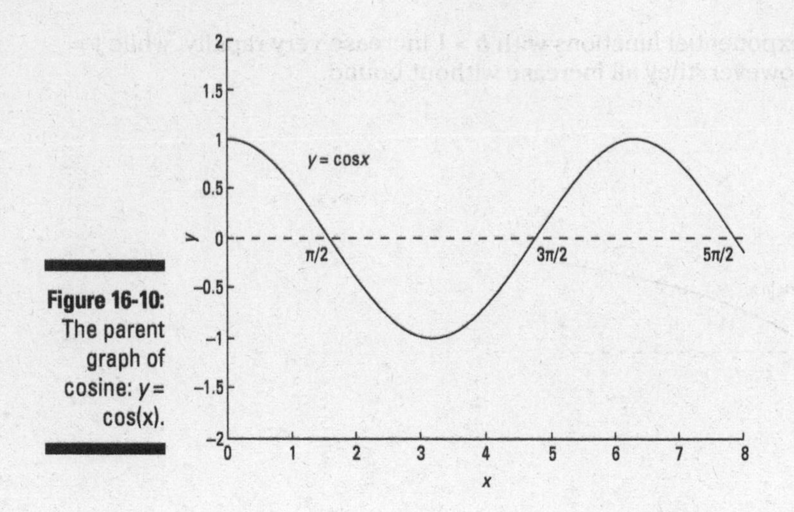

Figure 16-10: The parent graph of cosine: $y = \cos(x)$.

Covering Cosecant and Secant

Remember reciprocals? Well, cosecant is the reciprocal of sine, so the graph of cosecant reflects that. The parent sine graph and the parent cosecant graph are both depicted in Figure 16-11 so you can see the relationship. You can find specific graphing information in Chapter 7.

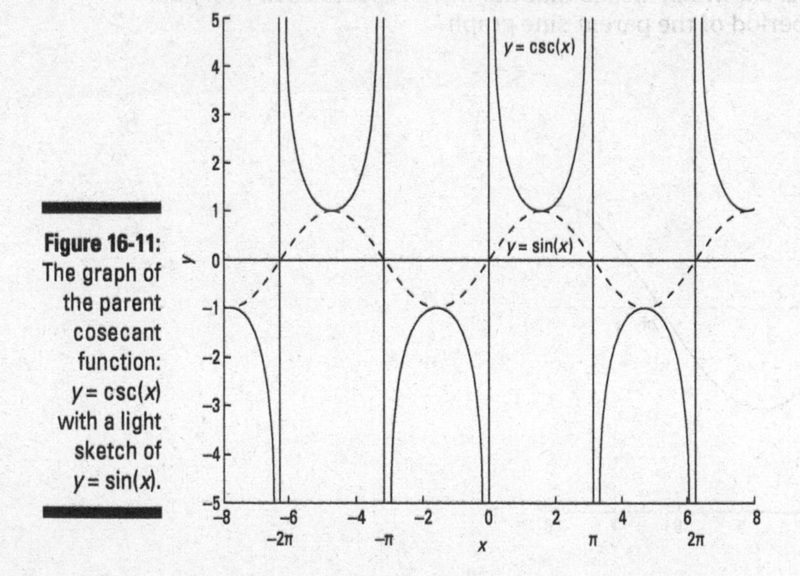

Figure 16-11: The graph of the parent cosecant function: $y = \csc(x)$ with a light sketch of $y = \sin(x)$.

Again, like cosecant, secant is the reciprocal of cosine, so the graph of secant is related to the graph of cosine. To picture this, we've lightly drawn the graph of cosine along with the parent graph of secant in Figure 16-12.

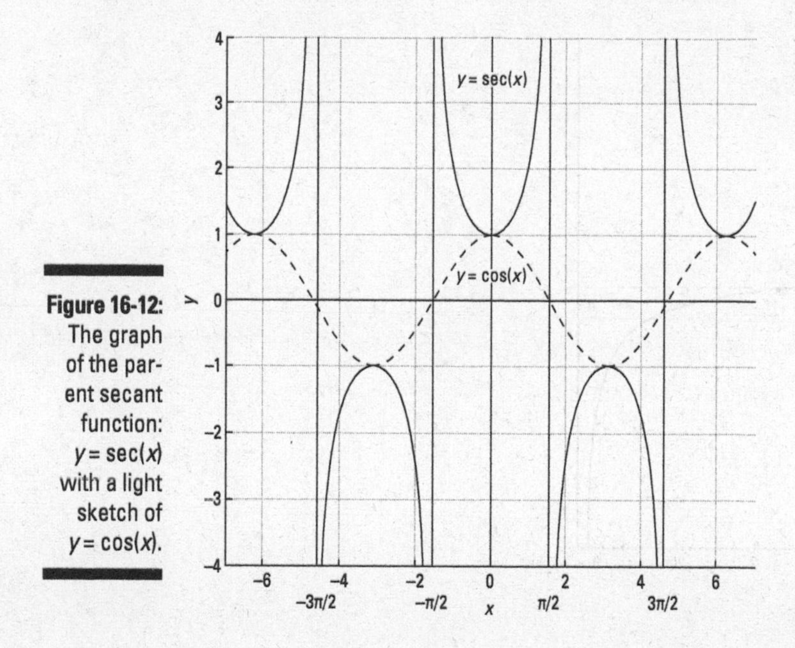

Figure 16-12: The graph of the parent secant function: $y = \sec(x)$ with a light sketch of $y = \cos(x)$.

Tripping over Tangent and Cotangent

One repeating pattern of the graph of tangent is its asymptotes, where the function is undefined. Like other trig graphs, a tangent graph has a period where it repeats itself. In this case, it's π. In Figure 16-13, we show you one period of the parent tangent graph. For more information about graphing tangents, turn to Chapter 7.

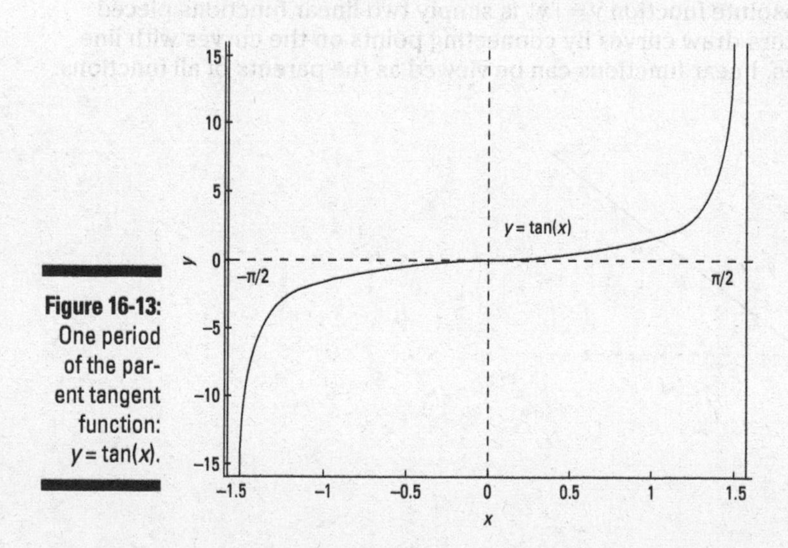

Figure 16-13: One period of the parent tangent function: $y = \tan(x)$.

Like tangents, the parent graph of cotangent has asymptotes at regular intervals. Also like tangents, the period of cotangent is π. In Figure 16-14, we show you one period of the parent cotangent graph. You can get more information about graphing cotangents in Chapter 7.

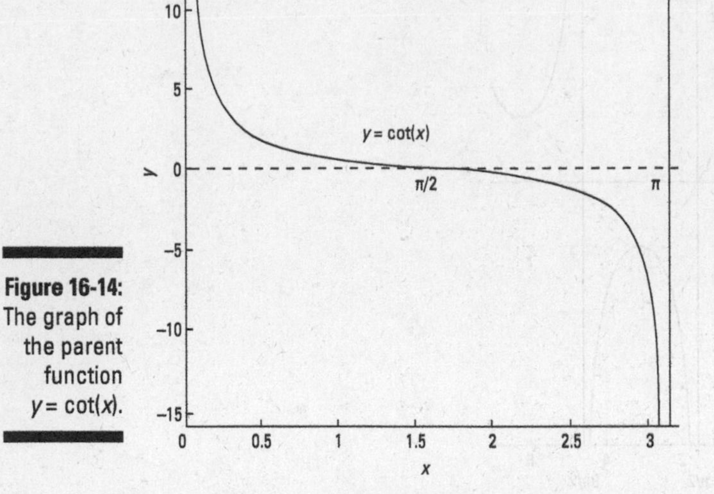

Figure 16-14: The graph of the parent function $y = \cot(x)$.

Lining Up and Going Straight with Lines

Straight lines are very invariant; you can perform all the transformations, translations, and reflections you want, but they still remain straight. These lines are graphs of linear functions of the form $y = ax + b$. Figure 16-15 depicts the line $y = x$. In applications, scientists often approximate those curvy (called *nonlinear*) functions by a collection of linear segments that represent linear functions. You'll see plenty of this when you take the calculus course. The V-shaped absolute function $y = |x|$ is simply two linear functions pieced together. Indeed, all computers draw curves by connecting points on the curves with line segments. So, in some senses, linear functions can be viewed as the parents of all functions.

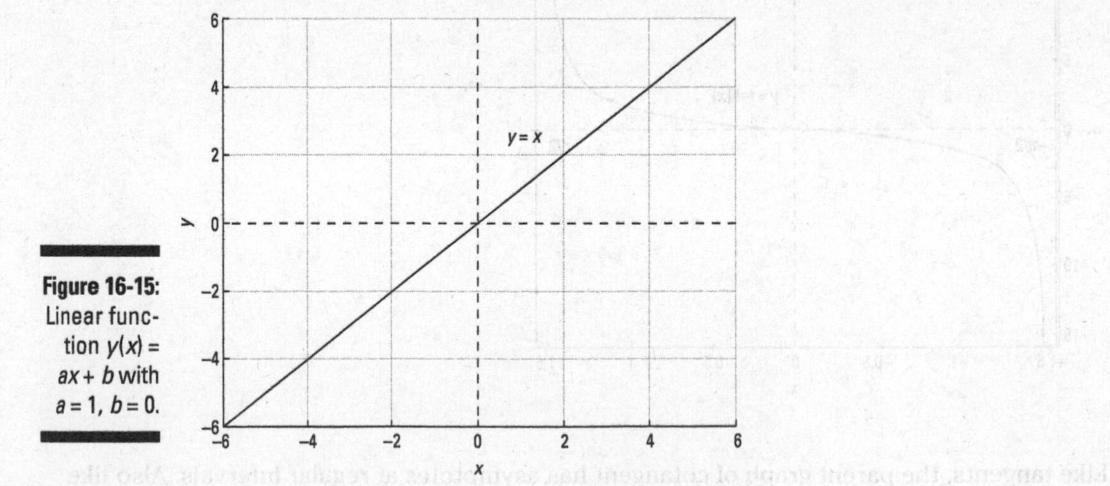

Figure 16-15: Linear function $y(x) = ax + b$ with $a = 1$, $b = 0$.

Chapter 17

Ten Missteps to Avoid in Pre-Calculus

In This Chapter

- Avoiding some common mistakes that pre-calculus students make
- Reviewing some relevant pre-calc rules

One of our favorite video games from childhood (we know this dates us, but . . .) featured a little guy with a square head running through the square jungle, swinging on square vines, and jumping over square alligators in square swamps. He was avoiding the pitfalls of the jungle. Consider this chapter the vine you can use to jump over the pitfalls that normally trip up the pre-calc student. And you don't have to be a video game geek to understand this chapter!

Going Out of Order (Of Operations)

Operations in an expression or an equation aren't meant to be done from left to right. For example, $3 - 7(x - 2)$ doesn't equal $-4(x - 2)$ or $-4x + 8$. You're supposed to do multiplication first, which means distributing the -7 first: $3 - 7x + 14$. Now combine like terms to get $-7x + 17$.

Remember your order of operations (PEMDAS) all the time, every time:

Parentheses (and other grouping devices)

Exponents

Multiplication and **D**ivision, from left to right in order as you find them

Addition and **S**ubtraction, also from left to right

To further review the order of operations, see Chapter 1.

FOILing Binomials Incorrectly

When multiplying binomials, always remember to multiply them in the correct order. You remember FOIL — First, Outside, Inside, Last. This includes squaring any binomial. The biggest mistake we see in these situations is something like: $(x - 4)^2 = x^2 + 16$. That's forgetting a whole lot of multiplying, though. It should look like this: $(x - 4)^2 = (x - 4)(x - 4) = x^2 - 4x - 4x + 16 = x^2 - 8x + 16$. You may use other orders, but it pays to be careful and consistent.

Breaking Denominators Incorrectly

Don't fall for a big trap and break a fraction up incorrectly. $\frac{5}{4x+3}$ doesn't equal $\frac{5}{4x}+\frac{5}{3}$. If you don't believe us, pick a value for x and plug it into both expressions and see whether you get the same answer twice. You won't, because it doesn't work. You're doing the order of operations right if you change $\frac{6}{3x-12}$ to $\frac{2}{x-4}$ (see the section "Going Out of Order (Of Operations)," earlier in this chapter). The division bar is a grouping symbol, and you have to simplify the numerator and denominator separately before doing the division: $\frac{6}{3x-12}=\frac{\cancel{3}\cdot 2}{\cancel{3}\cdot(x-4)}=\frac{2}{x-4}$

Combining Terms That Can't Be Combined

Yet another mistake we see frequently is students combining terms that aren't meant to be combined. $4x - 1$ suddenly becomes $3x$, which it's not. $4x - 1$ is simplified, meaning that it's an expression that doesn't contain any like terms. $3a^4b^5 + 2a^5b^4$ is also simplified. Those exponents are close, but close only counts in horseshoes and hand grenades. When counting in the real world (as opposed to the algebra one you probably feel stuck in now), you can't combine apples and bananas. Four apples plus three bananas is still four apples and three bananas. It's the same in algebra: $4a + 3b$ is simplified.

Forgetting to Flip the Fraction

When dealing with complex fractions, don't take all the rules that you've learned and throw them out the window. $\frac{\frac{3}{x+1}}{\frac{2}{x-2}}$ doesn't become $\frac{3}{x+1}\cdot\frac{2}{x-2}$. If you think it does, you're probably forgetting that a division bar is division.

$\frac{\frac{3}{x+1}}{\frac{2}{x-2}}=\frac{3}{x+1}\div\frac{2}{x-2}$. To divide a fraction, you must multiply by its reciprocal:

$\frac{3}{x+1}\div\frac{2}{x-2}=\frac{3}{x+1}\cdot\frac{x-2}{2}$

Losing the Negative (Sign)

We know that in life you're not supposed to be negative, but in math, don't throw away a negative sign — especially when subtracting polynomials.

$(4x^3 - 6x + 3) - (3x^3 - 2x + 4)$ isn't the same thing as $4x^3 - 6x + 3 - 3x^3 - 2x + 4$. If you do it that way, you're not subtracting the whole second polynomial, only its first term. The right way to do it is $4x^3 - 6x + 3 - 3x^3 + 2x - 4$, which simplifies to $x^3 - 4x - 1$. The issue here is a special case of the failure to correctly apply the distributive law; it frequently occurs with other coefficients as well (not just –1). Failure to write the parentheses often directly contributes to these errors.

Similarly, when subtracting rational functions, take care of that negative sign.

$$\frac{3x+5}{x-2}-\frac{x-6}{x-2}\neq\frac{3x+5-x-6}{x-2}$$

What happened? You forgot to subtract the whole second polynomial on the top. Instead, this is the way to do it: $\frac{3x+5}{x-2}-\frac{x-6}{x-2}=\frac{3x+5-(x-6)}{x-2}=\frac{3x+5-x+6}{x-2}=\frac{2x+11}{x-2}$

Oversimplifying Roots

When it comes to roots, we've seen all kinds of errors. For instance, $\sqrt{3}$ suddenly becomes 3 in a problem and loses the root altogether.

Don't add or subtract roots that aren't like terms, either. $\sqrt{7}+\sqrt{3}$ isn't $\sqrt{10}$, now or ever. They're not like terms, so you can't add them. $\sqrt{7}+\sqrt{3}$ is it — it's done.

Avoiding Exponent Errors

When multiplying monomials, don't multiply the exponents. We've seen students who've been dealing with exponents for a long time suddenly do something like this:

$x^4 \cdot x^3 = x^{12}$, when the right answer is actually $x^4 \cdot x^3 = x^{3+4} = x^7$. Also, when dealing with a power of a product, you must apply the power to everything. $(2x^5y)^3$ isn't $2x^5y^3$ or $2x^{15}y^3$. You must raise everything inside the parentheses to the third power, so the answer should be $8x^{15}y^3$.

Watch out when dealing with negatives and exponents in some calculators. -4^2 and $(-4)^2$ represent –16 and 16 respectively, so be sure you know which configuration you're looking for when you punch a number into your calculator.

Canceling Too Quickly

You can cancel terms when adding or subtracting if you have two terms that are exact opposites of each other. In multiplication and division, you can cancel terms if one common factor divides into all terms.

Here are the most common canceling mistakes we see:

- **Canceling constants:** If you see the rational expression $\frac{15x-2}{5}$, it doesn't equal $3x-2$. The 5 in the denominator has to divide into both terms on the top.

- **Canceling variables:** For the same reason as the one given in the preceding bullet, $\frac{4x^2-3x+2}{x}$ isn't $4x-3x+2$ or $4x-3+2$, because the denominator doesn't divide into everything on the top, only the first term.
- **Canceling everything:** To keep you aware, $\frac{4x^2-3x+2}{2x^2-6x+8}$ isn't $\frac{\not{4}^2x^2-\not{3}x+\not{2}}{\not{2}x^2-\not{6}_2x+\not{8}_4}=\frac{2}{2+4}=\frac{2}{6}=\frac{1}{3}$, which would be creating a constant out of two polynomials dividing.

Making Distribution Errors

When a polynomial is multiplied by a monomial, the process is known as distribution. Think of it like delivering the newspaper to every house on the block.

We've also seen students who don't distribute to every term, especially when the polynomial gets long.

$3(2x^5-6x^4+3x^3-x^2+7x-1)$ isn't equal to

$6x^5-6x^4+3x^3-x^2+7x-1$,

or $6x^5-18x^4+3x^3-x^2+7x-1$,

or even $6x^5-18x^4+9x^3-x^2+7x-1$,

or anything other than $6x^5-18x^4+9x^3-3x^2+21x-3$.

Index

• D •

• Q •

• R •

• S •

• T •